Herbicides – Environmental Impact Studies and Management Approaches
Edited by Ruben Alvarez-Fernandez

Published by InTech

Janeza Trdine 9, 51000 Rijeka, Croatia

Publishing Process Manager Petra Nenadic

Technical Editor Teodora Smiljanic

Cover Designer InTech Design Team

Additional hard copies can be obtained from orders@intechopen.com

Herbicides – Environmental Impact Studies and Management Approaches,
Edited by Ruben Alvarez-Fernandez
p. cm.
ISBN 978-953-307-892-2

For EU product safety concerns:
IN TECH d.o.o., Prolaz Marije Krucifikse Kozulić 3, 51000 Rijeka, Croatia,
info@intechopen.com or visit our website at intechopen.com.

Meet the editor

Dr Rubén Álvarez Fernández was born in Spain in 1978. Interested in botany and genetics, he obtained his degree in Biology in 2000 at the University of Oviedo. He specialized in Biotechnology and pursued his interests in plants with a PhD project on plant genetic transformation at the same university. Dr Álvarez obtained his PhD with Honors in 2006 and co-authored the winning project on conifer propagation in an entrepreneurs competition. He undertook a post-doc on plant physiology and response to drought before moving to the University of Cambridge (UK) to work on plant molecular evolution. Next he moved into scientific publishing as Scientific Curator on the Methods Navigator project - a cookbook on molecular biology protocols. Currently, with his interests on applied science, he is studying the genetics of plant responses to light stress and production of secondary metabolites in the University of Essex (UK).

Contents

Preface

Weeds severely affect crop quality and yield. Therefore, successful farming relied for a very long time on the combination of manual weeding and management approaches, including crop rotation, fire, tillage, and fallow systems.

A new farming era began during the Second World War with the discovery of two synthetic auxinic herbicides. In 1942, 2,4-dichlorophenoxyacetic acid (2,4-D)was first identified, which was soon followed by 2-methyl-4-chlorophenoxyacetic acid (MCPA) in 1945 and 1946, respectively. These two herbicides were then marketed and used to selectively control broad leaf weeds in cereal crops.

The use of herbicides on a wide scale started in the 1950's and expanded quickly until the 1970's. Thousands of new compounds and herbicides were identified, allowing for a qualitative increase in crop yield. However, this "easy chemistry" soon became fully exploited. Toxicities to the environment and humans were found, and consequently regulations tightened. The development of herbicide-resistant crops forced a change in the scope of herbicide development. And by 1980, the growth of this industry began to decline and today it remains in a static situation.

This book blends review chapters with scientific studies, providing the reader with an insight into some of the current trends in the field of herbicides. Included are environmental case studies on the toxicity and impact of herbicides on natural populations, as well as methods to reduce herbicidal inputs and the use of bioherbicides as natural alternatives. Importantly, this book describes ecological and economical aspects of herbicide application. On the one hand it shows the need for controlling or abandoning herbicide applications to reduce contamination, and on the other it shows how increased herbicide dosage does not necessarily imply higher crop yield and can actually lower the economic gain. Indeed, in some cases, herbicide savings could reach 80% with appropriate management and decision support systems - an approach that the farmer usually dismisses because he may think that killing the weed is always necessary. Consequently, the use of Integrated Weed Management approaches and Decision Support Systems is encouraged. These will assist in choosing optimal parameters such as herbicide type, application time, area, and mode aiming at obtaining the highest yield and crop quality with the minimum environmental impact.

Finally, as Editor of this book I would like to acknowledge the effort of the authors, their prompt response to comments, and their openness to discuss suggestions to improve the chapters. I would also like to thank my Publishing Process Manager, Ms. Petra Nenadic, for her kindness, willingness to help, and professionalism during the editorial process.

Dr Rubén Álvarez-Fernandéz
University of Cambridge,
Department of Plant Sciences,
UK

1

Evaluation of Toxicities of Herbicides Using Short-Term Chronic Tests of Alga, Daphnid and Fish

Norihisa Tatarazako[1] and Taisen Iguchi[2]
[1]*Environmental Quality Measurement Section, Research Center for Environmental Risk, National Institute for Environmental Studies, Tsukuba, Ibaraki,*
[2]*Department of Bio-Environmental Science, Okazaki Institute for Integrative Bioscience, National Institute for Basic Biology, National Institutes of Natural Sciences, Myodaiji, Okazaki, Aichi,*
Japan

1. Introduction

Agriculture plays a significant role around the world in order to achieve sustainable development, as it is the major source of food supply, and plays an important role in building the social structure of rural areas (Oerke et al., 2004; Cooper et al., 2007; Damalas et al., 2009; Wilson et al., 2001). Depending on the region/country, these aspects weigh differently. Japan's agricultural scenario, although not very massive, is sometimes termed as an outstanding contributing sector to the country's economy (1.5% of GDP) (Ministry of Agriculture, Forestry and Fisheries, 2010). It is characterized by high production and secured subsidies, owing to its socio-cultural background and environmental characteristics.

Japan, an archipelago of many islands with a northeast-southwest arc, stretches for approximately 2,400 km through the western Pacific Ocean. The country has a total land area of 377,887 square km. About 70-80% of the country is mountainous, leaving 46,280 square km available for profitable cultivation. Several agricultural crops are highly associated with rich social history of Japan. Similar to the other Asian countries, paddy is one of the major and oldest cultivated crops in Japan. Paddy cultivation in Japan is most favored because of its relatively wet climate, which prevails all round the year with intermittent rains and associated freshwater supplies. Approximately, 54.3% (2,516,000 ha) of total cultivable land in the country is used for paddy cultivation (Fig. 1) (Ministry of Agriculture, Forestry and Fisheries, 2011). Unfortunately, several factors, like ever-growing population, intense industrialization, and their associated demands for space, have reduced the agriculturally cultivable lands. This has not only resulted in the total land shrinkage, but the average percentage of agriculturist has also gone down drastically. For instance, in the year 2005, 3,353,000 people were engaged in agriculture but by 2011 it had declined to 2,606,000 people (▼22.3%) (Ministry of Agriculture, Forestry and Fisheries, 2005, 2010).

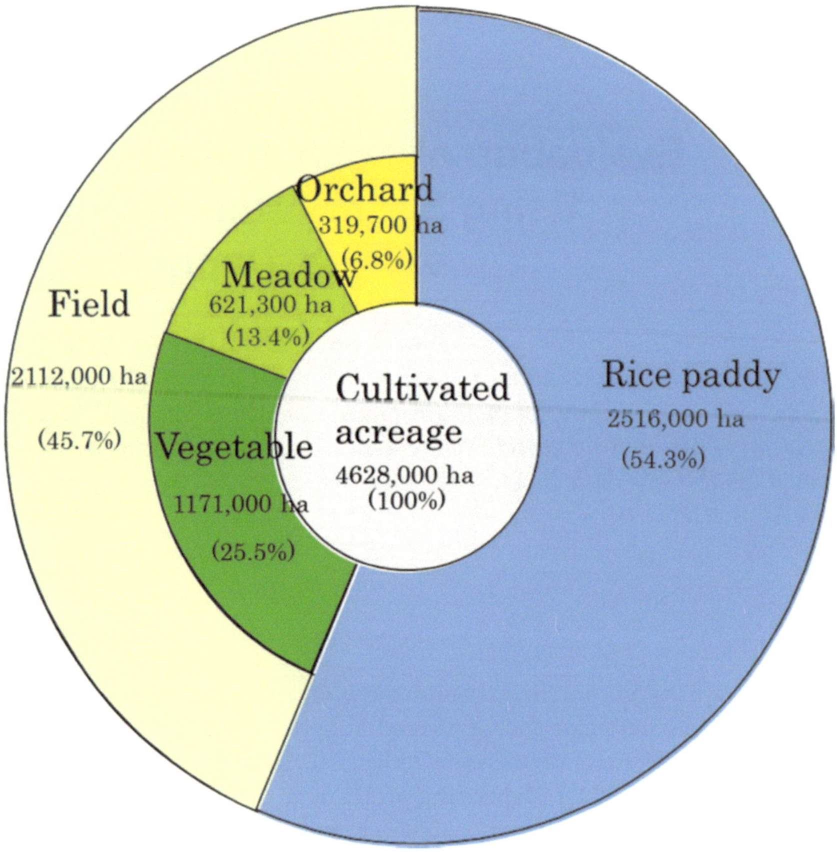

Fig. 1. Ratio of area according to the arable land.

Japanese agricultural farmers have very small land holding, unlike their western counterparts (Fig. 2) (Ministry of Agriculture, Forestry and Fisheries, 2011). Although the cultivable area in Japan is approximately 3,632,000 ha, the average management cultivated area per management constitution (farming family) is about 2.2 ha (in Hokkaido: 23.5 ha and other prefectures: 1.6 ha). This statistics therefore shows that the farmers are left with no other choice, other than use pesticides and fertilizers, in order to harvest the maximum yield from the available land resource. According to OECD (Organisation for Economic Co-operation and Development), Japan uses maximum pesticide (1.5 tons/km^2) followed by Korea (1.29 tons/km^2) Netherlands (1.06 tons/km^2), Belgium (0.92 tons/km^2) (OECD, 2002).

Majority of the Japanese farming communities (70% or 1,740,000 farmers) are engaged in paddy culture (Ministry of Agriculture, Forestry and Fisheries, 2011), thereby making its culture methods a monopoly amongst the farmers. This leads to a situation where the farmers are at a liberty to freely choose the pesticide as well as its dosage and application schedule. These practices result in the release of many different kinds of herbicides and insecticides (pesticides) in the environment during a stipulated period of time (Japan Plant Protection Association, 2010; Añasco et al., 2010; Sudo et al., 2002). Therefore, it is extremely difficult to analyze the effect and fate of each pesticide being used in a particular agricultural field as a whole.

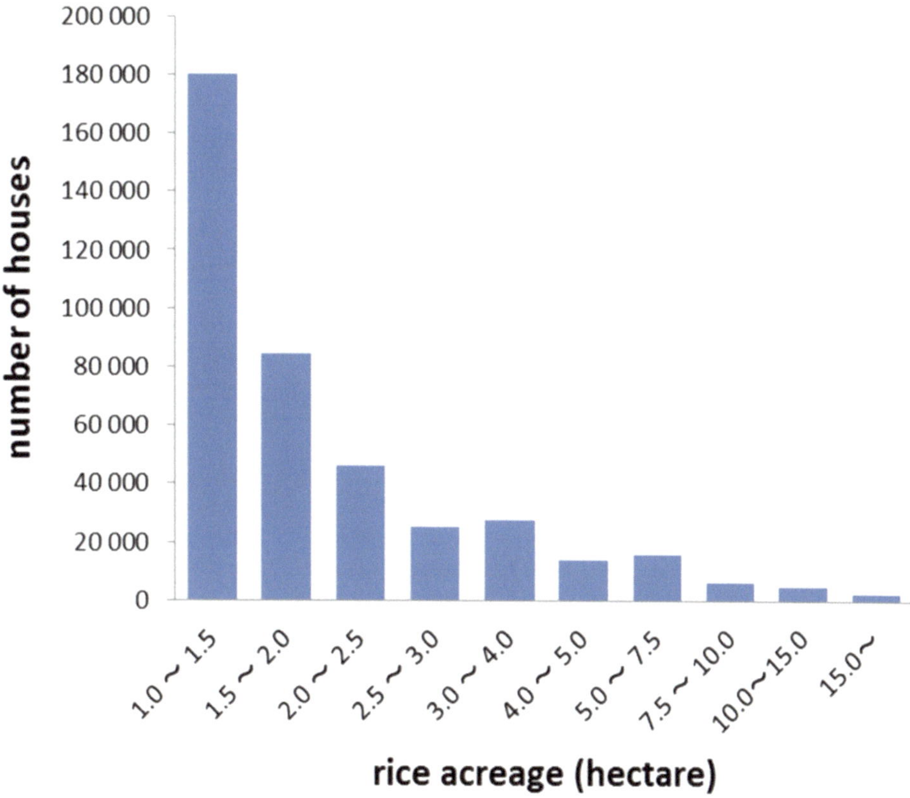

Fig. 2. Area under cultivation in Japan.

The mainstay of the Japanese agriculture scenario is the wet-rice farming. Therefore, the effect of these abundantly used herbicides in the paddy farming and on the aquatic environment cannot be ignored. The herbicides spread through the water, logged in the field, which later diffuse into the soil, thereby giving it a weeding effect. Therefore, Japanese paddy culturists maintain several days (preferably one week) of long static water environment after the application of herbicide in the field (Vu, 2006). Such conventional practice not only increases the weeding effect but also prevents the outflow of herbicides to a large extent. Despite such practices, traces of herbicides are detected in rivers and wetlands (Tanabe et al., 2001; Sudo et al., 2002; Nakano et al., 2004; Numaba et al., 2006; Miyashita et al., 2009; Tsuda et al., 2011). In order to reduce such environmental contamination, farmers may need to pay more attention regarding their use especially the amount of pesticide per square meter to be used.

As a statutory measure, Japanese government has established the application criteria and examination for different pesticides in order to reduce the residual concentration in the crops, soil and water and hence minimize the aquatic pollution (Nagayama, 2010). Few years back, in 2006, only 799 pesticides had been listed for their residual effect in food (Akiyama et al., 2005; Maitani, 2005, 2007; Saito, 2007; Hirahara et al., 2007). However, the criterion about remaining in environment did not exist. As for the pesticide registration reservation criterion about prevention of toxic damage to the fisheries animals and plants, the standard value was established only in consideration of the acute toxicity of the carp about a pesticide used conventionally in a rice field. But, by criterion revision of March, 2003, the criterion is set in consideration of effect for a fish, a crustacean, algae and the forecast concentration in public waters and postpones the pesticide registration of the criterion that it is inadequate. However, a major problem concerning this registration is the test organism and negligence about environmental exposure levels. For example, till now, the degree of adversity is calculated based on several acute toxicity tests, like LC50 at 96h for fish, EC50 (effective concentration 50) at 48h for water flea swimming inhibition and EC50 at 72h for green algae growth inhibition (Wei et al., 2008; Kim et al., 2008). But few chronic tests are being conducted to address the environmental health in the long term (Sakai, 2002; Kang et al., 2009; Marques et al., 2011). Information about the usage of herbicides and the levels of herbicides in the environment must be monitored to calculate the possible adverse effect of these pesticides on our ecosystem.

Although herbicides have a relatively shorter half-life their application window is more important. Majority of aquatic insects, fish, and amphibians in Japan, spawns/hatch in early spring, the time when weedcides are sprayed in field. This makes the newborns most vulnerable and more prone to death. The bioassay using the aquatic species is effective in predicting the effect of herbicides on the ecosystem. Therefore, it is necessary to carry out the herbicidal short-term chronic toxicity examinations (or sub-chronic toxicity examinations) as well as acute toxicity tests using the aquatic species (Matthews, 2006). It is also important to perform the chronic toxicity evaluations on the plural endpoints like breeding and/or growth of the aquatic species. In this regard, since April 2005, the Ministry of the Environment has initiated research to re-evaluate the predicted concentration of pesticides in the public waters, along with their toxic concentration for fish, crustacean and algae, in order to prevent the adverse effects of pesticides on wildlife and ecosystem. However, information about chronic toxicity tests using fish, crustacean and alga, which might help in eco-toxicological assessment of different herbicides, is incredibly little in Japan. Therefore, we have carried out bioassays for chronic toxicity to check the ecological effect of pesticides and also investigated the agriculture drainage sample, collected from several Japanese environments (Ministry of the Environment, 2006, 2007, 2008, 2009, 2010). In this present investigation, we collected water samples directly from rice field drainage just after 7-14 days of herbicide application and performed chemical analysis by GC/MS (Gas chromatography-mass spectroscopy). We also carried out short term chronic toxicity tests i.e. Algal growth inhibition test (OECD TG201;OECD test guideline No.201, 2002), Daphnia reproduction test (Canada Ministry of the Environment: Test of Reproduction and Survival Using the Cladoceran, *Ceriodaphnia dubia*) (Environment Canada, 2007; EPA Biological Test Method, 2007), and Fish short-term toxicity test on embryo and sac-fry stages (OECD TG212, 1998) using an alga (*Pseudokirchneriella subcapitata*), a crustacean (*Ceriodaphnia dubia*), and fish (*Danio rerio*) respectively, for the 10 major kinds of herbicide formulation.

2. Research methodologies

2.1 Sampling

Present investigation was carried out in 14 sampling points of 9 prefectures of Japan from 2009 to 2010 (Ministry of the Environment, 2009, 2010). The field samplings were conducted during May and June, generally after 1-2 weeks from herbicide application, from paddy field drainage system. Three water samples were collected in glass bottles, and then these samples were filtrated with a glass fiber of 0.7 μm and kept at 4°C till further analysis.

In order to keep correct information about dispersion period, respective owner/ farmer was interviewed. However, the effect from target outskirts farmhouse and the weather conjugation was not controlled. To gain some idea about the risk factors associated with uncertainties, water from one sampling station was sampled every week during one month from a day of the herbicide dispersal.

2.2 Chemicals

Dimethametryn (CAS: 22936-75-0), pretilachlor (CAS: 51218-49-6), bromobutide (CAS: 74712-19-9), carbetamide (CAS: 16118-49-3), bendiocarb (CAS: 22781-23-3), triazine (CAS: 101-05-3), cyanazine (CAS: 21725-46-2), simetryn (CAS: 1014-70-6), esprocarb (CAS: 85785-20-2), and mefenacet (CAS: 73250-68-7) were purchased from Wako Pure Chemicals Industries, Ltd (Osaka, Japan). Chemicals were dissolved at a concentration of 100ppm in the test medium for 48 h at room temperature. Chemicals with low solubility were dissolved in the test water at concentration of aqueous solubility (100% solution) for 48 h at room temperature, and filtrated with 1.2 μm glass fiber.

2.3 Analysis of chemical composition by GC-MS

The water samples were subjected to Sep-pacC18 matrix solid-phase dispersion clean up followed by simultaneous analysis of 917 different kinds of herbicides/pesticides using gas chromatography-mass spectrometry-choice ion monitoring system (GC-MS-SIM; Shimazu QP-2010, Kyoto, Japan). Seven substances (naphthalene-d8, acenaphthene-d10, phenanthrene-d10, chrysene-d12, perylene-d12, 4-chlorotoluene-d4, 1,4-dichlorobenzene-d4) were used as internal standard. In addition, liquid chromatograph mass spectrometry (LC/MS) was applied to analyse a pesticide which we failed to characterized by GC-MS (Kadokami et al., 2005).

The results were analyzed according to "GC/MS Database Software Ver.2 for Environmental Simultaneous Analysis" which was developed by Shimazu Corporation and Kitakyusyu Municipal University and was purchased from Shimazu Corporation (Tokyo, Japan).

2.4 Short-term chronic toxicity tests

2.4.1 Fish toxicity test

Fishtoxicity test was performed using zebrafish (*Danio rerio*) according to the OECD test guideline No. 212 (Fish short-term toxicity test on embryo and sac-fry stages). Zebrafish were obtained from National Institute for Environmental Studies, Japan. Eggs were collected within 4h of fertilization and distributed at 20 eggs/test jar (80 ml glass cup). Four replicates of each control and exposure group were incubated in 50 ml of filtered water samples and test medium respectively for 9 days. 100% media replacement was carried out every 2 days. The water samples were diluted (0-80%) using test water when required. The experiment was conducted at 26 ± 1°C (water temperature) and photoperiod of 16h : 8h (light : dark). The hatchability and survivability were recorded daily.

2.4.2 Daphnia toxicity test

Ceriodaphnia dubia has been acclimatized for several years in NIES, Japan to breed at hardness of 70 mg/l (similar to average hardness of Japanese water). We used water flea subacute toxicity test method "Test of Reproduction and Survival using the Cladoceran *C. dubia*" (Environment Canada, 2007). Changes (decreases) in the number of offspring per female in the definite period (7-8 days) upon exposure were used as an index of the effect. A fixed quantity of green unicellular alga (*Chlorella vulgaris*, Chlorella Industry Co., Ltd., Tokyo, Japan) and YCT (yeast, Cerophyll and trout chow) (Marinco Bioassay Laboratory, Inc., FL, USA) were fed every day. The cultures were performed at a water temperature of 26 ± 1°C under 16:8h of light and dark cycle. Homogenous populations of female offspring born within 24 h from mature individuals (age >7 days) were used for a test. *C. dubia* was exposed in semi-static condition for 7-8 days with three times water exchange per week. The filtered water was subjected to 2 fold serial dilution to form 6-7 different concentrations and 10 replicates/ concentration were tested for this study. The environmental water sample were examined without pH adjustment in the case of 6.5-8.5. Those falling out of range were adjusted with HCl or NaOH 1N. The lethality of the parent and the number of the neonates were counted every day and documented. When more than 60% of control population had lain their third batch of neonates, the test was terminated and added up number of offspring in litter size/ mother.

2.4.3 Green alga growth inhibition test

A stock (NIES -35) of Green unicellular alga *Pseudokirchneriella subcapitata* (former name *Selenastrum capricornutum* Printz) was used for the test. OECD-TG201 test that has been used in the Japanese "Act on the Evaluation of Chemical Substances and Regulation of Their Manufacture" was used to measure the algal growth inhibition. A test was conducted using a standard medium of the OECD. The filtered water samples were sterilized with 0.22 µm filter prior to incorporation in the medium. pH was also measured and adjusted to the range of 7.5-8.5 with NaOH or HCl.

The tests were conducted using following conditions: the initial cell density, 5×10^3 cells/ml; test temperature, 23±2 degrees Celsius; the light intensity of the flask surface neighborhood, 60-120 µmol/m^2/s; pH, 7.8-8.0; and 72 h concussion culture (100 rpm). The test concentration sets 5 steps or more in dilution ratio 2 or less, because these concentrations are necessary to detect NOEC (No Observed Effect Concentration)/LOEC (Lowest Observed Effect Concentration). The cell numbers were estimated using a particle counter for all test containers at 24 h interval. In case of abnormal cell count, the cells were critically assessed under microscope.

2.4.4 Statistics

All the data were analysed by one-way ANOVA (Analysis of Variance) at 95% confidence limit. Dunnett's test was performed to calculate the significance between different groups. When data showed nonequivalent dispersibility, an analysis of variance was carried out if dispersibility was satisfied by log tranformation. When equal dispersibility was not satisfied by the variable conversion, statistical test independ on distribution (e.g. Kruskal-Wallis or Dunn's test) was carried out. We also analyzed the statistical significance using the software JMP (Ver. 6.0.3, SAS Institute, Inc. North Carolina, USA).

3. The ecological effect of the pesticide

3.1 The bioassay using the environmental samples

Because the environmental water is already contaminated with a large number of herbicides, newly introduced pesticides may have little ecological effect; on the other hand, some may even become moderately to deleteriously toxic due to the factors such as the composition effect between pesticides and/or humic substances in the environment. Therefore, low detection level and relatively small ecological toxicity of herbicides does not always imply an ecologically undisturbed environment. Hence, further collection of data to understand the effects of herbicides on the environment.

For two years, from 2009 through 2010, chemical analysis and short-term chronic bioassays using an alga, a water flea, and a fish were performed with the agricultural drainage water collected from 14 different spots (A-N) of Japan.

The result of the chronic toxicity test in each species is shown in Table 1. The algal test showed evidence of effect in 9 spots. Of these, 3 spots, D, L, and N, showed TU (Toxicity Units) value as 10 or more. TU here is shown as 100/NOEC, and is directly proportional to ecological outcome. That is, if a value of TU is larger, it signifies that ecological effect is larger. The effect on water flea was observed in two spots, and only spot D had TU greater than 10; Spot D showed strong effects in both alga and water flea tests. The effect on hatching, survival rate of fish was checked as well, but no significant toxicity was found in any spots examined.

	NOEC (%)*			TU (=100/NOEC)		
	P. subcapitata	*C. dubia*	*D. rerio*	*P. subcapitata*	*C. dubia*	*D. rerio*
A	25	100	No data	4	-	No data
B	12.5	100		8	-	
C	50	100		2	-	
D	6.25	6.25		16	16	
E	25	100		4	-	
F	25	100		4	-	
G	25	100		4	-	
H	80	80	80	-	-	-
I	80	80	80	-	-	-
J	80	80	80	-	-	-
K	80	80	80	-	-	-
L	2.5	80	80	40	-	-
M	80	80	80	-	-	-
N	2.5	40	80	40	2.5	-
N (late 1 month)	80	80	80	-	-	-

* Sample water is considered to be 100 percent and shows the concentration of NOEC at percentage when it was diluted with unaffected water (e.g. culture water).

Table 1. Results of chronic toxicity test using the environmental sample.

3.2 Short-term chronic toxicity tests using candidate herbicides

Ten candidate herbicides, which were detected in the water samples, are typical herbicides used in Japan. Short-term chronic toxicity tests using alga, water flea and fish were carried out. As a note, it is important to collect data on each chemical substance accumulated in the environment before conducting mixed exposure of herbicides.

The results of the chronic toxicity tests in each species are shown in Table 2. In this study, the strongest effect was obtained in the alga test. Toxic effects on the alga were found in pretilachlor, cyanazine and simetryn at 1-10 mg/L.

The effects of herbicides on water flea were weaker than on alga, with the only exception of bendiocarb, which showed adverse effect on daphnid at 10 times lower concentrations compared to an alga. However, the effects of herbicides were generally found in order of alga> water flea> fish. In addition, bromobutide showed no adverse effects in all species tested.

	NOEC (mg/L)		
	P. subcapitata	*C. dubia*	*D. rerio*
Dimethametryn	0.005	0.63	2.5
Pretilachlor	0.001	1.3	1.5
Bromobutide	1	1.5	3
Carbetamide	0.063	2	10
Bendiocarb	0.078	0.006	12.5
Triazine	0.21	0.31	100<
Cyanazine	0.005	3.85	20.7
Simetryn	0.004	5.59	2.4
Esprocarb	0.028	0.27	0.37
Mefenacet	0.026	0.45	0.84

Table 2. Chronic toxicity tests of herbicides using the aquatic species.

3.3 Concentrations of herbicides in water samples in Japan

Simultaneous analysis of the GC-MS of the water samples from the bioassay was carried out. Ninety-two different herbicides were detected in water samples in this study, equivalent to approximately one-tenth of 917 kinds of analyzable pesticides in this analytical method. Actual concentrations of 10 typical herbicides are shown in Table 3a,b and Fig 3.

Bromobutide was detected at high concentration in each spot. Dimethametryn and pretilachlor were also detected from more than half of the samples. In addition, other herbicides were also detected at high concentration at spots D, G, K and N. At point N, water samples were collected again one month later, and were checked for the decrement of pesticides in the environment. Results showed a significant decrease in concentration of bromobutide, below detection limit value of other substances.

According to the resultant data, the herbicide including Bromobutide seemed to disappear immediately by disintegration or proliferation in the environment.

2009	MEC (μg/L)						
	A	*B*	*C*	*D*	*E*	*F*	*G*
Dimethametryn	0.65	0.71	0.64	3.33	0.11	0.17	0.17
Pretilachlor	0.40	0.56	0.13	1.66	5.98	-	3.23
Bromobutide	7.78	0.94	2.99	38.60	7.27	5.28	81.40
Carbetamide	0.85	-	-	1.35	-	-	3.56
Bendiocarb	0.16	-	0.26	5.68	-	-	-
Triazine	-	-	-	-	-	-	-
Cyanazine	-	-	-	-	-	-	-
Simetryn	-	-	-	-	-	-	-
Esprocarb	0.27	0.32	-	-	0.11	-	1.85
Mefenacet	0.92	-	1.89	-	1.32	0.21	0.65

-;not detected in the field.

(a)

2010	MEC (μg/L)							
	H	*I*	*J*	*K*	*L*	*M*	*N*	*N (30d after)*
Dimetha -metryn	0.05	0.04	0.13	0.03	-	-	0.03	-
Pretilachlor	-	-	-	-	-	-	-	-
Bromobutide	0.35	0.55	47.29	54.84	0.54	0.18	53.88	0.13
Carbetamide	-	-	0.83	1.66	-	-	1.64	-
Bendiocarb	-	-	-	-	-	-	-	-
Triazine	-	-	-	-	-	-	-	-
Cyanazine	-	-	-	-	-	-	-	-
Simetryn	0.27	0.17	-	-	-	-	-	-
Esprocarb	-	-	-	-	-	-	-	-
Mefenacet	-	-	-	-	-	-	0.03	-

-;not detected in the field.

(b)

Table 3. Actual concentrations of herbicides in the water samples by the simultaneous analysis (MEC; Measured Environmental Concentrations) in 2009 (a) and 2010 (b).

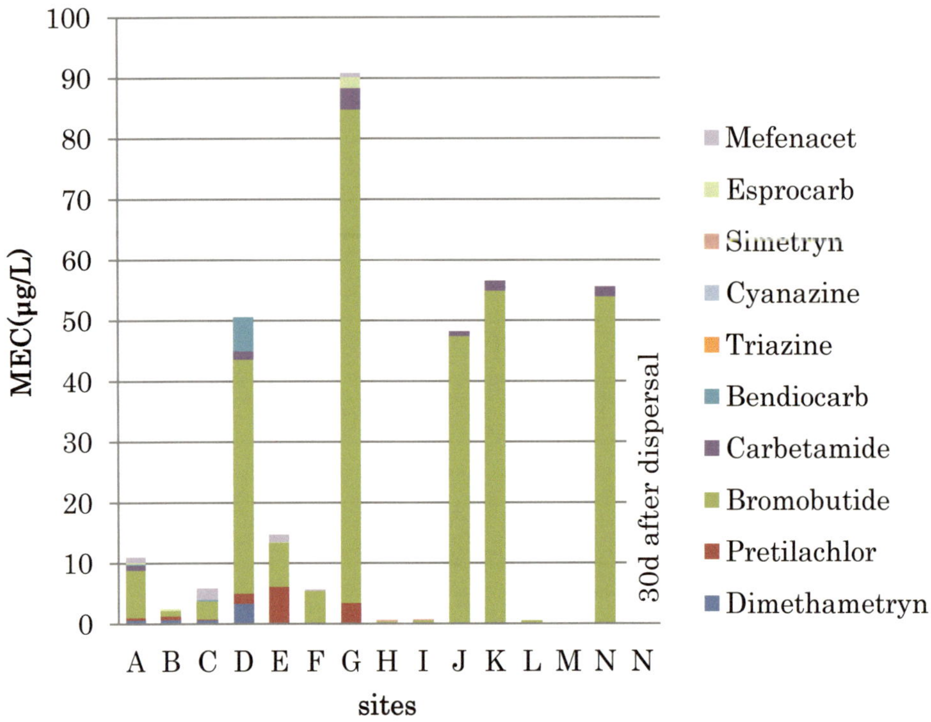

Fig. 3. Actual concentrations of herbicides in each sampling spot.

3.4 The risk evaluation of each spot based on the field measurements

Applying ecological toxicity data of 10 herbicides, a risk evaluation based on the MEC of each spot was performed. MEC/NOEC was calculated with the species that showed lowest NOEC for each target substance used in the present study (shown in Table 4 and Fig 4).

When the MEC/NOEC ratios of each herbicide are simply tallied, the total sums exceeded 1 in three spots (D, E and G). Yet from the results of Table 1, ecological effects at nine spots, A, B, C, D, E, F, G, L and N, are reported to be present. From the results of Table 4, at least at spots D, E, and G the ecological effect of the pesticide, which was measured in this study, is suspected to be present. Because Σ (MEC/NOEC) measured in this exposure was less than 1 in eleven spots of A, B, C, F, H, I, J, K, L, M and N, it is suggested that the observed effect may be attributed to a wholly different chemical substance, perhaps a herbicide that is unaccounted for in Table 4, or a non-pesticide chemical.

2009	MEC/NOEC						
	A	*B*	*C*	*D*	*E*	*F*	*G*
Dimetha -metryn	0.129	0.143	0.128	0.666	0.023	0.034	0.034
Pretilachlor	0.398	0.558	0.133	1.660	5.980	-	3.230
Bromobutide	0.008	0.001	0.003	0.039	0.007	0.005	0.081
Carbetamide	0.014	-	-	0.022	-	-	0.057
Bendiocarb	0.025	-	0.042	0.902	-	-	-
Triazine	-	-	-	-	-	-	-
Cyanazine	-	-	-	-	-	-	-
Simetryn	-	-	-	-	-	-	-
Esprocarb	0.010	0.013	-	-	0.004	-	0.066
Mefenacet	0.036	-	0.073	-	0.051	0.008	0.025
Σ(MEC/NOEC)	0.619	0.714	0.378	3.288	6.065	0.047	3.493

(a)

2010	MEC/NOEC							
	H	*I*	*J*	*K*	*L*	*M*	*N*	*N (30d after)*
Dimetha -metryn	0.01	0.008	0.026	0.006	-	-	0.006	-
Pretilachlor	-	-	-	-	-	-	-	-
Bromobutide	0.0004	0.0006	0.0473	0.0548	0.0005	0.0002	0.0539	0.0001
Carbetamide	-	-	0.013	0.027	-	-	0.026	-
Bendiocarb	-	-	-	-	-	-	-	-
Triazine	-	-	-	-	-	-	-	-
Cyanazine	-	-	-	-	-	-	-	-
Simetryn	0.075	0.047	-	-	-	-	-	-
Esprocarb	-	-	-	-	-	-	-	-
Mefenacet	-	-	-	-	-	-	0.001	-
Σ (MEC/NOEC)	0.085	0.056	0.087	0.087	0.001	0.0002	0.087	0.0001

(b)

Table 4. A risk evaluation based on the measurement value (MEC/NOEC) in 2009 (a) and 2010 (b).

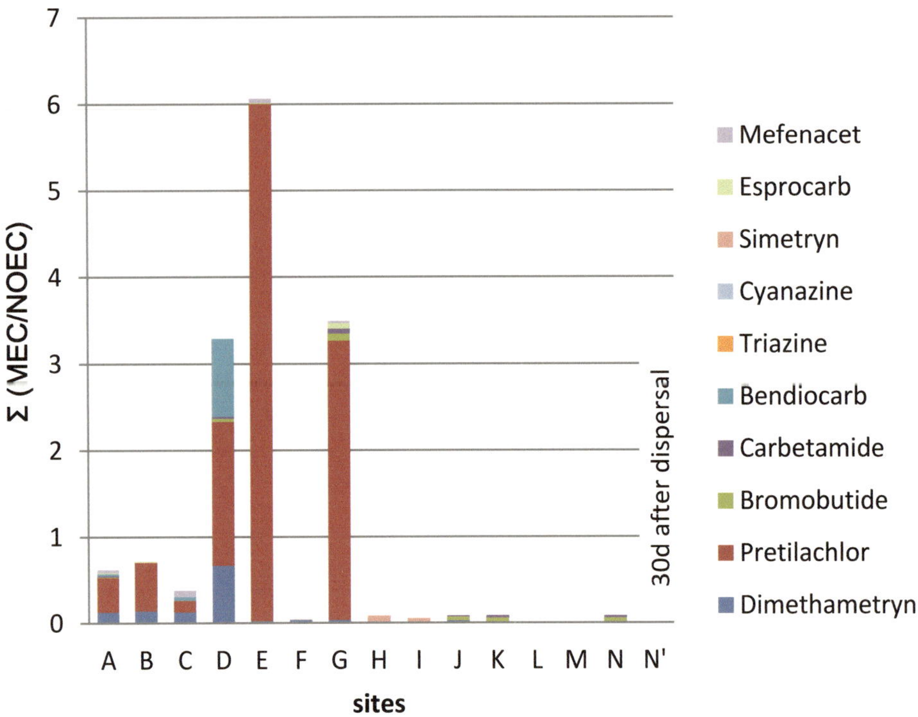

Fig. 4. Σ (MEC/ NOEC) of each sampling spot.

Three aquatic species were used for toxicity tests using dimethametryn, pretilachlor, cyanazine and simetryn, all of which showed growth inhibition of alga even at concentrations lower than 10 µg/L. The aquatic species most strongly affected by these herbicides was the alga in the present study. Separately, in bendiocarb the highest toxicity was encountered in the crustacean, decreasing the number of offspring at 12.5 µg/L and was 10 times more sensitive compared to alga. Daphnia had the highest sensitivity to bendiocarb. In summary, 100-1000 times differences in toxicity of various herbicides were encountered. The fish were far less sensitive to toxicity of herbicides than alga, similarly or less sensitive than daphnid in this test. Though fish were shown to be less sensitive, the pesticide dispersion period in Japanese farm occurs during same time as spawning and/or hatching period in wildlife. Therefore, accumulation of toxicity data including fish is needed to perform a more detailed evaluation of ecological risk of herbicides. In addition, accumulation of chronic data of herbicides using the aquatic species is also needed to protect wildlife and the ecosystem.

4. A green alga and a blue-green alga

Relying solely upon green alga for risk evaluation and analysis of herbicide effect is not only insufficient for proper analysis, but may also lead to bias and error. For example, the effect of a chemical substance on germination and rooting cannot be evaluated because the green alga is a unicellular organism. Furthermore, different toxicity for various organism species has been reported in some herbicides (Suárez-Serrano et al., 2010; Roubeix et al., 2011; Pereira et al., 2009). In other words, a herbicide may have a selective property; imposing no effect on growth of agricultural crops, yet able to effectively inhibit weeds growth e.g., the ineffective to rice and effective to wild millet. *Lemna* sp. Growth Inhibition Test (OECD TG221, 2006) can be used in addition to green alga toxicity test; however, herbicides toxicity data using duckweed are limited at present.

Blue-green alga (*Synechococcus leopoliensis*) has been used as a test species in addition to the green alga (*P. subcapitata*), and compared for herbicide toxicity. Because the blue-green alga is also a single cell organism, it can only serve as a biological reference to show species specific difference (Kaur et al., 2002; Vaishampayan et al., 1984; Lehmann-Kirk et al., 1979). Differences in toxicity effect between the green alga and blue-green alga using eight kinds of pesticides are shown in Figure 5.

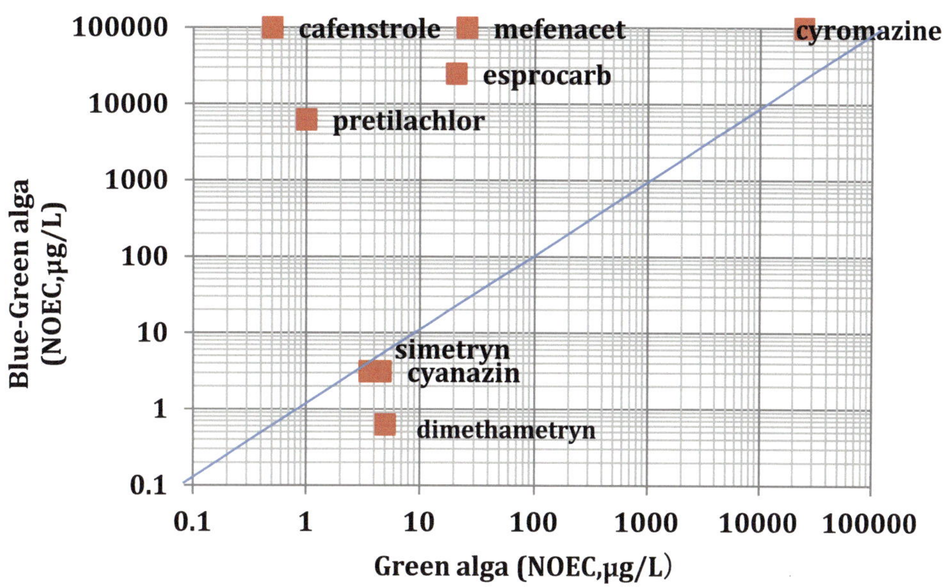

Fig. 5. Comparison of herbicide toxicity using green alga and blue-green alga.

Correlation of herbicide toxicity was hardly shown between green alga and blue-green alga (Fig. 3). However, the green alga and blue-green alga displayed approximately similar sensitivity to simetryn, cyanazin, and cyromazine. The green alga showed susceptible sensitivity in the toxicity other than dimethametryn. The green alga has been commonly used for ecological risk evaluation of chemicals including herbicides; however, it is also necessary to accumulate the test data using multicellular plants such as floating weeds in the future.

5. Conclusions

Fate of herbicides after their release into the environment is extremely difficult to grasp precisely. Regarding the adverse effects of herbicides on the environment (water, soil and air contamination from leaching, runoff, and spray drift, as well as the detrimental effects on wildlife, fish, plants, and other non-target organisms), the well being of resulting environmental state depends on the toxicity of the herbicides themselves(Monaco et al., 2002; Eleftherihorinos, 2008). Detailed information will be needed concerning measurements of exposure levels of herbicides during their application, the dosage applied, the adsorption on soil colloids, the weather conditions prevailing after application, and pesticide persistence in the environment.

As for the risk assessment of the impact of herbicides on the environment, a simple and precise process does not exist (Commission of the European Communities, 1991; EPA, 2009; FAO, 2002; Abrantes et al., 2009). Various examples point to multivariate ecological effect based on various environments, and the ecological risk changes on a case-by-case basis. Hence, we need to instead depend upon data gained through exposure periods and exposure levels, toxicity and the durability of applied herbicides, as well as taking in account the local environmental characteristics for proper risk evaluation of herbicides.

It has been recognized, however, that an impact on the environment of herbicides included in the agriculture drainage could be estimated to some extent by performing short-term chronic toxicity tests (Cantelli-Forti et al., 1993). The ecological toxicity tests may detect the effect of not only herbicides but also the chemical substances used for daily life and sewage effluents. For consideration of environmental risk of chemicals in general, synergistic effects with herbicides and other substances should be detected. The monitoring of the environmental water using the aquatic species will become an important index for the chemical safety and control of environmental chemicals including herbicides.

6. Acknowledgments

Part of the data used here was carried out as a government-funded research sponsored by the Agricultural Chemicals Control Office of the Ministry of the Environment. We thank Dr. Tapas Chakraborty, National Institute for Basic Biology, Japan for his critical reading of this manuscript.

7. References

Abrantes N, Pereira R, de Figueiredo DR, Marques CR, Pereira MJ, Gonçalves F. (2009). A whole sample toxicity assessment to evaluate the sub-lethal toxicity of water and sediment elutriates from a lake exposed to diffuse pollution. Environ Toxicol. 24(3): 259-70.

Akiyama Y, Yoshioka N, Ichihashi K. (2005). Study of pesticide residues in agricultural products for the "Positive List" system. Shokuhin Eiseigaku Zasshi 46(6): 305-18.

Añasco N, Uno S, Koyama J, Matsuoka T, Kuwahara N. (2010). Assessment of pesticide residues in freshwater areas affected by rice paddy effluents in Southern Japan. Environ Monit Assess. 160(1-4): 371-83.

Cantelli-Forti G, Paolini M, Hrelia P. (1993). Multiple end point procedure to evaluate risk from pesticides. Environ Health Perspect. 101(Suppl 3): 15–20.

Commission of the European Communities. (1991). Council Directive 91/414/EEC of 15 July 1991 Concerning the Placing of Plant Protection Products on the Market; Official Journal L 230; Commission of the European Communities: Brussels, Belgium.

Cooper, J.; Dobson, H. (2007). The benefits of pesticides to mankind and the environment. Crop Prot., 26: 1337-1348.

Damalas, C.A. (2009). Understanding benefits and risks of pesticide use. Sci. Res. Essays, 4: 945-949.

Eleftherohorinos, I.G. Weed Science. (2008). Weeds, Herbicides, Environment, and Methods for Weed Management; AgroTypos: Athens, Greece.

Environment Canada (2007). Biological Test Method: Test of Reproduction and Survival Using the Cladoceran *Ceriodaphnia dubia*. 74pp.

EPA. Registering Pesticides (2009). Available online: http://www.epa.gov/pesticides/regulating/ re-gistering/index.htm (accessed on 1 Sep. 2011).

EPA Biological Test Method (2007): Test of Reproduction and Survival Using the Cladoceran ('*Ceriodaphnia dubia*') EPS1/RM/21, http://www.ec.gc.ca/Publications/AB93350E-9791-487E-81DB-E784433B2419/21--NO-HL.pdf (accessed on 1 Sep. 2011).

FAO. (2002). Manual on the Submission and Evaluation of Pesticide Residues Data for the Estimation of Maximum Residue Levels in Food and Feed; Food and Agriculture Organization: Rome, Italy.

Hirahara Y. (2007). Status of inspection of imported foods after introduction of the positive list system for agricultural chemical residues in foods mainly about analysis of pesticide residue in foods. Shokuhin Eiseigaku Zasshi. 48(4): J299-302.

Japan Plant Protection Association. (2010). Annual Inventory of Registered Pesticides and Their Use (in Japanese).

Kadokami K, Tanada K, Taneda K, Nakagawa K. (2005). Novel gas chromatography-mass spectrometry database for automatic identification and quantification of micropollutants. J Chromatogr A. 1089 (1-2): 219-226.

Kang HS, Park CJ, Gye MC. (2009). Effects of molinate on survival and development of *Bombina orientalis* (Boulenger) embryos. Bull Environ Contam Toxicol. 82(3): 305-309.

Kim Y, Jung J, Oh S, Choi K. (2008). Aquatic toxicity of cartap and cypermethrin to different life stages of *Daphnia magna* and *Oryzias latipes*. J Environ Sci Health B. 43(1): 56-64.

Lehmann-Kirk U, Bader KP, Schmid GH, Radunz A. (1979). Inhibition of photosynthetic electron transport in tobacco chloroplasts and thylakoids of the blue green alga *Oscillatoria chalybea* by an antiserum to synthetic zeaxanthin. Z Naturforsch C. 34(12): 1218-1221.

Maitani T. (2005). Introduction of the "Positive List" system for agricultural chemicals in foods and development of analytical methods --recent amendment of regulation on pesticides by the Ministry of Health, Labour and Welfare, the Ministry of Agriculture, Forestry and Fisheries, and the Ministry of the Environment. Shokuhin Eiseigaku Zasshi. 46(6): J327-334.

Maitani T. (2007). Notification of the "positive list system for agricultural chemicals in foods" and subsequent management. Shokuhin Eiseigaku Zasshi. 48(6): J402-410.

Marques CR, Pereira R, Antunes SC, Cachada A, Duarte AC, Gonçalves F. (2011). In situ aquatic bioassessment of pesticides applied on rice fields using a microalga and daphnids. Sci Total Environ. 409(18) : 3375-3385.

Matthews, G.A. (2006). *Pesticides: Health, Safety and the Environment*; Blackwell Publishing: Oxford, UK.

Ministry of Agriculture, Forestry and Fisheries. (2006). The Census of Agriculture and Forestry 2005.

Ministry of Agriculture, Forestry and Fisheries. (2011). The Census of Agriculture and Forestry 2010., http://www.maff.go.jp/j/tokei/census/afc/about/pdf/kakutei_zentai.pdf (accessed on 1 Sep. 2011).

Ministry of Agriculture, Forestry and Fisheries. (2010). Statistics about the GDP (gross domestic product), http://www.maff.go.jp/j/tokei/sihyo/data/01.html (accessed on 1 Sep. 2011).

Ministry of the Environment. (2006). Actual situation research report for the aquatic effect of the pesticide, (an independent administrative agency National Institute for Environmental Studies version) (in Japanese).

Ministry of the Environment. (2007). Actual situation research report for the aquatic effect of the pesticide, (an independent administrative agency National Institute for Environmental Studies version. (in Japanese).

Ministry of the Environment. (2008). Actual situation research report for the aquatic effect of the pesticide, (an independent administrative agency National Institute for Environmental Studies version). (in Japanese).

Ministry of the Environment. (2009). Actual situation research report for the aquatic effect of the pesticide, (an independent administrative agency National Institute for Environmental Studies version). (in Japanese).

Ministry of the Environment. (2010). Actual situation research report for the aquatic effect of the pesticide, (an independent administrative agency National Institute for Environmental Studies version). (in Japanese).

Miyashita S, Shimoya M, Kamidate Y, Kuroiwa T, Shikino O, Fujiwara S, Francesconi KA, Kaise T. (2009). Rapid determination of arsenic species in freshwater organisms from the arsenic-rich Hayakawa River in Japan using HPLC-ICP-MS. Chemosphere 75(8): 1065-1073.

Monaco, J.T.; Weller, S.C.; Ashton, F.M. (2002). Herbicide registration and environmental impact. In Weed Science: Principles and Practices, 4th ed.; Monaco, T.J., Weller, S.C., Ashton, F.M., Eds.; John Wiley & Sons: New York, NY, USA.

Nagayama T. (2010). Regulation of the pesticide residues in foods and the positive list system. Shokuhin Eiseigaku Zasshi 51(6): 340-348.

Nakano Y, Miyazaki A, Yoshida T, Ono K, Inoue T., (2004). A study on pesticide runoff from paddy fields to a river in rural region--1: field survey of pesticide runoff in the Kozakura River, Japan. Water Res. 38(13): 3017-3022.

Numabe A, Nagahora S. (2006). Estimation of pesticide runoff from paddy fields to rural rivers. Water Sci Technol. 53(2): 139-146.

OECD Environmental Performance Reviews: Japan (2002). Organisation for Economic Co-Operation and Development, Organization for Economic (2002/05).

OECD TG 201, (2002), OECD Guideline No. 201. (2002). Freshwater Alga and Cyanobacteria, Growth Inhibition Test, OECD Guidelines http://www.oecd.org/dataoecd/58/60/1946914.pdf (accessed on 1 Sep. 2011).

OECD TG 221, (2006). OECD Guideline No. 221. (2006). OECD Lemna sp. Growth Inhibition Test.

Oerke, E.C.; Dehne, H.W. (2004). Safeguarding production-losses in major crops and the role of crop protection. Crop Prot. 23: 275-285.

Pereira JL, Antunes SC, Castro BB, Marques CR, Gonçalves AM, Gonçalves F, Pereira R. (2009). Toxicity evaluation of three pesticides on non-target aquatic and soil organisms: commercial formulation versus active ingredient. Ecotoxicology18(4): 455-463.

Roubeix V, Mazzella N, Schouler L, Fauvelle V, Morin S, Coste M, Delmas F, Margoum C. (2011). Variations of periphytic diatom sensitivity to the herbicide diuron and relation to species distribution in a contamination gradient: implications for biomonitoring. J Environ Monit. 13(6): 1768-1774.

Saito I . (2007). The viewpoint and approach for regulation of Japanese positive list system for agricultural chemical residues in foods as one cooperative federation. Shokuhin Eiseigaku Zasshi 48(4): J291-295.

Sakai M. (2002). Use of chronic tests with *Daphnia magna* for examination of diluted river water. Ecotoxicol Environ Saf. 53(3): 376-381.

Suárez-Serrano A, Ibáñez C, Lacorte S, Barata C. (2010). Ecotoxicological effects of rice field waters on selected planktonic species: comparison between conventional and organic farming. Ecotoxicology 19(8): 1523-1535.

Sudo M, Kunimatsu T, Okubo T. (2002). Concentration and loading of pesticide residues in Lake Biwa basin (Japan). Water Res., 36(1): 315-329.

Tanabe A, Mitobe H, Kawata K, Yasuhara A, Shibamoto T. (2001). Seasonal and spatial studies on pesticide residues in surface waters of the Shinano river in Japan. J Agric Food Chem. 49(8): 3847-3852.

Tsuda T, Igawa T, Tanaka K, Hirota D. (2011). Changes of concentrations, shipment amounts and ecological risk of pesticides in river water flowing into lake Biwa. Bull Environ Contam Toxicol., 87(3): 307-311.

Vaishampayan A. (1984). Powerful mutagenicity of a bipyridylium herbicide in a nitrogen-fixing blue-green alga *Nostoc muscorum*. Mutat Res., 138(1): 39-46.

Vu SH, Ishihara S, Watanabe H. (2006). Exposure risk assessment and evaluation of the best management practice for controlling pesticide runoff from paddy fields. Part 1: Paddy watershed monitoring. Pest Manag Sci. 62(12): 1193-1206.

Wei D, Lin Z, Kameya T, Urano K, Du Y. (2008). Application of biological safety index in two Japanese watersheds using a bioassay battery. Chemosphere 72(9): 1303-1308.

Wilson, C.; Tisdell, C. (2001). Why farmers continue to use pesticides despite environmental, health and sustainability costs. Ecol. Econ. 39: 449-462.

WHO (2010). International Code of Conduct on the Distribution and Use of Pesticides: Guidelines for the Registration of Pesticides; World Health Organization: Rome, Italy.

2

Herbicide Phytotoxicity and Resistance to Herbicides in Legume Plants

Agnieszka I. Piotrowicz-Cieślak[1] and Barbara Adomas[2]
[1]*Department of Plant Physiology and Biotechnology,*
[2]*Department of Air Protection and Environmental Toxicology,*
University of Warmia and Mazury in Olsztyn
Poland

1. Introduction

Active substances in herbicides, just like in other pesticides, are chemical compounds synthesized in order to kill organisms which are harmful for cultivated plants. Therefore, they are toxins introduced on purpose by man into the environment. From the perspective of environmental protection, it is very significant that herbicides are most often applied directly into the soil to manage weeds. Since DDT and chloro-organic herbicides such as 2,4,5-T were withdrawn (in the 1970s) and since the EU regulations were unified for all its member countries, plant protection techniques have advanced considerably. Yet, pesticides, thus herbicides as well, continue to be a big group of xenobiotics periodically occurring at high levels in agroecosystems. These compounds infiltrate into related biocenoses from air, soil, water and food (Allinson & Morita, 1995; Kolpin et al., 1998; Adomas at al., 2008). Soil may become a reservoir of various pollutants, including herbicides. Herbicides remain active in soil for different periods. Paraquat has a relatively long half-life in soil (estimated at about 1000 days). The half-life of glyphosate in soil is only 10 to 100 days, and according to Monsanto the average half-life of this herbicide is 32 days (Hornsby et al., 1996; Monsanto, 2005). Remainders of persistent herbicides (e.g. atrazine, metribusin, and trifluralin) can stay in soil and destroy subsequent plantations a year or more after herbicides had been used. Herbicides from soil leach into surface water and ground water. The assessment of herbicides content in the aquifers in Iowa shows that 75% of herbicides (Kolpin et al., 1998), despite degradation, are still detected. From soil, water or air, herbicides get into crops (Adomas et al., 2008). When pesticides are applied, acceptable remainders of active substances (MRL) can often be detected in cultivated plants. Depending on physicochemical properties of the active substances of pesticides and the ways of their detoxification, some of these pollutants tend to increase concentration while passing through organisms of higher trophic levels. It can lead to a significant bioaccumulation of toxins in the food chains (Allinson & Morita, 1995; Dinis-Oliveira et al., 2006). No doubt therefore, monitoring of herbicide (including desiccant) residues in cultivated plants is needed, so that people and environment can be safe. Moreover, application of herbicide desiccants modifies physiological properties of seeds and may thus lead to delayed problems, becoming evident long after the treatment.

2. Physiological changes after preharvest desiccation of plants

Preharvest application of herbicides (desiccants, e.g. glufosinate ammonium (Basta 200 SL) or diquat (Reglone Turbo 200 SL)), is aimed at removing excessive water from plant cells and balancing maturation. Dehydration is best tolerated by plant reproductive organs/cells: seeds, pollens, spores, aestive buds and somatic embryos. In the developmental cycle, the majority of seeds go through the stage of natural, physiological drying, yet not all the seeds dry. Based on their tolerance to desiccation, seeds have been divided into three groups: orthodox (which can be chemically desiccated, Fig. 1), recalcitrant and intermediate (Piotrowicz-Cieślak et al., 2007). Orthodox seeds tolerate dehydration relatively well and may retain their viability for a long time (Murthy et al., 2003).

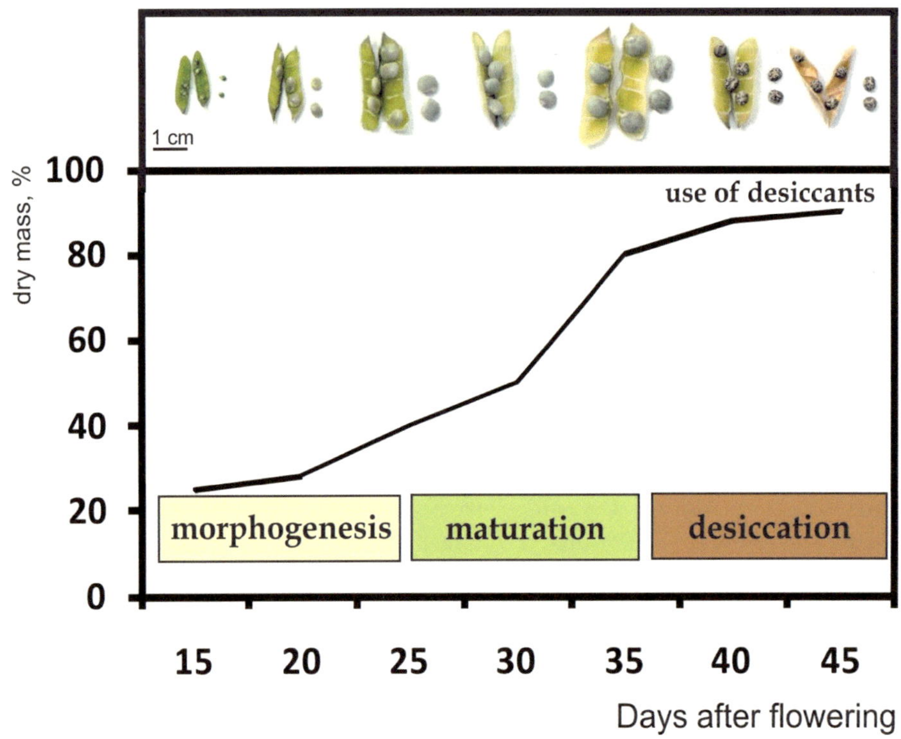

Fig. 1. Basic stages of seed ontogenesis and dry mass accumulation in yellow lupin seeds. Chemical desiccation is carried out during natural dehydration of seeds.

They are resistant to water loss below 5% of the initial water content. During desiccation, orthodox seeds can lose over 90% of water, and as a result slow down their metabolism and become dormant (Adams & Rinne, 1980). Seeds of this type are produced by most monocots and dicots of the temperate climate zone (Walters & Towill, 2004). Seed desiccation is a prerequisite for completion of their life cycle and it can be chemically forced. Recalcitrant seeds perish as a result of drying below a critical level of water, which is usually quite high and amounts to 40% (Roberts, 1973). Desiccation cannot be applied in this group of plants which includes tropical trees, e.g. avocado, mango, litchi, cocoa (Farrant et al., 1993) and some grasses (Probert & Longley, 1989). The third group - i.e. intermediate (suborthodox) seeds are tolerant to dehydration if it stops at seed water level 6-10% (Ellis et al., 1991). It seems that chemical desiccation can be successfully applied in this group, however mild (with small dessicant doses). Moreover, intermediate seeds do not tolerate cold while being dehydrated (Hong & Ellis, 1996). Chemical desiccation of plants is performed during the period when orthodox seeds start their natural drying. Yet, precise determination of this period depends on the species, the pace of water loss and its expected final content. Many processes contribute to preservation of seed vigour and viability during both natural and accelerated (chemical) drying.

2.1 Biochemical changes in seeds resulting from desiccant application

The rate of water loss during seed desiccation has a profound effect on seed condition at cellular and subcellular levels. Organelles in the cytoplasm become compacted together, cytoplasm viscosity increases and a number of degenerative reactions intensify. In order for the cell to protect itself against dehydration, it has to accumulate protective compounds. A wide group of such substances was detected in seeds, for example proline, glutamine, fructans, poliols, trehalose, carnitine and others. A lot of plants respond to desiccation caused dehydration by accumulating so-called 'compatible solvents' (i.e. molecules not interfering with the structure and functions of the cell), including one kind or more of the above given substances. Unfortunately, concentration of these hydrophilic compounds is often not sufficient to bind water properly (Hare et al., 1998). Yet, probably it is one of the most important mechanisms protecting macromolecules under conditions of limited water loss. Molecules of compatible compounds form a protective layer around proteins, thus preventing protein deformations (Crowe et al., 1990). During desiccation, changes occur at all levels of plant cell functions. Maintaining the integrity of genetic material and keeping the DNA repair mechanisms functional during dehydration are the most important for seed survival. As a result of water loss during seeds desiccation the genetic material undergoes conformational changes, depending on its nucleotide sequence and interactions with specific DNA-binding proteins (Osborne & Boubriak, 1994). A higher stability of seed DNA structure probably indicates increased seed tolerance to desiccation. At the early stages of seed ontogenesis, desiccation leads to frequent chemical modifications of DNA bases, resulting in a modified DNA methylation level. The later chemical desiccation is applied in the ontogenetic development of seeds, the fewer epigenetic changes are thus caused.

2.1.1 Soluble carbohydrates and their derivatives content in seeds

Yellow lupin plants (*Lupinus luteus* L. cv. Taper and Mister) were grown in 10-L pots in greenhouse (Fig. 2) with a 12-h photoperiod at 20°C day/18°C night and 140 μmoles photons m^{-2} s^{-1} irradiace. Mixture of peat, garden soil and sand (1:1:1, v:v:v) was used as substrate for plant growth. On the day of flowering the plants were labelled and divided into three groups: a control and two treatments. Basta 200 SL (producer Bayer Poland, active substance (a.i.) glufosinate ammonium) was applied in the amount of 4,1 μg a.i. per pot (the preparation was diluted in 15 ml of distilled water). Roundup Ultra 360 SL (producer Monsanto Poland, a.i. glyphosate) was applied in the amount of 4,1 μg a.i. per pot (here the preparation was diluted in 15 ml of distilled water too). Herbicide levels applied in this experiment corresponded to the field doses 2.5 and 3 L/ha, respectively.

Fig. 2. Lupin plants and seeds during and two weeks after desiccation treatment.

Seeds were collected in five-day intervals, starting from 15 days after flowering (DAF) until full physiological seeds maturity. Soluble carbohydrates content in seeds were analysed by GC chromatography according to Piotrowicz-Cieślak (2005). Seeds (30 - 60 mg fresh mass) were homogenised in ethanol: water, 1:1 (v/v) containing 300 µg phenyl-α-D-glucose as internal standard. The homogenate and the wash were combined in a 1.5 ml microfuge tube, heated at 75°C for 30 min to inactivate endogenous enzymes and centrifuged at 15 000 g for 20 min. The supernatant was passed through a 10 000 MW cut-off filter (Lida, Kenosha, WI USA). Aliquots of 0.3 ml filtrate were transferred to silylation vials and evaporated to dryness (Eppendorf Concentrator 5301). Dry residues were derived with 300 µl of silylation mixture (trimethylsilylimidazole : pyridine, 1:1, v/v) in silylation vials (Thermo Scientific) at 70°C for 30 min, and then cooled at room temperature. One µl carbohydrate extract was injected into a split-mode injector of a Thermo Scientific gas chromatograph equipped with flame ionisation detector. Soluble carbohydrates were analysed on a DB-1 capillary column (15 m length, 0.25 mm ID, 0.25 µm film thickness, J&W Scientific). Soluble carbohydrates were identified with internal standards as available, and concentrations were calculated from the ratios of peak area, for each analysed carbohydrate, to the peak area of respective internal standard. Quantities of soluble carbohydrates were expressed as mean ± SD for 3-5 replications of each treatment.

2.1.2 The role and content of soluble carbohydrates in seed drying

The composition of carbohydrates, with particular regard to content of raffinose family oligosaccharides, has been measured in seeds of many species. Raffinose content exceeds 1 % d.m. in seeds of bean, soybean and pea. The high content of stachyose (over 4 % d.m.) is found in vetch and soybean seeds, whereas verbascose content over 2 % d.m. was measured in bean and pea (Horbowicz & Obendorf, 1994). Soluble carbohydrates content in lupin seeds varies from 6 to 15.6 % d.m. (Piotrowicz-Cieślak et al., 1999; Piotrowicz-Cieślak, 2005). Raffinose family oligosaccharides (RFO) content in the overall content of soluble carbohydrates in lupin seeds amounts to 46 to 76 %, with the prevalence of stachyose. A common characteristic of lupin seeds is that the RFO content in embryonic axes is two times higher than in cotyledons.

This feature, connected with higher metabolic activity of embryonic axes than cotyledons, was also observed in all other legume seeds studied (Horbowicz & Obendorf, 1994; Górecki et al., 1997). Oligosaccharide contents in seeds depend on the rate of seed maturation (Piotrowicz-Cieślak et al., 2003). In maturing seeds of lupin, Taper and Mister varieties, accumulation of fresh and dry mass was at first slow, and then it rapidly increased between the 25th and 35th day after flowering. The first symptoms of seed viability were observed on the 20th day after flowering. In the period of natural seed desiccation, the seed mass increased rapidly. In both lupin varieties, dry mass rapidly increased after applying desiccants, which resulted in an decrease in fresh mass. Quick desiccation also increased the electrical conductivity of seed exudates. Yet, such accelerated maturation does not have any significant impact on seed germination (Fig. 3). A properly conducted desiccation is of key importance for the maturing seeds to acquire vigour (Sanhewe & Ellis, 1996). Both these phenomena are dependent on biosynthesis of specific stress-related proteins, induced by ABA (Late Embryogenesis Proteins, Blackman et al., 1992) and accumulation of a considerable level of soluble carbohydrates in seeds. The assessment of soluble carbohydrates accumulation has been carried out in maturing seeds of *L. luteus* cv. Taper and Mister. The length of maturation period, from flowering to full maturity of seeds, was 45 days. Seeds accumulated monosugars (glucose, fructose, galactose) sucrose, cyclitols, raffinose family oligosaccharides and galactosyl cyclitols.

In the initial phase of seeds development (15-20 days after flowering) the monosugars content (fructose, glucose and galactose) was high (Fig. 4) and gradually growing, reaching the maximum on the 30th day after flowering (DAF). At full seed maturity monosugars were present in negligible amounts. Upon chemical drying the content of monosugars decreased rapidly, particularly in the first five days after desiccants application. The rate of the decrease in monosugars content was similar when Basta and Roundup were applied (Fig. 4).

The chemical drying of seeds facilitated increase of cyclitols content, particularly *myo*-inositol. The content of sucrose in the initial phase of seeds development was high. At the beginning of desiccation the sucrose level decreased, reaching the minimum at full physiological seed maturity (45 DAF).

Chemical drying induced sucrose synthesis in seeds (Fig. 5). Among galactosyl cyclitols: galactinol, ciceritol and trigalactopinitol A, galactopinitol B and digalacto-*myo*-inositol were present in the highest amounts. In seeds which were not chemically dried the content of galactinol increased to reach the maximum at 30 DAF. Its maximum concentration in seeds desiccated with Basta and Roundup dropped five days after treatment. In the course of seeds drying up, the content of galactinol decresed, reaching the minimum level in the naturally growing, fully mature seeds. In the chemically dried seeds the content of galactinol was higher; also, more intense synthesis of ciceritol and trigalactopinitol A was found after chemical drying of seeds (Fig. 5).

The dominant reserve carbohydrates in lupin seeds were raffinose family oligosaccharides (Fig. 6). The level of these metabolites gradually increased in the course of seeds desiccation, with particular intensity in chemically dried seeds. Stachyose was the member of raffinose family oligosaccharides that occurred in the highest concentration. In mature, non-desiccated seeds its level was 43.17 mg/g d.m. In the chemically dried seeds, after the application of both Basta and Roundup, the level of stachyose was higher and persisted until the end of the physiological maturity of seeds (Fig. 6).

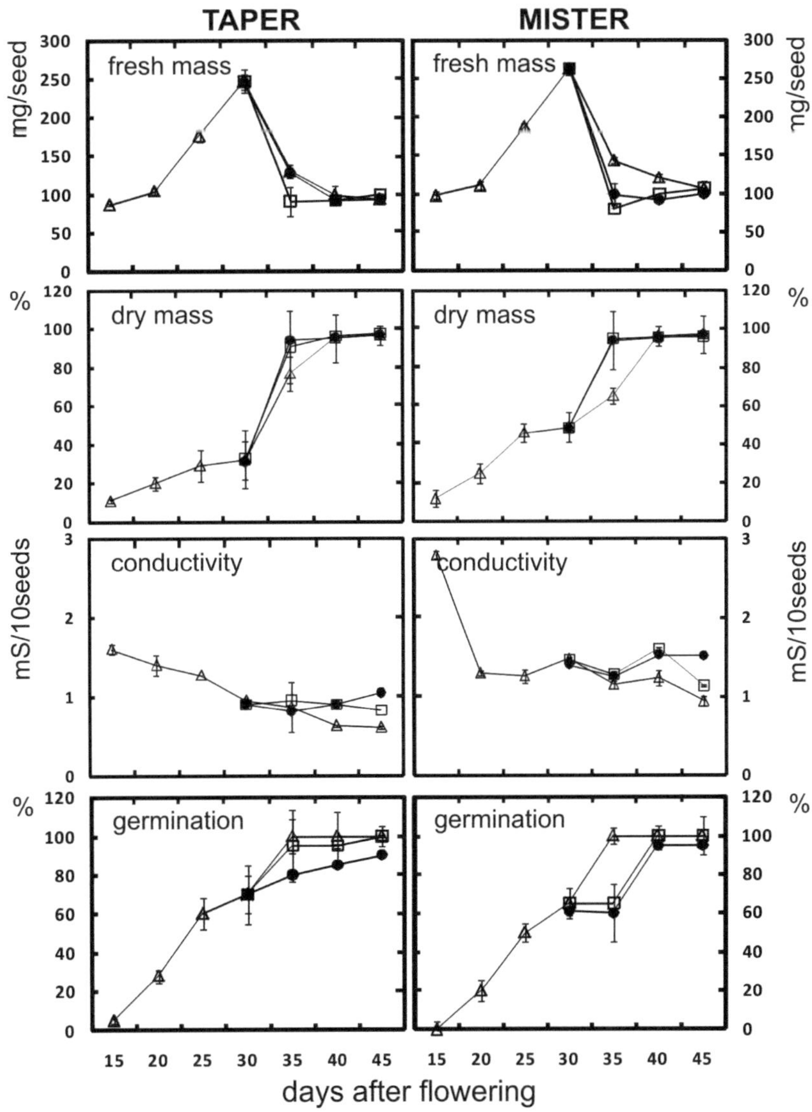

Fig. 3. Seed dry and fresh mass, conductivity and germination during maturation and after application of herbicides: Basta 200 SL and Roundup Ultra 360 SL. Data points represented the mean ± SD for fifteen replicate samples - control (Δ), Basta treatments (□), Roundup treatments (•).

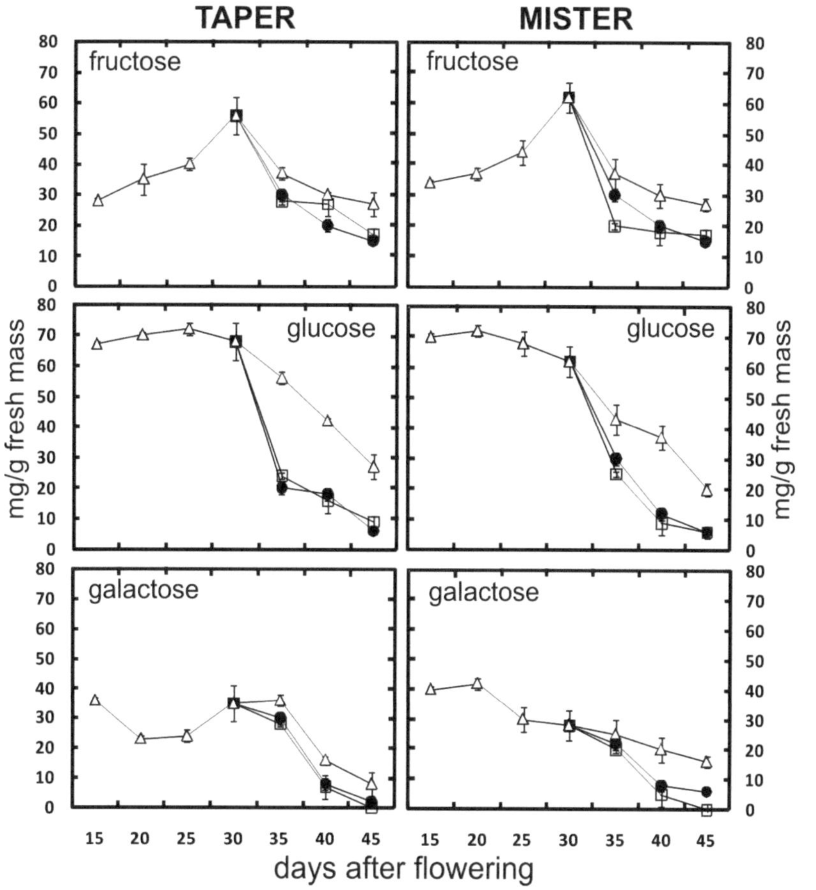

Fig. 4. Monosugars content in lupin seeds during maturation and after application of herbicides: Basta 200 SL and Roundup Ultra 360 SL. Data points represented the mean ± SD for fifteen replicate samples - control (Δ), Basta treatments (□), Roundup treatments (•).

The authors of the papers published so far on the accumulation of RFO in seeds come to unequivocal conclusion that the drying induces the accumulation of RFO in embryonic tissues. The increases in the RFO contents are paralleled by increasing seed resistance to desiccation (Obendorf, 1997). The protective action of RFOs results from their role in inducing the cytoplasm vitrification and stabilization of membranes and macromolecules in dehydrated cells (Bernal-Lugo &Leopold, 1985). In the course of seed drying (chemical or natural) accumulation of osmoprotective substances takes place, such as proline, betaine, oligosaccharides including RFO, cyclitols and their derivatives (Carpenter & Crowe, 1989; Ramanjulu & Bartels, 2002; Piotrowicz-Cieślak et al., 2007). Our research demonstrates that the dominant soluble carbohydrates in maturing embryos of *L. luteus* were raffinose family oligosaccharides and galactosyl cyclitols.

Seeds accumulated significant amounts of stachyose, verbascose, trigalactopinitol A and B. Soluble carbohydrates characterised by high molecular weight (stachyose, verbascose) belong to the main osmoprotectors in seeds of lupin. They also contain several hydroxyl groups (seven in the case of verbascose). Polyhydroxy compounds can substitute for water

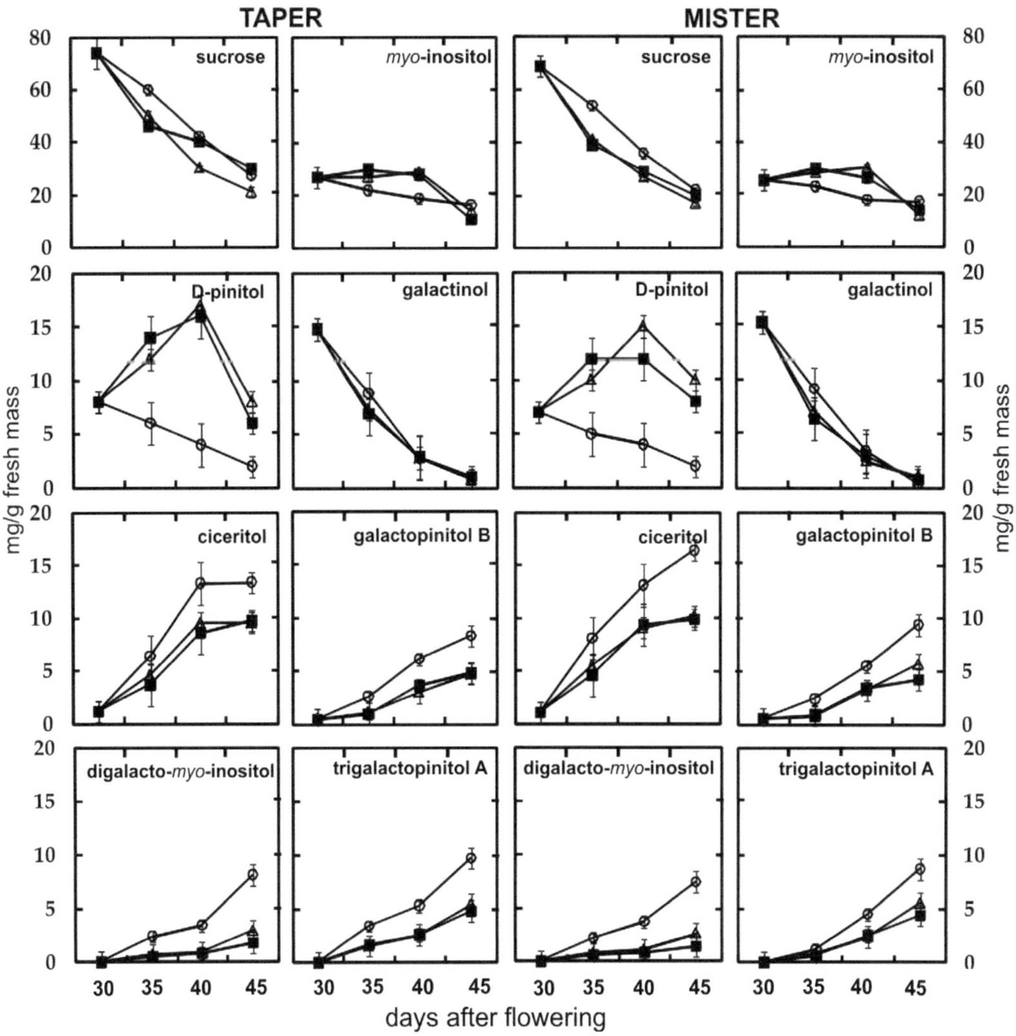

Fig. 5. Soluble carbohydrates content [mg/ g fresh mass] in lupin seeds during maturation and after application of herbicides: Basta 200 SL and Roundup Ultra 360 SL. Data points represented the mean ± SD for ten replicate samples - control (o), Basta treatments (Δ), Roundup treatments (■).

in stabilizing membrane structure in the dry state. Trehalose is indicated as the optimal osmoprotector; it contains 8 hydroxyl groups and has molecular weight 2.5 times lower (Crowe & Crowe, 1984) than verbascose. Lupin seeds do not accumulate trehalose, but similar functions can be performed in them by other soluble carbohydrates. Bianchi et al. (1993) point out that tissues should contain from 10 to 15 % of sucrose or 5g sucrose for every gram of lipid to tolerate the drying well. Lupin embryos contain less than 3 % of sucrose, but they tolerate chemical drying relatively well. Therefore, it is the total content of all soluble carbohydrates, including sucrose, that likely becomes significant. Chemical drying of seeds enhances the increase of soluble carbohydrates content in lupin embryos –

raffinose family oligosaccharides and, to a smaller extent, galactosyl cyclitols. Thus, galactosyl cyclitols are probably important osmoprotective agents in lupin seeds.

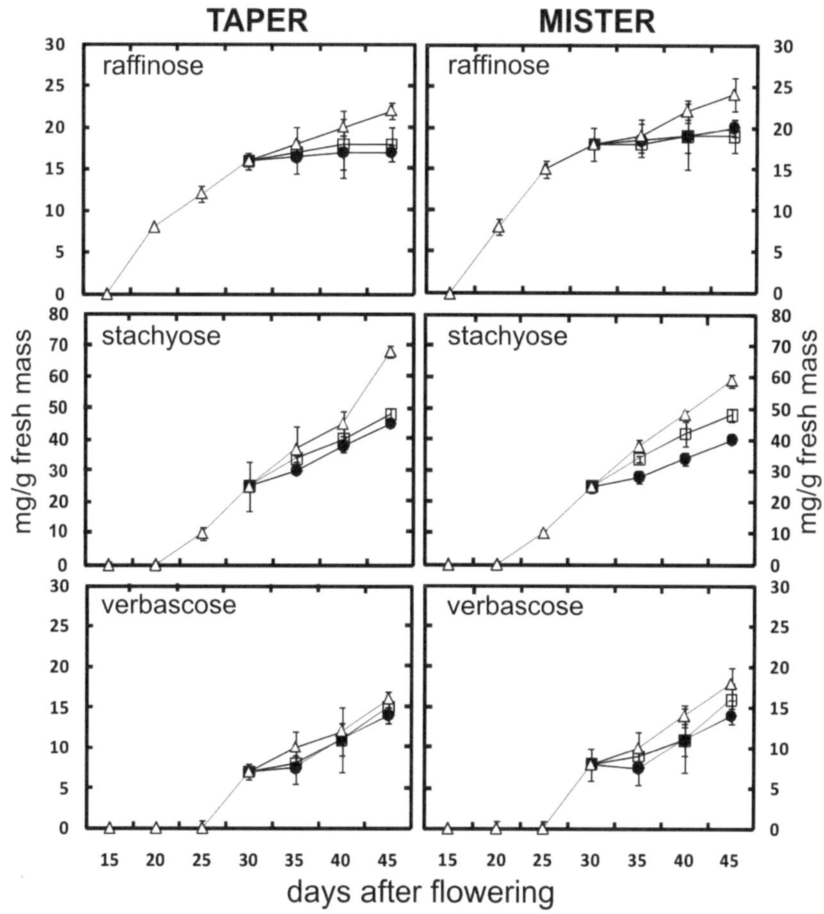

Fig. 6. Raffinose family oligosaccharides content in lupin seeds during maturation and after application of herbicides: Basta 200 SL and Roundup Ultra 360 SL. Data points represented the mean ± SD for fifteen replicate samples - control (Δ), Basta treatments (□), Roundup treatments (•).

This hypothesis is based on biochemical and physiological characteristics of galactosyl cyclitols, similar to those of RFO (Obendorf, 1997). It was shown that in buckwheat seeds galactosyl cyclitols indeed contribute to the development of seed resistance to desiccation (Horbowicz et al., 1998). These compounds are formed by attaching one or more galactosyl radicals to cyclitol (most commonly inositol). Galactose is attached to cyclitol by α-(1→2) or α-(1→3) bindings. Galactosyl cyclitols series based on the form of cyclitol which is an acceptor or galactosyl radical. So far the following structures of galactosyl cyclitols have been discovered and defined: galactinol, galactopinitol, fagopyritol, galactoononitol. The accumulation of galactosyl cyclitols was observed in seeds of different plants (Piotrowicz-Cieślak, 2004). Particularly rich in galactosyl cyclitols are the seeds of buckwheat, castor oil plant and lentil (Horbowicz & Obendorf, 1994). Lupin seeds accumulate up to 2 % d.s. of galactosyl cyclitols (Piotrowicz-Cieślak et al., 2003). Under desiccation seeds of legume plants (soybean, yellow lupin) accumulate mainly RFO, despite the fact that during the natural drying these seeds also form galacto-D-pinitol and D-*chiro*-inositol (Górecki et al., 1997). Water stress, resulting from soil drought, low or high temperature, may have a significant influence on prompting the biosynthesis of galactinol and RFO in vegetative tissues and maturing seeds. In leguminous plants exposed to water stress an intense accumulation of α-D-galactosides is found (Streeter et al., 2001). Water stress resulting from soil drought and chill induces the accumulation of galactinol and raffinose also in vegetative tissues of alfalfa and *Arabidopsis* (Taji et al., 2002; Cunningham et al., 2003; Zuther et al., 2004) and tomato germinating seeds (Downie et al, 2003). RFO level depends on the activity of galactinol synthase. The activity of this enzyme is related to temperature (Panikulangara et al., 2004) and RFO total amount in leaves and seeds (Castillo et al, 1990). In the genome of *Arabidopsis* seven genes have been identified as responsible for the activity of galactinol synthase. Two of them are activated by drought and salinity, whereas one is activated by low temperature (Taji et al, 2002). The level of stachyose and verbascose depends on the level of initial substrates, including *myo*-inositol (Hitz et al., 2002) and sucrose. Herbicides modify the content of soluble carbohydrates in seeds and remain in soil after having been applied. Toxicological tests are a simple, inexpensive, and quick method to assess their impact on subsequent plants.

3. Toxicological tests in the environmental assessment

A wide range of analytical approaches are used to asses the effect of contaminants on environment at all levels of its complexity - from studying the biochemical changes in single cells to changes measured on ecosystem levels. The holistic, system approach so characteristic of modern science, combining efforts of experts in many fields, increasingly becomes applied also in environmental toxicology studies. Yet, it is not possible to precisely characterize all possible pollutants due to their big number, different concentrations, molecular weights, or high reactivity of widely spread low-molecular-weight substances (Kahru et al., 2008). The key task of toxicology and ecotoxicology is a direct evaluation of risks resulting from environmental contamination and refers to, among other issues, creating classification systems based on increasing levels of toxicity (Persoone et al., 2003). Until recently, physical-chemical methods were considered the basic way to diagnose the condition of environment and its specific elements. Although these analyses greatly facilitate elimination of some toxic substances (Wolska et al., 2007), they do not fully characterize the biological activity of any substance on affected organisms. They only inform us about the level of contamination, but do not predict its biological consequences. The precise analytical methodologies are mostly worked out for those compounds only which are subject to strict legal regulations. Moreover, such chemical analyses sometimes blur the real environmental threats (Manusadžianas et al., 2003; Persoone et al., 2003; Wolska et al., 2007). A valuable alternative supplementing this purely chemical approach was worked out, based on the following principle: measure the exposure and analyze the accumulation and metabolism of contaminant in living organisms. Relatively inexpensive and biologically founded biotests are being developed for this purpose.

A biotest is a standard procedure indicating the impact of contamination or their mixture on a biological system, i.e. a plant or animal organism, a system of organisms or a fragment of an ecosystem (a mesosystem, a microsystem), at the same time defining make-up of substances and their possible interactions (Kratasyuk et al., 2001; Wang et al., 2003; Persoone et al., 2003). Biotests show an interdisciplinary approach to organisms' response to the ocurrence of particular chemical substances. In order to determine the impact of these chemical substances, biotests employ the methods of physiology and biochemistry. The obtained results unambiguously determine the toxicity of a given sample or lack of its toxicity, as compared to the response of an organism not exposed to the activity of all chemical substances present in the analysed sample (Simeonov et al., 2007). Methods used in toxicology and ecotoxicology based on morphological and/or physiological disturbances at the cell level or organism level, and sometimes as a consequence death of a given organism, aim at protecting the environment, which indirectly results in protecting the life and health of human population (Celik et al., 1996; Dinis-Oliveira et al., 2006). The current form and shape of these tests and recommendations for their proper usage are a result of more than twenty years of research. In consequence, conditions of conducting such tests became standardized, costs related to their application decreased, the test cultures became widely available (Kaza et al., 2007), and measurements unified and comparable.

3.1 Biotests and their classification

Interdisciplinary analytical methods based on living material illustrate potential threats posed by contaminants or their mixture in diversified environmental matrices. The multitude of methods causes many problems with classifying them (Persoone et al., 2003). Primary classifications referred to information about the level of contamination in a given element of the environment, considering the place of performed analysis. There are laboratory tests conducted, which are based on modelling samples in controlled conditions, results of which determine the level of toxicity for real samples. Moreover, samples retrieved from specific components of the environment (e.g. water, soil, wastewater) are analysed and compared with standard samples. In comparative studies of water environment, it is best to apply various methods and to test the same samples with various organisms in order to determine the level of test sensitivity and eliminate errors resulting from application of only one test (Kratasyuk et al., 2001). Another way to best account for the specific environmental conditions is to perform an *in situ* analysis, which utilises responses of organisms living in the natural environment (Anderson et al., 2004; Mc William & Baird, 2002). Such an analysis enables continuous replacement of the medium, mainly water, when studies are performed on fish or plants as indicator organisms, and it is classified to be a dynamic biotest (Sundt et al., 2009). If replacement takes place in set time periods, such biotest is defined as a half-static one (Blanck et al., 2003; De Liguoro et al., 2009). In a static test the medium is not replaced until the end of analysis. While analysing toxicity, the tested pollutant at a certain concentration is only once placed on the medium, i.e. water, sediment, or soil.

Studies based on the assessment of various substances' impact on morphological and physiological changes in the indicator organism pose many difficulties in classifying them to a specific group of toxicological tests due to a wide spectrum of methods they employ. The active element of the test - i.e. an organism used in the test - may be considered as a basic criterion in classification of toxicological and ecotoxicological biotests. Various plants, bacteria and animals are used as active elements (Nałęcz-Jawecki & Persoone, 2006; Adomas & Piotrowicz-Cieślak, 2008; Nałęcz -Jawecki et al., 2010). Nowadays, microbiotests called 'toxkits' are available. Sets of microbiotests - TOXKIT - used to determine acute and short-term chronic toxicity have been developed with the use of many test organisms which belong to many phylogenetic groups and trophic levels. They are used in order to assess land and fresh water, as well as coastal and sea environments. Among tests used to assess fresh water there are:

- ALGALTOXKIT F™ using *Selenastrum capriconrnutum*
- DAPHTOXKIT F ™ magna using *Daphnia magna*
- DAPHTOXKITF ™ pulex using *Daphnia pulex*
- CERIODAPHTOXKIT F ™ with the aid of *Ceriodaphnia dubia*
- THAMNOTOXKIT F ™ with *Thamnocephalus platyurus*
- ROTOXKIT F ™ acute using *Brachionus calyciflorus*
- ROTOXKIT F ™ short - chronic with *Brachionus calyciflorus*
- PROTOXKIT F ™ with the aid of *Tetrahymena thermophila*
- MARA biotest with 11 microorganisms individually lyophilized in microplate wells.

Tests used to assess brackish and saline waters include, among others:

- ROTOXKIT M™ rotifers *Brachionus plicatilis* are the test organism;
- ARTOXKIT M™ with sea crustacean *Artemia salina;*
- MARINE ALGALTOXKIT ™ uses *Phaeodactylum tricornutum.*

Sediments and soil are assessed with the use of:

- PHYTOTOXKIT - seeds of mono- and dicotyledonous plants are the phytobioindicators;
- OSTRACODTOXKIT F™ uses *Heterocypris incongruens;*
- PHYTOTOXKIT (Fig. 7) is a microbiotest of phytotoxicity of soil, sludge, compost sewage used for watering, chemical substances and biocides.

Biotests based on the analysis of changes within a plant organism and called phytotests provide information concerning organisms of key importance to a given ecosystem, thanks to which it is possible to determine its condition and disturbances in matter flow or substance circulation. Phytotoxic substances disturb absorbance and transport of essential micro- and macro-elements in plants, which results in delayed seed germination and plant sprouting as well as deformations and underdevelopment of certain plant elements (Jin et al., 2009). Thanks to phytotests, a solid knowledge of the impact of an environmental factor is acquired. Phytotests are used to assess soil (Piotrowicz-Cieślak et al., 2010 a) and water (Fernández-Alba et al., 2002; Drzewicz et al., 2004) contaminated with, among others, pesticides. Biotests used in biomonitoring (Holgado et al., 2004) are applied in order to assess the condition of ecosystems, to establish the capacity of ecosystems to absorb pesticides, including herbicides, as well as to assess interactions among pesticides, and between pesticides and the environment. In these studies many kinds of plants are used; algae from the class of *Chlorophyta: Selenastrum capricornutum, Scenedesmus quadricauda, S. subiscatus* are used most often to evaluate fresh waters (Küster et al., 2003; Simeonov et al., 2007; Wadhia & Thompson, 2007), while the class of *Bacillariophyta: Phadeodactylum tricornutum, Skeletonema constatum* is used to evaluate brackish and saline waters (Nendza, 2002; Wadhia & Thompson, 2007). Duckweed is widely used in laboratory studies of substances and their accumulation (OECD, 221), *Lemna minor* and *L. gibba* are the organisms most frequently applied as indicators in biotests (Lemna test). Phytobioindicators are chosen also from among tracheophytes, among which *Sorghum saccharatum, Lepidium sativum* or *Sinapis alba* can be distinguished. They have very small seeds, and the test is performed for three days (OECD, 208). To evaluate soil contaminated with glyphosate yellow lupin was applied, with seeds 20 times bigger than the size recommended by the PHYTOTOXKIT™ producer (MicroBio Test Inc., Belgium) (Adomas & Piotrowicz-Cieślak, 2008).

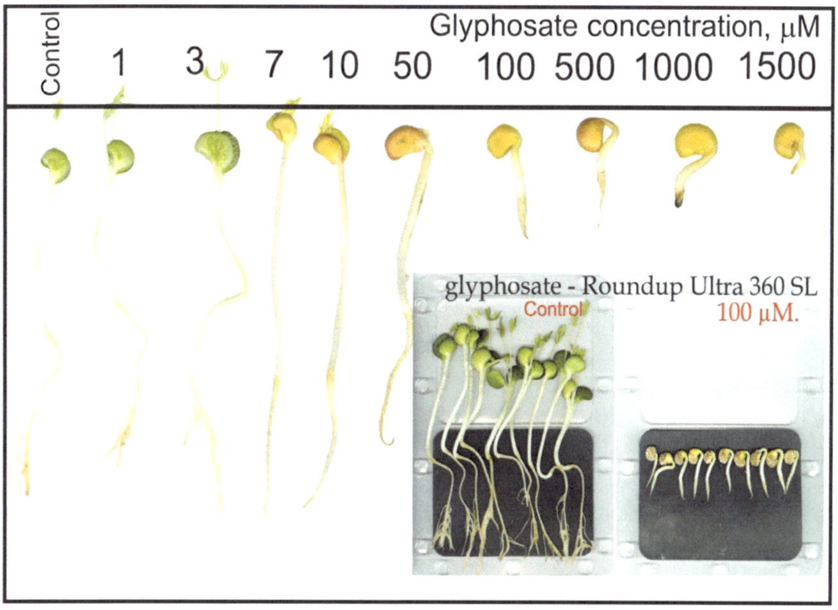

Fig. 7. PHYTOTOXKIT is used to determine phytotoxicity of soil contaminated different glyphosate concentration.

In soil environment assessment, apart from assessing the germination of grasses, crucifers, leguminous plants and grains, also macrophytic plants are taken into consideration. However, a complicated process of growing them as well as a long time of their growth and their requirements with reference to space precludes their common use in biotests.

The choice of plant organisms should characterize a given ecosystem. Plant species are chosen considering their availability, methods of growing, simplicity of conducting studies and biological sensitivity to a compound or a group of compounds, confirmed in a number of repeated tests (Eberius et al., 2002). In such kind of research, the opportunity to assess toxicity in a multigenerational cycle is extremely important. Thus the majority of applied phytotests are of chronic character, i.e. they last longer than 1/3 of pre-productive period. Nowadays, as a result of long-lasting studies in the field of environmental assessment, scientists tend to use available plant cultures in their experiments, which can be used in laboratories. They grow their own algae and tracheophytes or use commercially available selected cultures in the cryptobiotic or lyophilized form, available with microbiotests, e.g. Algaltoxkit F™, Phytotoxkit™ (Persoone et al., 2003; Wolska et al., 2007).These are kits equipped with accessories essential in determining the level of toxicity in samples. Every microbiotest contains an active element, i.e. a living organism, ready to be activated at any moment. Toxicological assessment with the use of such tests makes it possible to compare results between continents. Moreover, thanks to cultures attached to the test, there is no need to grow any organisms. Assessments performed with the use of these tests are characterized by lower research costs than in the case of the conventional ones; the time of organism response is shortened, and the study can be performed for samples of lower volume. Additionally, a researcher can work on a few samples at the same time (Persoone et. al., 2003; Kaza et al., 2007; Wolska, et al., 2007), which enables planning the experiment as a battery of tests. A battery of tests is a toxicological study which encompasses more than just one species of indicator organisms. It is used in order to widen the scope of research conducted on a given chemical substance onto a bigger portion of the ecosystem, considering also other organisms which live there on different trophic levels, in the following order: producer-consumer-reducer (Kaza et al., 2007). Toxicological analyses as well as the obtained results are specifically diversified. They often aim at determining the very lethality or inhibition effects with reference to different bioindicators, after particular exposure times. Lethal Concentration (LC) or Lethal Dose (LD) is often determined taking into account Lowest Observed Effect Concentration (LOEC) or Observable Effect Concentration (NOEC). Typically, Effect Concentration (EC) is established for rotifers and crustaceans, and denotes inhibition of a particular physiological or biochemical activity expressed as a percentage in a given test group. In growth tests, inhibition of growth of plants, fungi as well as biomass of algae is considered. Apart from the above parameters, due to the development of biotechnology, also enzymatic tests (Budantsev, 2005) and tests of genotoxicity (Yamamoto et al., 2001; Küster et al., 2003; Jha, 2008) have become important in toxicological studies. The first group determines inhibition of activity of an enzyme or a group of enzymes which catalyze a given biochemical reaction, while the second group determines genetic changes brought about by a toxicant. Sometimes toxicologists analyze parameters relating to absorption and storage of certain substances in tissues and organs depending on the time of exposure. These tests are referred to as bioaccumulation tests (Kahle & Zauke, 2002). Sensitivity of leguminous plants to herbicides has been successfully used in our research.

3.2 Estimation of herbicites phytotoxicity to legumes

Lupinus luteus, Glycine max, Pium sativum, Lens esculenta, Vigna angularis and *Medicago sativa* were used to estimation soil contamination with glufosinate ammonium, glyphosate and diquat. Seeds of yellow lupin (*Lupinus luteus*), soybean (*Glycine max*), adzuki bean (*Vigna angularis*), pea (*Pium sativum*), lentil (*Lens esculenta*),) and alfalfa (*Medicago sativa*) were germinated for nine days using PHYTOTOXKIT™ (MicroBio Test Inc., Belgium). Germination was carried out in controlled climatic conditions, 25°C and 90 % RH humidity, in darkness. Ninety ml of soil (sand, vermiculite, peat, 1:0.3 :1, v:v:v) were placed in plastic microbiotest plates. The soil was covered with Whatman No. 1 filter-paper and watered with 27 ml distilled water supplemented with different glyphosate (Roundup Ultra 360 SL), glufosinate ammonium (Basta 200 SL), and diquat (Reglone Turbo 200 SL) final concentrations: 1, 3, 7, 10, 50, 100, 500, 1000, 1500 μM. The control plants were watered with pure distilled water. The root length was estimated using Image Tool for Windows. The effective concentration causing a 50% response (EC_{50}) was calculated for inhibition of root growth and no observed effect concentration (NOEC) that did not affect germination was noted. Electroconductivity of seedlings soak water (4×10 seedlings, each in 100 ml deionised water for 24 h) was measured using Hanna conductivity meter HI 4321. Dry and fresh mass and *myo*-inositol content (see 2.1.1) were determined.

3.3 Phytotest used in herbicides study

Yellow lupin, soybean, adzuki bean, pea, lentil and alfalfa seedlings responded in a similar manner to herbicides (glyphosate, glufosinate ammonium and diquat). All seeds, irrespectively of the concentration of herbicides, germinated with frequency between 65% and 95% (Fig. 8). Seeds were considered germinated when the radicle penetrated the seed coat. With increasing concentration of herbicides an inhibition of root elongation was observed in all plants (Fig. 9).

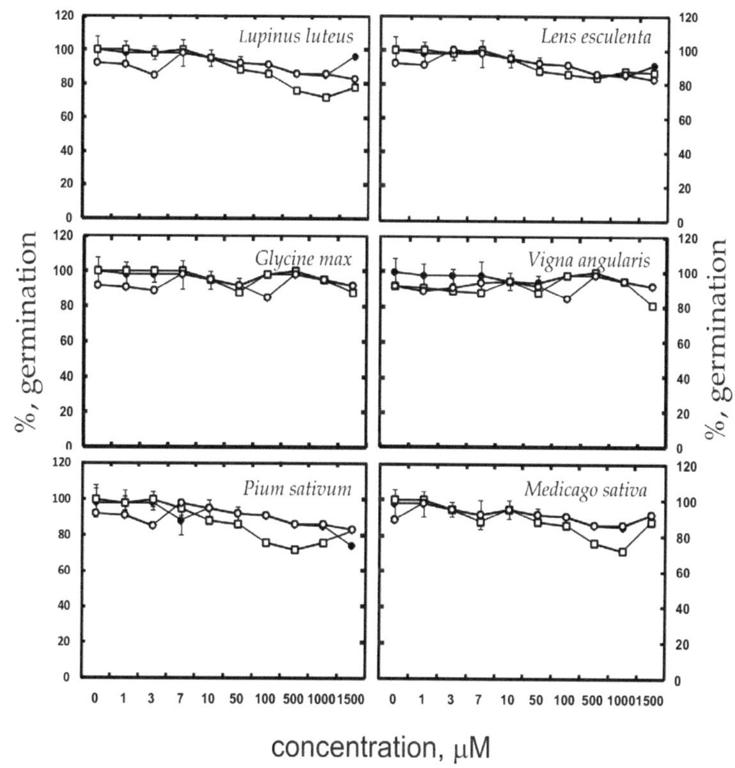

Fig. 8. Yellow lupin (*Lupinus luteus*), soybean (*Glycine max*), adzuki bean *(Vigna angularis)*, pea *(Pium sativum)*, lentil (*Lens esculenta*) and alfalfa (*Medicago sativa*) seedlings germination [%] after application of herbicides: glyphosate, glufosinate ammonium and diquat. Data points represented the mean ± SD for fifteen replicate samples - glyphosate (o), glufosinate ammonium (□) and diquat treatments (•)

Active substances of herbicides, depending on their chemical structure, remain in the soil solution and they are absorbed by roots of germinating seeds. When the weeds are removed, within one vegetation season these substances should decompose to compounds which are natural in the environment (e.g. nitrogen and carbon dioxide). Yet, among herbicides there are chemical compounds which have varying half-lives. For instance, the half-life of glyphosate in soil is only 10 to 100 days (47 days on average) according to Hornsby et al., (1996) and according to Monsanto, (2005) the average half-life is 32 days. Paraquat has a relatively long half-life in soil (estimated at about 1000 days). The residue of persistent herbicides (e.g. atrazine, metribuzin, trifluralin) may stay in soil and negatively affect subsequent crop even more than a year after application. It pertains especially to active substances used year by year on the same field (e.g. atrazine) (Sheets & Shaw, 1963). Thus for both agriculture and environmental protection it is important to check what happens with the active substances of herbicides in soil (their translocation on different levels of soil and water, and their degradation) and how they are absorbed by plants (Beckie & McKercher, 1990).

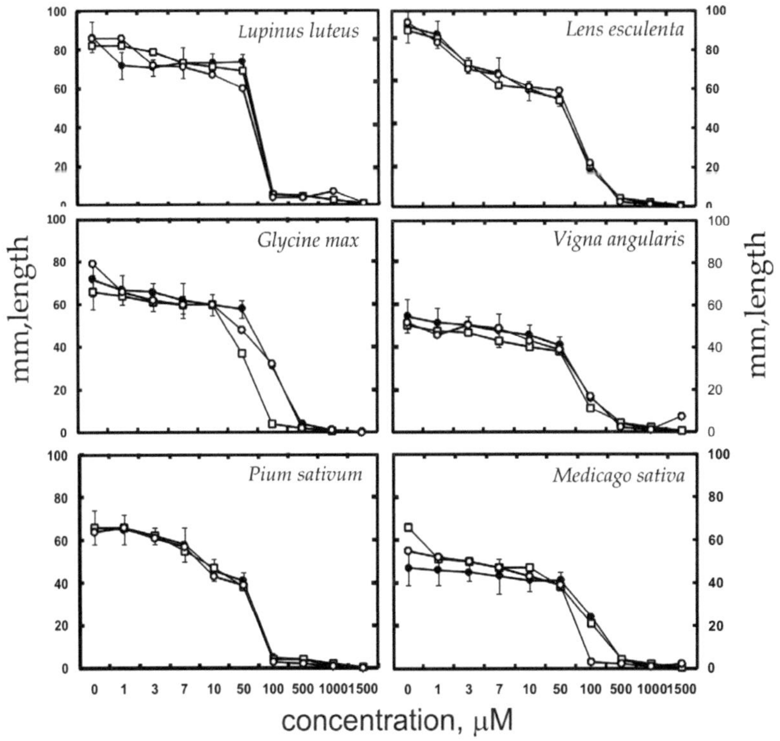

Fig. 9. Yellow lupin (*Lupinus luteus*), soybean (*Glycine max*), adzuki bean *(Vigna angularis)*, pea *(Pium sativum)*, lentil (*Lens esculenta*) and alfalfa (*Medicago sativa*) seedlings length [mm] after application of herbicides: glyphosate, glufosinate ammonium and diquat. Data points represented the mean ± SD for fifteen replicate samples - glyphosate (o), glufosinate ammonium (□) and diquat treatments (•).

In such research, phytotests in which different plants are phytobioindicators are successfully applied. Plant sensitivity to environment contamination is often used to estimate the degree of environment degradation. Plants respond in different manner to many kinds of toxic substances. The symptoms include morphological deformations (e.g. seed germination or the length of roots and shoots) and changes in plant biochemistry (osmoprotectors content) (Pardo, 2010; Piotrowicz-Cieślak et al., 2010a). The phytotoxic effect is a result of an interaction between the compound and the plant in given environmental conditions. Environmental toxicity is usually determined with the use of phytotests according to OECD norms (2006) (OECD, 221), mainly with reference to pesticide (Stork & Hannah, 1996) and veterinary medicine contamination (Piotrowicz-Cieślak et al., 2010b). Biotests, in contrast to instrumental (chemical) methods, enable simple and inexpensive estimation of very low levels of active substances in soil which can be phytotoxic to crop plants (Wolska et al., 2007).

Phytotoxkit has been successfully applied to estimate phytotoxicity of glyphosate used all over the world to fight weeds and desiccate crop plants (leguminous plants and rape seed). Soil contaminated with increasing concentrations of glyphosate from 0 to 2000 µM was assessed with the use of leguminous plants (*Lupinus luteus*), crucifers (*Brassica napus, Sinapis alba, Lepidium sativum*), grains (*Avena sativa*) and a plant from the *Poaceae* family (*Sorghum saccharatum*) (Piotrowicz-Cieślak et al., 2010a). Glyphosate concentrations highter than 40 µm inhibited root growth.

In order to see a complete picture of herbicide phytotoxicity, one has to assess also key cell metabolites, among others *myo*-inositol, in seedlings which grow in contaminated soil. *myo*-Inositol is an aliphatic alcohol derived from glucose-6-phosphate, it has six OH groups on its ring. It is the most commonly spread cyclitol in the environment and a precursor of optical and methyl derivatives. *myo*-Inositol is localized in cytosol and plastids in small quantities (Paul & Cockbourn 1989), it is easily incorporated into cell metabolism and thus its (galactosyl) derivatives are said to have storage functions (Piotrowicz-Cieślak et al., 2008). *myo*-Inositol constitutes a key component of cell membranes (poli-phosphatidylinositols take part in receiving and transducing signals). This compound occurs in considerable quantities in the form of an ester of phytic acid, being an easily accessible form of phosphate ions (Loewus et al., 1990). In dehydration stress, *myo*-inositol plays the role of an osmoprotectant (Nelson et al., 1998; Piotrowicz-Cieślak et al., 2007), first of all limiting destructive changes in the biological membranes which were induced by the stress factor.

The level of *myo*-inositol in plants treated with glyphosate, glufosinate ammonium and diquat indeed increased when the herbicides were used at very high concentrations (Fig. 10).

Phytotoxkit was also used to assess the residue of herbicides applied to control weed in winter wheat: chlorsulfuron, nicosulfuron, 2,4 DP and dicamba within the range of 0.025-1.2 mg/kg of soil. Seeds of *Sinapis alba, Fagopyrum esculentum* and *Cucumis sativus* were phytobioindicators here. On the basis of root elongation in 5-day seedlings, it was shown that *Sinapis alba* was most effective in detecting chlorsulfuron and nicosulfuron residues in soil. *Cucumis sativus* seedlings were the most sensitive plants to 2,4 DP residue in soil, as the highest of the analysed concentrations reduced root growth by 99%. The roots of *Fagopyrum esculentum* were most suitable for the detection of dicamba residue. The research demonstrated that very sensitive plants are able to detect the residue of herbicides in soil at the level of 0.0015 mg/kg (Sekutowski & Sadowski, 2009).

While applying herbicides, e.g. in agrotechnical activities and to destroy plants in water reservoirs (e.g. glyphosate), xenobiotics are introduced into natural aquatic systems. The mechanism of toxic activity of herbicides inhibiting photosynthesis in plants in PI and PII system was used to assess phytotoxicity of paraquat, atrazine, metribuzinan and diuron towards *Scenedesmus obliquus* green algae. The F684/F735 chlorophyll fluorescence ratio in *S. obliquus* algae can be a quick and sensitive measurement method of contamination levels in water reservoirs when they were contaminated with active substances of herbicides chosen for the research (Eullaffroy & Vernet, 2003).

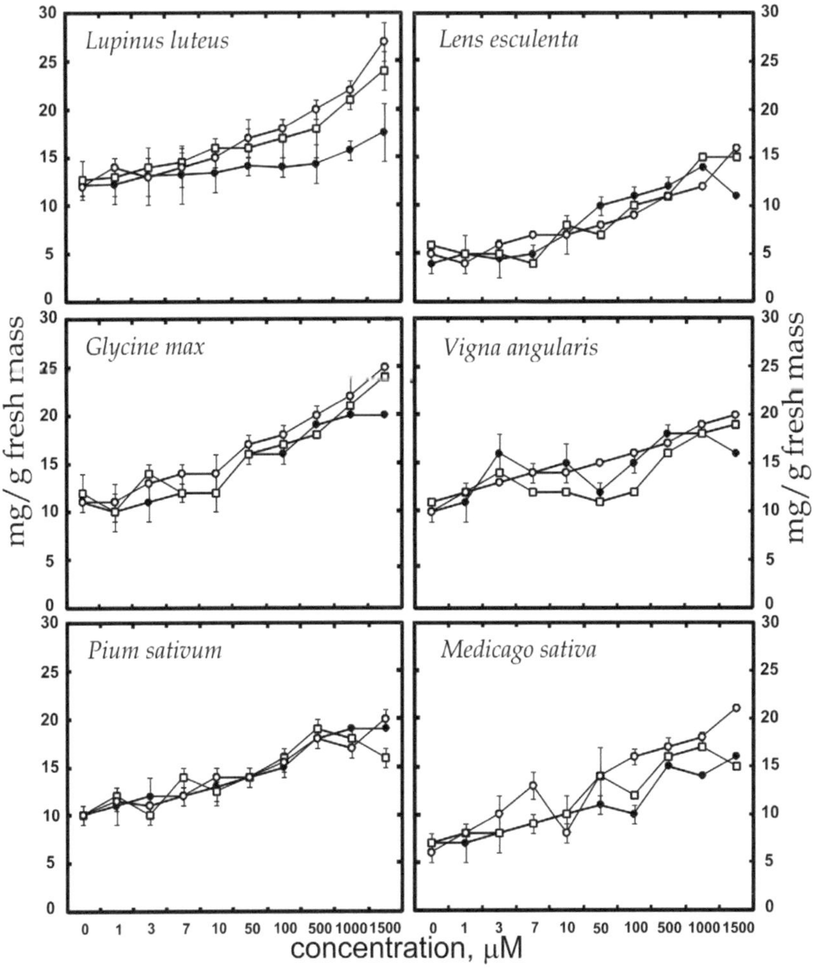

Fig. 10. *myo*-Inositol content [mg/g fresh mass] in yellow lupin (*Lupinus luteus*), soybean (*Glycine max*), adzuki bean *(Vigna angularis)*, pea *(Pium sativum)*, lentil (*Lens esculenta*) and alfalfa (*Medicago sativa*) seedlings after application of herbicides: glyphosate, glufosinate ammonium and diquat. Data points represented the mean ± SD for fifteen replicate samples - glyphosate (o), glufosinate ammonium (□) and diquat treatments (•).

The unicellular algae *Chlamydomonas reinhardtii* were used to detect 16 herbicides, belonging to 11 different chemical groups, in water. Different reactions (sensitive or no effect) of one species of algae were shown for acifluorfen, chlorpropham, diclofop-methyl (DFM), glyphosate, isoxaben, pinnacle, and trifluralin, dichlorobenzonitrile (DCB), 2,4-dichlorophenoxyacetic acid (2,4-D), metobromuron, 2-ethyl-4-chlorophenoxyacetic acid (MCPA), metribuzin, atrazine, hexazinone, norflurazon and terbacil - after 3 days from their application. The reaction of algae depended on the quantity of substances and the mechanism of toxic activity of particular chemical groups of herbicides. Such unicellular organisms can be used as phytobioindicators for a quick and easy detection of different active substances of herbicides in water (Li et al., 2008).

The results of toxicity assessment of active substances in herbicides may vary, depending on the sensitivity of the species used for biotests. These relations were demonstrated through the analysis of pesticides (42 insecticides and 45 herbicides) transported in 1985-2004 into the rivers in Mississippi River basin, which then were entering Chesapeake Bay. The U.S. EPA estimates that approximately 75% of all pesticide usage in the United States is agricultural. The other 25% is for home and garden use, industrial, commercial and government sectors. It has been found that of the 45 tested herbicides the ones most often utilized in 1985-2004 were atrazine, glyphosate, and metalachlor. Atrazine and metalachlor have been identified as more toxic to nontarget plant species than glyphosate. On the other hand, insecticides were more toxic to animals and plants used in the study (trout, bluegill, daphnia, selenastrum, skeletonema, lemna) than herbicides. In the years 1985 and 2004, 12 herbicides inhibiting weed roots were examined in Maryland. The lowest amounts of these herbicides (0.0054, 0.0145, 0.0345 mg/L) were phytotoxic for selenastrum, lemna, and skeletonema that had been used as bioindicators section (Hartwell, 2011).

Pea (*Pisum sativum* L.) and lupin (*Lupinus angustifolius* L.) roots were used seven days after sowing for assessment of soil contaminated with herbicide from sulfonylurea class (chlorsulfuron, triasulfuron and metsulfuron-methyl) and from the sulfonanylide class (flumetsulam and metosulam). The sensitivities of the species were similar in chlorsulfuron and flumetsulam trials and their response range varied with soil type and herbicide, e.g. between 0.75 and 6.0 ng triasulfuron g^{-1} in the Wimmera grey clay and between 0.125 and 8.0 ng chlorsulfuron g^{-1} soil in the Mallee sand (Stork & Hannah, 1996). Sensitivity of leguminous plants to herbicides has been successfully used in our research.

4. Conclusion

Research into pesticides toxicity to animal and plant organisms has been conducted for many years and has developed as production and registration of active substances of these xenobiotics have increased. Most toxicological studies concern animal organisms, yet embryophytes - and especially crop plants - have been used rather more rarely as bioindicators for the environment contaminated with pesticides. Over the recent years the ecotoxicological research has proved that active substances of herbicides present in water and soil are, even in very low quantities (examples), phytotoxic. Yet, these are very low quantities (e.g. over 10 μM of glyphosate) (Piotrowicz-Cieślak et al, 2010a) in soil and even lower in water - 0.08 mg/L of atrazine for *Lemna* (Harwell, 2011).

On the other hand, before it is registered, every active substance is analysed in order to determine its dosage phytotoxic to standardised mono- and dicot assay plants. Our studies have proved that glyphosate, glufosinate ammonium and diquat applied in doses recommended by producers are not phytotoxic to leguminous plants. Only if a 50 μM dose is exceeded, morphological and biochemical changes are apparent in these plants. Applying desiccants prior to harvest (to immature seeds) induces changes in their metabolism of carbohydrates. We also found that in order to determine phytotoxicity it is better to use a root elongation test than a test of seed germination in soil contaminated with herbicides. The need to conduct research into the natural environment assessment with the use of up to-date toxicity assessment methods has lead to many various legal modernisations in the field of environmental protection of many countries (European Commission, 76/464/EEC; European Commission, 2455/2001/EC). Thanks to this, a lot of concentrations of pollutants so far believed to be safe have been found unsuitable, and appropriate regulations require their elimination or lowering their content in the environment. Biotests - as compared with conventional methods - constitute a relatively quick and inexpensive analysis method. Combining two methods (the biological and chemical ones) enables to provide an exhaustive and wide-ranging toxicological description. As a consequence, the natural environment protection, including human population, is becoming more and more effective and preventive steps are taken much earlier than they used to be.

5. References

Adams, C.A. & Rinne, R.W. (1980). Moisture content as controlling factor in seed development and germination, *International Review of Cytology*, Vol.68, pp. 1- 8, ISSN 0074-7696

Adomas, B. & Piotrowicz-Cieślak, A.I. (2008). Yellow lupin is a good bioindicator of soil contamination with sulfamethazin, Lupin for Health and Wealth, *Proceedings of the 12th International Lupin Conference*, pp. 362-367, ISBN 0-86476-153-8, Fremantle, Western Australia, September 14-18, 2008

Adomas, B.; Piotrowicz-Cieślak, A.I.; Michalczyk, D.J. & Sadowski, J. (2008). Residues of dimethipin in pods and seed of narrow-leaved (*Lupinus angustifolius* L.) and yellow (*Lupinus luteus* L.) lupin, *Fresenius Environmental Bulletin*, Vol.17, No.9, pp. 1288-1293, ISSN 1018-4619

Allinson, G. & Morita, M. (1995). Bioaccumulation and toxic effects of elevated levels of 3,3',4,4'-tetrachloroazobenzene (33'44'-TCAB) towards aquatic organisms. I: A simple method for the rapid extraction, detection and determination of 33'44'-TCAB in multiple biological samples, *Chemosphere*, Vol.30, No.2, pp. 215-221, ISSN 0045-6525

Anderson, B.; Hunt, J.; Phillips, B.; Nicely, P.; Tjeerdema, R. & Martin, M. (2004). A Comparison of In Situ and Laboratory Toxicity Tests with the Estuarine Amphipod *Eohaustorius estuaries*, *Archives of Environmental Contamination and Toxicolog*, Vol..46, No.1, pp. 52-60, ISSN 0090-4341

Beckie, H.J. & McKercher, R.B. (1990). Mobility of two sulfonylurea herbicides in soil, *Journal of Agricultural and Food Chemistry*, Vol.38, pp. 310-315, ISSN 0021-8561

Bernal - Lugo, I. & Leopold, A.C. (1995). Seed stability during storage: Raffinose content and seed glassy state, *Seed Science Research*, Vol.5, pp. 75-80, ISSN 1475-2735

Bianchi, G.; Gamba, A.; Limiroli, R.; Pozzi, N.; Elster, R.; Salamini, F. & Bartels, D. (1993). The unusual sugar composition in leaves of the resurrection plant *Myrothamnus flabellifolia*, *Physiologia Plantarum*, Vol.87, pp. 223–226, ISSN 0031-9317

Blackman, S.A.; Obendorf, R.L. & Leopold, A.C. (1992). Maturation proteins and sugars in desiccation tolerance of developing soybean seeds, *Plant Physiology*, Vol.100, pp. 225-230, ISSN 0032-0889

Blanck, H.; Porsbring, T. & Scholze, M. (2003). (MOP/74) Predictability of the toxicity of metal(oid) mixtures to marine periphyton communities. *The Society of Environmental Toxicology and Chemistry*, Europe 2003, ISSN 0730-7268

Budantsev, A.Y. (2005). A Biotest Based on an Acetylcholinoesterase Tissue Preparation Immobilized on Paper, *Journal of Analytical Chemistry*, Vol.60, No.12, pp. 1155-1158, ISSN 1061-9348

Carpenter, J.F. & Crowe, J.H. (1989). An infrared spectroscopic study of the interactions of carbohydrates with dried proteins, *Biochemistry*, Vol.28, pp. 3916-3922, ISSN 1522-1059

Castillo, E.A.; de Lumen, B.O.; Reyes, P.S. & de Lumen, H.Z. (1990). Raffinose synthase and galactinol synthase in developing seeds and leaves of legumes, *Journal of Agriculture and Food Chemistry*, Vol.38, pp. 351-355, ISSN 0021-8561

Celik, I.; Camas, H.; Arslan, O.; Yegin, E. & Kufrevioglu, O. I. (1996). The effects of some pesticides on the activity of liver and erythrocyte enzymes (in Vitro), *Journal of Environmental Science and Health*, Part A, Vol.31, No.7,pp.1645-1649, ISSN 1093-4529

Crowe, B.L. & Crowe, L.M. (1984). Preservation of membranes in anhydrobiotic organisms: the role of trehalose, *Science*, Vol.223, pp. 701-703, ISSN 0036-8075

Crowe, J.H.; Carpenter, J.F; Crowe, L.M. & Anchordoguy, T.J. (1990). Are freezing and dehydration similar stress vectors? A comparison of modes of interaction of stabilizing solutes with biomolecules, *Cryobiology*, Vol.27, pp. 219-231, ISSN 0011-2240

Cunningham, S.M.; Nadeau, P.; Castonguay, Y.; Laberge, S. & Volence, J.J. (2003). Raffinose and stachyose accumulation, galactinol synthase expression, and winter injury of contrasting alfalfa germplasms, *Crop Science*, Vol.43, pp. 562-570, ISSN 1835-2693

De Liguoro, M.; Fioretto, B.; Poltronieri, C. & Gallina, G. (2009). The toxicity of sulfamethazine to *Daphnia magna* and its additivity to other veterinary sulfonamides and trimetroprim, *Chemosphere*, Vol.75, pp. 1519-1524, ISSN 0045-6535

Dinis-Oliveira, R.J.; Remião, F.; Carmo, H.; Duarte, J.A.; Navarro, A.S.; Bastos, M.L. & Carvalho, F. (2006). Paraquat exposure as an etiological factor of Parkinson's disease, *NeuroToxicology*, Vol.27, No.6, pp. 1110-1122, ISSN 0161-813X

Downie, B.; Gurusinghe, S.; Dahal, P.; Thacker, R.R.; Snyder, J.C.; Nonogaki, H.; Yim, K.; Fukanaga, K.; Alvarado, V. & Bradford, K.J. (2003). Expression of a galactinol synthase gene in tomato seeds is up-requlated before maturation desiccation and again after imbibition whenever radicle protrusion is prevented, *Plant Physiology*, Vol.131, pp. 1347-1359, ISSN 0032-0889

Drzewicz, P.; Nałęcz-Jawecki, G.; Gryz, M.; Sawicki, J.; Bojanowska-Czajka, A.; Głuszewski, W.; Kulisa, K.; Wołkowicz, S. & Trojanowicz, M. (2004). Monitoring of toxicity turing degradation of selected pesticides using ionizing radiation, *Chemosphere*, Vol.57, pp. 135-145, ISSN 0045-6535

Eberius, M.; Mennicken, G.; Reuter, I. & Vandenhirtz, J. (2002). Sensitivity of different growth inhibition Test-Just a question of mathematical calculation. Theory and practice for algae and duckweed, *Ecotoxicology*, Vol.11, pp. 293-297, ISSN 0963-9292

Ellis, R. H.; Hong, T. D. & Roberts, E. H. (1991). An intermediate category of seed storage behaviour? II. Effects of provenance, immaturity, and imbibition on desiccation-tolerance in coffee, *Journal of Experimental Botany*, Vol.42, No.5, pp. 653-657, ISSN 0022-0957

Eullaffroy, P. & Vernet, G. (2003). The F684/F735 chlorophyll fluorescence ratio: a potential tool for rapid detection and determination of herbicide phytotoxicity in algae, *Water Research*, Vol.37, pp. 1983-1990, ISSN 0043-1354

European Commission, Council Directive of 4 May 1976 on pollution caused by certain dangerous substances discharged into the aquatic environment of the Community (76/464/EEC), *Official Journal of the European Community*, L129 (1976) 23

European Commission, Decision No 2455/2001/EC of the European Parliament and of the Council of 20 November 2001 establishing the list of priority substances in the field of water policy and amending Directive 2000/60/EC, *Official Journal of the European Community,* L331 (2001) 1

Farrant, J.M.; Pammenter, N.W. & Berjak, P. (1993) Seed development in relation to desiccation tolerance: A comparison between desiccation-sensitive (recalcitrant) seeds of *Avicennia marina* and desiccation-tolerant types, *Seed Science Research,* Vol.3, pp. 1–13, ISSN 1475-2735

Fernández-Alba, A.R.; Hernando, M.D.; Piedra, L. & Chisti, Y. (2002). Toxicity evaluation of single and mixed biocides measured with acute toxicity bioassays, *Analytica Chimica Acta,* Vol.456, pp. 303-312, ISSN 0003-2670

Górecki, R.J.; Piotrowicz-Cieślak, A.I.; Lahuta, L. & Obendorf, R.L. (1997). Soluble carbohydrates in desiccation tolerance of yellow lupin seeds during maturation and germination, *Seed Science Research,* Vol.7, pp. 107-115, ISSN 1475-2735

Hare, P.D.; Cress, W.A. & Van Staden, J. (1998). Dissecting the roles of osmolyte accumulation during stress, *Plant Cell and Environment,* Vol.21, pp. 535–553, ISSN 0140-7791

Hartwell, S. I. (2011). Chesapeake Bay Watershed Pesticide Use Declines But Toxicity Increases, *Environmental Toxicology and Chemistry,* Vol.30, No.5, pp. 1223–1231, ISSN 0730-7268

Hitz, W.D.; Carlson, T.J.; Kerr, P.S. & Sebastian, S.A. (2002). Biochemical and molecular characterization of a mutation that confers a decreased raffinosaccharide and phytic acid phenotype on soybean seeds, *Plant Physiology,* Vol.128, pp. 650–660, ISSN 0032-0889

Holgado, R.; Rowe, J.; Andersson, S. & Magnusson, C. (2004). Electrophoresis biotest studies on some population of cereal cyst nematode, *Heterodera spp.* (Tylenchida: *Heteroidae*), *Nematology,* Vol.6, No.6, pp. 857-865, ISSN 1388-5545

Hong, T.D. & Ellis, R.H. (1996). A protocol to determine seed storage behavior, *IPGRI Technical Bulletin N° 1. International Plant Genetic Resources Institute,* pp. 4-62, ISBN 92-9043-279-9

Horbowicz, M. & Obendorf, R.L. (1994). Seed desiccation tolerance and storability: dependence on flatulence-producing oligosaccharides and cyclitols - review and survey, *Seed Science Research,* Vol.7, pp. 385-405, ISSN 1475-2735

Horbowicz, M.; Brenac, P. & Obendorf, R.L. (1998). Fagopyritol B1, CHx-D-galactopyranosyl-(1-2)-D-*chiro*-inositol, a galactosyl cyclitol in maturing buckwheat seeds associated with desiccation tolerance, *Planta,* Vol.205, pp. 1-11, ISSN 0032-0935

Hornsby, AG.; Wauchope, R.D. & Herner, A.E. (1996). Pesticide properties in the environment, *Springer-Verlag,* New York, Berlin, Heidelberg, ISBN 0-387-94353-6

Jha, A. (2008). Ecotoxicological applications and significance of the comet assay, *Mutagenesis,* Vol.23, No.3, pp. 207-221, ISSN 0267-8357

Jin, C.; Chen, Q.; Sun, R.; Zhou, Q. & Liu, J. (2009). Eco-toxic effects of sulfadiazine sodium, sulfamonomethoxine sodium and enrofloxacin on wheat, Chinese cabbage and tomato, *Ecotoxicology,* Vol.18, No.7, pp. 878-885, ISSN 0963-9292

Kahle, J. & Zauke, G. (2002). Bioaccumulation of trace metals in the copepod *Calanoides acutus* from the Weddell Sea (Antarctica): comparison of two-compartment and hyperbolic toxicokinetic models, *Aquatic Toxicology,* Vol.59, No.1-2, pp.115-135, ISSN 0166-445X

Kahru, A.; Dubourgier, H-C.; Blinowa, I.; Ivask, A. & Kasemets, K. (2008). Biotests and biosensors for ecotoxicology of metal oxide nanoparticles, *Sensors,* Vol.8, No.8, pp. 5153-5170, ISSN 1424-8220

Kaza, M.; Mankiewicz-Boczek, J.; Izydorczyk, K. & Sawicki, J. (2007). Toxicity assessment of water samples from rivers in central Poland using a battery of microbiotests - a pilot study. *Polish Journal of Environmental Studies,* Vol.16, No.1, pp. 81-89, ISSN 1230-1485

Kolpin, D.W.; Barbash, J.E. & Gilliom, R.J. (1998). Occurrence of pesticides in shallow groundwater of the United States: Initial results from the National Water-Quality Assessment Program, *Environmental Science & Technology,* Vol.32, pp. 558-566, ISSN 0013-936X

Kratasyuk, V.; Esimbekova, E.; Gladyshev, M.; Khromichek, E.; Kuznetsov, A. & Ivanova, E. (2001). The use bioluminescent biotests for study of natural and laboratory aquatic ecosystems, *Chemosphere,* Vol.42, pp. 909-915, ISSN 0045-6535

Küster, E.; Dorusch, F.; Meiβner, B.; Weiss H.; Schüürmann, G. & Altenburger, R. (2003). Toxizitätsreduktion durch (Grundwasser-) Sanierung? Erfolgskontrolle von In-situ-Grundwasser-Sanierungsverfahren mithilfe von kontinuierlichen und diskontinuierlichen Biotest, *Grundwasser. Zeitschrift der Fachsektion Hydrogeologie,* Vol.1, pp.32-40, ISSN 1430-483X

Li, X-Q.; Ng, A.; King, R. & Durnford, D.G. (2008). A Rapid and Simple Bioassay Method for Herbicide Detection, *Biomarker Insights,* Vol.3, pp. 287–291, ISSN 1177-2719

Loewus, F.A.; Everard, J.D. & Young, K.A. (1990). Inositol metabolism: precursor role and breakdown. In: *Inositol Metabolism in Plants,* D.J. Morre, W.F. Boss, F.A. Loewus, (Eds.), 21-45, Wiley-Liss Inc, New York

Manusadžianas, L.; Balkelytė, L.; Sadauskas, K.; Blinova, I.; Põllumaa, L. & Kahru, A. (2003). Ecotoxicological study of Lithuanian and Estonian wastewaters: selection of the biotests, and correspondence between toxicity and chemical-based indices, *Aquatic Toxicology,* Vol.63, pp. 27-41, ISSN 0166-445X

McWilliam, R. & Baird, D. (2002). Postexposure feeding depression: a new toxicity endpoint for use in laboratory studies with *Daphnia magna, Environmental Toxicology & Chemistry,*Vol.21, No.6, pp.1198–1205, ISSN 0730-7268.

Monsanto (2005). Available from http://www.monsanto.com/monsanto/content/products/productivity/roundup/gly_halflife_bkg.pdf

Murthy, U.M.N.; Kumar, P.P. & Sun, W.Q. (2003). Mechanisms of seed aging under different storage conditions for *Vigna radiata* (L.) Wilczek: Lipid peroxidation, sugar hydrolysis, Maillard reactions and their relationship to glass state transition, *Journal of Experimental Botany,* Vol.54, pp. 1057–1067, ISSN 0022-0957

Nałęcz-Jawecki, G. & Persoone, G. (2006). Toxicity of selected pharmaceuticals to the anostracan crustacean *Thamnocephalus platyurus* comparison of sublethal and lethal effect levels with the 1h Rapidtoxkit and the 24h Thamnotoxkit microbiotests, *Environmental Science and Pollution Research,* Vol.13, pp. 22-27, ISSN 0944-1344

Nałęcz-Jawecki G.; Wadhia K.; Adomas B.; Piotrowicz-Cieślak, A.I. & Sawicki, J. (2010). Application of microbial assay for risk assessment biotest in evaluation of toxicity of human and veterinary antibiotics, *Environmental Toxicology*, Vol.25, pp. 487-494, ISSN 1520-4081

Nelson, D. E.; Rammesmayer, G. & Bohnert, H. J. (1998). Regulation of cell-specific inositol metabolism and transport in plant salinity tolerance, *Plant Cell*, Vol.10, No.5, 753-764., ISSN 1040-4651

Nendza, M. (2002). Inventory of marine biotest methods for the evaluation of dredged material and sediments, *Chemosphere*, Vol.48, pp. 865-883, ISSN 0045-6535

Obendorf, R.L. (1997). Oligosaccharides and galactosyl cyclitols in seed desiccation tolerance, *Seed Science Research*, Vol.7, pp. 63-74, ISSN 1475-2735

OECD 208: Terrestrial Plant,: Growth Test, Guideline for testing of chemicals, OECD, *Organization for Economic Cooperation and Development*, Paris, 1984

OECD 221: *Lemna sp.* Growth Inhibition Test. Guideline for the testing of chemicals, OECD, *Organization for Economic Cooperation and Development*, Paris, 2006

Osborne, D.J. & Boubriak, I.I. (1994). DNA and desiccation tolerance, *Seed Science Research*, Vol.4, pp. 175–185, ISSN 1475-2735

Panikulangara, T.J.; Eggers-Schumacher, G.; Wunderlich, M.; Stransky, H. & Schöffl, F. (2004). Galactinol synthase 1. A novel heat shock factor target gene responsible for heat-induced synthesis of raffinose family oligosaccharides in *Arabidopsis*, *Plant Physiology*, Vol.136, pp. 3148-3158, ISSN 0032-0889

Pardo, J. (2010). Biotechnology of water and salinity stress tolerance, *Current Opinion in Biotechnology*, Vol.21, No.2; pp. 185-196, ISSN 0958-1669

Paul, M.J. & Cockburn, W. (1989). Pinitol, a compatible solute in *Mesembryanthemum erystallinum* L, *Journal of Experimental Botany*, Vol.47, pp. 1093-1098, ISSN 0022-0957

Persoone, G.; Marsalek, B.; Blinowa, I.; Törökne, A.; Zarina, D.; Manusadžianas, L.; Nałęcz-Jawecki, G.; Tofan, L.; Stepanowa, N.; Tothova, L. & Kolar, B. (2003). A practical and user-friendly toxicity classification system with microbiotests for natural waters and wastewaters, *Environmental. Toxicology.*, Vol.18, No.6, pp. 395-402, ISSN 1520-4081

Piotrowicz-Cieślak, A.I.; Górecki, R.J. & Adomas, B. (1999). The content and composition of soluble carbohydrates in lupin seeds of different species and cultivars. *Plant Breeding and Seed Science*, Vol.43, No.2, pp. 29-37, ISSN 1429-3862

Piotrowicz-Cieślak, I.A.; Gracia-Lopez, P.M. & Gulewicz, K. (2003). Cyclitols, galactosyl cyclitols and raffinose family oligosaccharides in Mexican wild lupin seeds, *Acta Societatis Botanicorum Poloniae*, Vol.72, No.2, pp. 109-114, ISSN 0001-6977

Piotrowicz-Cieślak, A.I. (2004). Flatulence-producing galactosyl cyclitols. D-*chiro*-inositol fraction in maturing yellow lupin seed, In: *Recent advances of research in antinutritional factors in legume seeds and oilseeds*, Eds: M. Muzquiz,; G.D. Hill,; C. Cuadrado,; M.M. Pedros,; C. Burbano, (Eds.), 69-72, EAAP Publication 110, ISSN 0071-2477, Wageningen Academic Publishers

Piotrowicz-Cieślak, A.I. (2005). Changes in soluble carbohydrates in yellow lupin seed under prolonged storage, *Seed Science and Technology*, Vol. 33, pp. 141-145, ISSN 0251-0952

Piotrowicz-Cieślak, A.I.; Michalczyk, D.J.; Adomas, B. & Górecki, R.J. (2007). Different effects of soil drought on soluble carbohydrates of developing *Lupinus pilosus* and

Lupinus luteus embryos, *Acta Societatis Botanicorum Poloniae*, Vol.76, No.2, pp. 119–125, ISSN 0001-6977

Piotrowicz-Cieślak, A.I.; Rybiński, W. & Michalczyk, D.J. (2008). Mutations modulate soluble carbohydrates composition in seeds of *Lathyrus sativus*, *Acta Societatis Botanicorum Poloniae*, Vol.77, No.4, pp. 281–287, ISSN 0001-6977

Piotrowicz-Cieślak, A.I.; Adomas, B. & Michalczyk, D. J. (2010)a. Different glyphosate phytotoxicity to seeds and seedlings to selected plant species, *Polish Journal of Environmental Studies*, Vol.19, No.1, pp. 123-129, ISSN 1230-1485

Piotrowicz-Cieślak, A.I.; Adomas, B.; Nałęcz-Jawecki, G. & Michalczyk, D. J. (2010)b. Phytotoxicity of sulfamethazine soil pollutant to six legume plant species, *Journal of Toxicology & Environmental Heath, Part A*, Vol.73, No.17-18, pp.1220-1229, ISSN 1528-7394

Probert, R.J. & Longley, P.L. (1989). Recalcitrant seed storage physiology in three aquatic grasses (*Zizania palustris*, *Spartina angelica* and *Porteresia coarctata*), *Annals of Botany*, Vol.63, pp. 53–63, ISSN 0305-7364

Ramanjulu, S., Bartels, D. (2002). Drought- and desiccation-induced modulation of gene expression in plants, *Plant, Cell & Environment*, Vol.25, pp. 141–151, ISSN 0140-7791

Roberts, E.H. (1973). Predicting the storage life of seeds, *Seed Science and Technology*, Vol.1, pp. 499-514, ISSN 0251-0952

Sanhewe, A.J. & Ellis, R.H. (1996). Seed development and maturation in *Phaseolus vulgaris* I. Ability to germinate and to tolerate desiccation, *Journal of Experimental Botany*, Vol.47, No. 300, pp. 949-958, ISSN 0022-0957.

Sekutowski, T. & Sadowski, J. (2009). Phytotoxkit ™ microbiotest used in detecting herbicide residue in soil, *Environment Protection Engineering*, Vol.35, No.1, pp. 105-1010, ISSN 0324-8828

Sheets, T.J. & Shaw, W.C. (1963). Herbicidal properties and persistence in soil of s-triazines. *Weed Science*, Vol..11, pp. 15-21, ISSN 0043-1745

Simeonov, V.; Wolska, L.; Kuczyńska, A.; Gurwin, J.; Tsakovski, S. & Namieśnik, J. (2007). Chemo metric estimation of natural water samples using toxicity tests and physicochemical parameters, *Critical Reviews in Analytical Chemistry*, Vol.37, No.2, pp. 81-90, ISSN 1040-8347

Stork, P. & Hannah, M.C. (1996). Bioassay method for formulation testing and residue studies of sulfonylurea and sulfonanylide herbicides. *Weed Research*, Vol.36, No.3, pp. 271-281, ISSN 0043-1737

Streeter, J.G., Lohnes, D.G. & Fioritto, R.J. (2001). Patterns of pinitol accumulation in soybean plants and relationships to drought tolerance, *Plant Cell & Environment*, Vol.24, pp. 429–438, ISSN 0140-7791

Sundt, R.C.; Meier, S.; Jonsson, G.; Sanni, S. & Beyer, J. (2009). Development of a laboratory exposure system using marine fish to carry out realistic effect studies with produced water discharged from offshore production, *Marine Pollution Bulletin*, Vol.58, No.9, pp. 1382-1388, ISSN 0025-326X

Taji, T.; Ohsumi, C.; Iuchi, S.; Seki, M.; Kasuga, M.; Kobayashi, M.; Yamaguchi-Shinozaki, K. & Shinozaki, K. (2002). Important roles of drought - and cold-inducible genes for galactinol synthase in stress tolerance in *Arabidopsis thaliana*, *The Plant Journal*, Vol.29, pp. 417-426, ISSN 0960-7412

Wadhia, K., & Thompson, C.K. (2007). Low-cost ecotoxicity testing of environmental samples using microbiotest for potential implementation of the Water Framework Directive, *Trends in Analytical Chemistry*, Vol.26, No.4, pp. 300-307, ISSN 0165-9936

Walters, C. & Towill, L. (2004). Seeds and Pollen, Available from http://www.ba.ars.usda.gov/hb66/153seeds.pdf

Wang, C.; Wang, Y.; Kiefer, F.; Yediler, A., Wang, Z. & Kettrup, A. (2003). Ecotoxicological and chemical characterization of selected treatment process effluents of municipal sewage treatment plant, *Ecotoxicology and Environmental Safety*, Vol.56, No.2, pp. 211-217, ISSN 0147-6513

Wolska, L.; Sagajdakow, A.; Kuczyńska, A. & Namieśnik, J. (2007). Application of ecotoxicological studies in integrated environmental monitoring: Possibilities and problems, *Trends in Analytical Chemistry*, Vol.26, No.4, pp. 332-344, ISSN 0165-9936

Yamamoto, Y.; Rajbhandari, N.; Xiaohong, L.; Bergmann, B.; Nishimura, Y.; Stomp, A-M. (2001). Genetic transformation of duckweed *Lemna gibba* and *Lemna minor*, *In Vitro Cellular & Developmental Biology - Plant*, Vol.37, No.3, pp. 349-353, ISSN 1054-5476

Zuther, E.; Büchel, K.; Hundertmark, M.; Stitt, M.; Hincha, D.K. & Heyer, A.G. (2004). The role of raffinose in the cold acclimation of *Arabidopsis thaliana*, *FEBS Lettres*, Vol.576, No.1, pp. 169–173, ISSN 0014-5793

3

Use of Sugar Beet as a Bioindicator Plant for Detection of Flucarbazone and Sulfentrazone Herbicides in Soil

Anna M. Szmigielski[1], Jeff J. Schoenau[1] and Eric N. Johnson[2]
[1]*Soil Science Department, University of Saskatchewan, Saskatoon, SK*
[2]*Agriculture and AgriFood Canada, Scott, SK*
Canada

1. Introduction

Determination of herbicide residues can be challenging due to the very low herbicide concentrations that can persist and remain bioactive in soil. Detection of residual herbicides is of great importance since these miniscule herbicide amounts may cause injury to sensitive rotational crops. Plant bioassays are a valuable alternative to instrumental procedures for determination of herbicides in soil. Instrumental methods such as gas chromatography or high performance liquid chromatography require solvent or solid phase extractions before sample analysis, and these highly efficient extractions enable the determination of total amount of herbicide in soil. In contrast, bioavailable herbicide is determined by bioassay procedures because plant response varies with soil type and generally decreases in soils of high organic matter and clay contents and low soil pH (Thirunarayanan et al. 1985; Renner et al. 1988; Che et al. 1992; Wang & Liu 1999; Wehtje et al. 1987; Grey et al. 1997; Szmigielski et al. 2009). Typically bioassay detection of herbicides that belong to different groups with different modes of action requires use of different plant species and/or measuring different plant parameters. Use of herbicides with different modes of action applied either in rotation or as pre-mixed combinations has become a common practice in farming to combat weed resistance problems. Thus performing more than one bioassay may be necessary for assessment of herbicide residues in soil after field applications of herbicides with different modes of action.

Flucarbazone is used in western Canada for control of certain grass and broadleaf weeds in wheat (*Triticum aestivum* L.) and its recommended application rate is 20 g ai ha^{-1}. Flucarbazone belongs to an acetolactate synthase (ALS) group of herbicide; these herbicides inhibit the biosynthesis of branched amino acids (valine, leucine and isoleucine) and affect primarily root growth of susceptible plants through inhibition of cell division at the root tips. Flucarbazone is a weak acid (pKa = 1.9) and therefore it is present mostly in the anionic form at environmentally relevant pH levels (Senseman 2007). Flucarbazone dissipation rate in soil is fast and the flucarbazone half-life in different soil types has been reported to range from 6 to 110 days (Eliason et al. 2004). However, as is the case with other ALS-inhibiting herbicides (Goetz et al. 1990; Anderson & Humburg 1987; Anderson & Barrett 1985; Loux & Reese 1992; Walker & Brown 1983), flucarbazone may persist in soil particularly under conditions of low moisture and cool temperature. Residual activity of a herbicide in soil is desirable in providing weed control late in the season; however, if the herbicide persists to the following year, it may damage rotational crops as has been reported for various sensitive crops seeded one year after an ALS-herbicide application including canola (*Brassica napus* L.), flax (*Linum usitatissimum* L.), lentil (*Lens culinaris* Medic), oriental mustard (*Brassica juncea* L.), corn (*Zea mays* L.), and sugar beet (*Beta vulgaris* L.) (Bresnahan et al. 2000; Moyer et al. 1990; Moyer & Esau 1996; Moyer & Hamman 2001).

Sulfentrazone is a soil applied herbicide and is registered in western Canada for control of grass and broadleaf weeds in chickpea (*Cicer arietinum* L.), field pea (*Pisum sativum* L.), and flax at application rates of 105 to 140 g ai ha^{-1}. It is a protox herbicide and its mode of action is the inhibition of protoporphyrinogen oxidase that leads to the disruption of lipid cell membranes and consequently causes shoot desiccation after plants emerge from soil and are exposed to light. Sulfentrazone is a weak acid with a pKa of 6.56; therefore, it exists predominantly in ionized form in soils with a pH higher than the pKa (Senseman 2007). Sulfentrazone is relatively persistent in soil, with a half-life reported in the range of 24 to 302 days (FMC Corporation 1999; Martinez et al. 2008; Ohmes et al. 2000). Because of slow sulfentrazone dissipation in some soils especially under conditions of drought and cool weather, a potential risk of carry-over to rotational crops is of concern. Injury to cotton (*Gossypium hirsutum* L.) (Ohmes et al. 2000; Main et al. 2004; Pekarek et al. 2010), sugar beet, and sorghum (*Sorghum bicolor* L.) (FMC Corporation 1999) has been reported one year after sulfentrazone application, and consequently extended recropping intervals are advised for these crops. In western Canada, lentil has exhibited sensitivity to sulfentrazone residues (Johnson E.N. unpublished data) and re-cropping intervals of 36 months are recommended.

This review presents our research on (1) the development of a sugar beet bioassay for detection of flucarbazone and sulfentrazone in soil, (2) the assessment of flucarbazone and sulfentrazone interactions in soil, (3) the evaluation of the N-fertilizer effect on detection of flucarbazone and sulfentrazone, and (4) the investigation of the landscape effect on flucarbazone and sulfentrazone bioactivity and dissipation in Canadian prairie soils.

2. Sugar beet bioassay

Selecting suitable plant species for a bioassay is critical, and the plant parameter measured in a bioassay has to be sensitive and correlate well with herbicide concentration. Typically, ALS-herbicides are detected using root inhibition bioassays, and various susceptible plant species including oriental mustard (Eliason et al. 2004; Szmigielski et al. 2008), corn (Mersi & Foy 1985; Hsiao & Smith 1983), red beet (Jourdan et al. 1998), and sunflower (*Helianthus annuus* L.) (Hernández-Sevillano et al. 2001; Günther et al. 1993) have been used. Protox-inhibiting herbicides influence mainly shoot development of sensitive plants, and cotton (Main et al. 2004; Grey et al. 2007) and sugar beet (Szmigielski et al. 2009; Blanco & Velini 2005) have been reported as a suitable species for sulfentrazone detection in soil.

We investigated the use of sugar beet (*cv.* Beta 1385) as a bioindicator plant for detection of both flucarbazone and sulfentrazone in one bioassay, by measuring both root and shoot length reduction of sugar beet in response to these two herbicides. This bioassay is performed in 4-oz Whirl-Pak™ plastic bags that are 16 cm long and 6 cm wide. A quantity of 100 g of soil is wetted to 100% field capacity, and then hand mixed in a plastic dish and transferred to a Whirl-Pak™ bag. The soil in the bag is gently packed to form a rectangular layer approximately 14 cm deep and 1 cm thick (Fig. 1a).

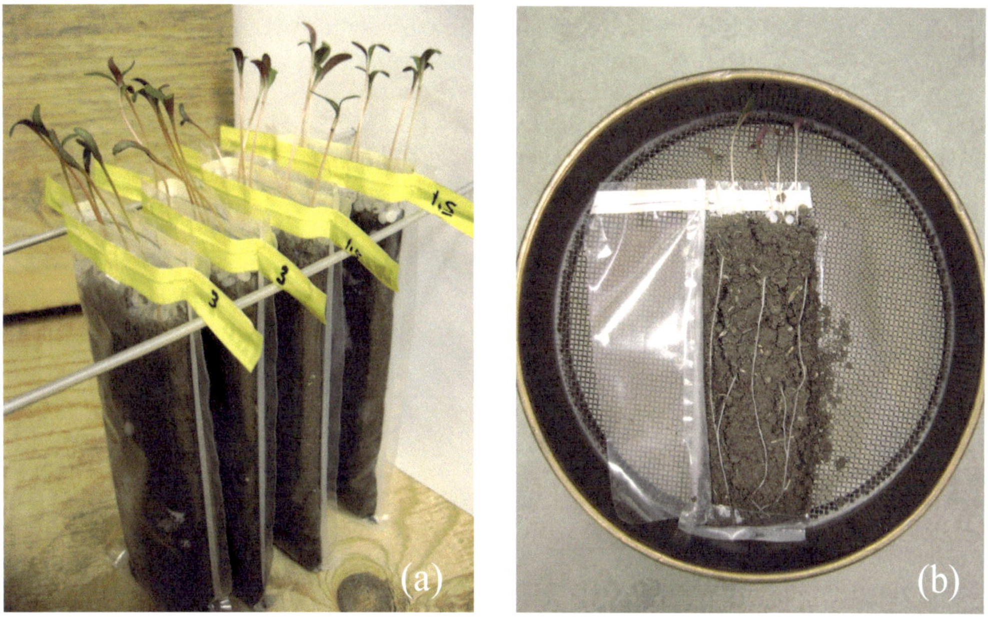

Fig. 1. (a) Sugar beet bioassay performed in WhirlPak™ bags; (b) Opened WhirlPak™ bag on sieve before plant removal from soil with water.

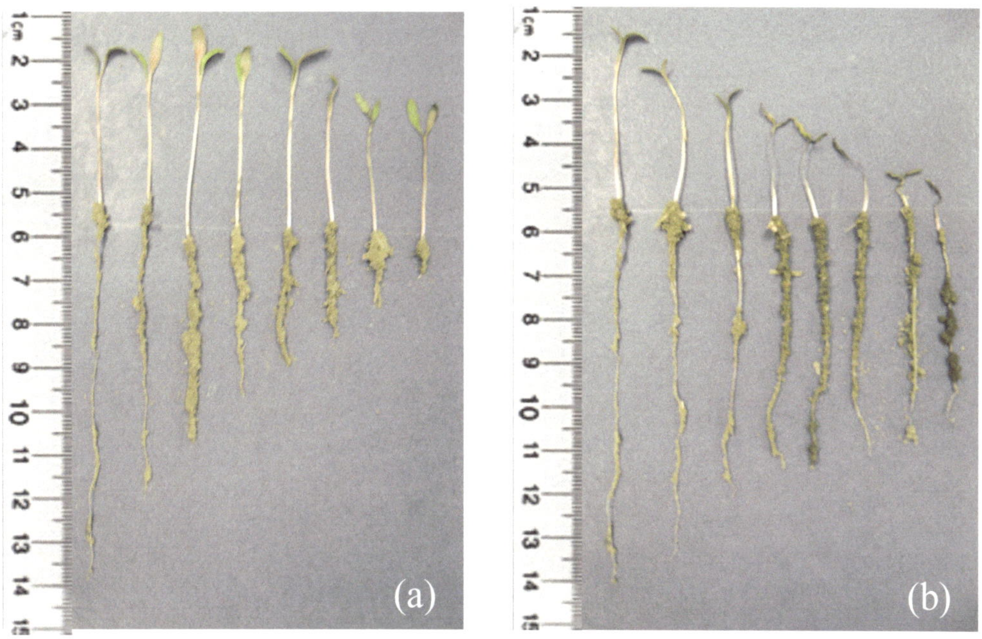

Fig. 2. Response of sugar beet plants to increasing concentration of (a) flucarbazone in the range from 0 to 15 ppb, and (b) sulfentrazone in the range from 0 to 200 ppb in soil.

Six sugar beet seeds are planted and the soil surface is covered with a layer of plastic beads to reduce water evaporation. Plants are then grown in a fluorescent canopy with light intensity of approximately 16 µmol $m^{-2} s^{-1}$ (Szmigielski et al. 2009) and watered to 100% field capacity daily by adding water to the predetermined weight. After a 6-day growth period, length of shoots is measured with a ruler from the soil level to the node where the cotyledons split from the stem. Next, plants are recovered from soil after the Whirl-Pak™ bag is cut open with scissors and placed on a sieve (Fig. 1b); soil is then washed away with water and root length measured with a ruler (Fig. 2a and 2b).

Sugar beet response to flucarbazone in the range from 0 to 15 ppb and to sulfentrazone in the range from 0 to 200 ppb was assessed, and root and shoot inhibition calculated using the formula (Beckie & McKercher 1989):

$$\text{Inhibition (\%)} = (1 - L_t/L_0) \times 100\% \quad (1)$$

where L_t is the root or shoot length in the herbicide-treated soil and L_0 is the root or shoot length in the untreated (control) soil. Dose-response curves can be constructed using a log-logistic regression model (Seefeldt et al. 1995):

$$\text{Inhibition (\%)} = 100\% - (C + [D - C]/[1 + \{x/I_{50}\}^b]) \quad (2)$$

where x is herbicide concentration, (100 - C) is the upper limit of the log-logistic inhibition curve, (100 - D) is the lower limit of the log-logistic inhibition curve, I_{50} is the concentration required for 50% plant growth inhibition, and b is the slope of the curve around the I_{50} value. Measuring both roots and shoots of sugar beet revealed that although flucarbazone primarily inhibits root length it also causes shoot reduction (Fig. 3a), and that while sulfentrazone primarily inhibits shoot length it also affects root development (Fig. 3b).

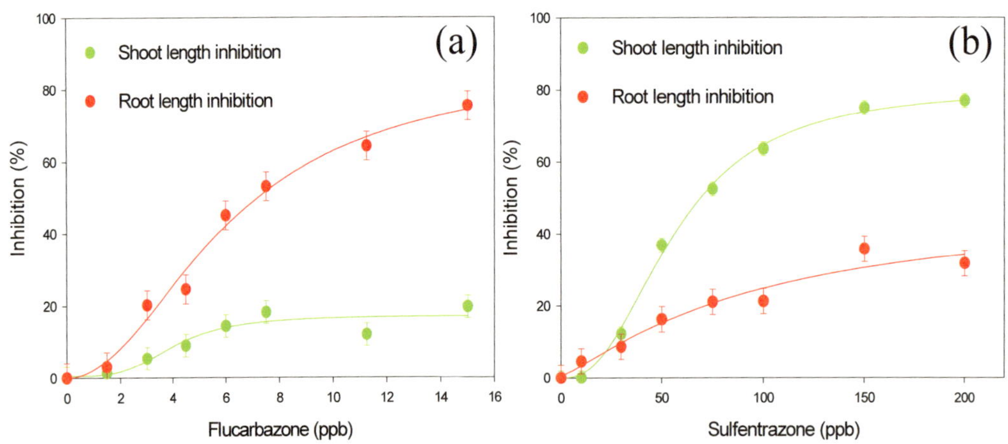

Fig. 3. Shoot and root length inhibition of sugar beet in response to (a) increasing concentration of flucarbazone, and (b) increasing concentration of sulfentrazone in soil.

This sugar beet bioassay is sensitive; concentration of approximately 2 ppb of flucarbazone based on root length measurements and of approximately 20 ppb of sulfentrazone using shoot length measurements was detected. However, bioassay detection limits vary with soil type and are generally lower in sandy soils of low organic matter content and high pH (Jourdan et al. 1998; Eliason et al. 2004; Szmigielski et al. 2009).

Since flucarbazone and sulfentrazone decrease both root and shoot length of sensitive plants such as sugar beet, sequential or simultaneous applications of these two herbicides could potentially result in herbicide interactions.

3. Flucarbazone and sulfentrazone interactions

Repeated applications of herbicides with the same mode of action have resulted in weeds developing resistance (Vencill et al. 2011; Colborn & Short 1999; Whitcomb 1999). Using herbicides with different mode of action either applied as pre-mixed combinations or applied in rotation reduces problems related to weed resistance and consequently improves weed control. However, combinations of herbicides are generally chosen to improve the spectrum of weed control without prior knowledge of the possible consequences of the interactions between herbicides (Zhang et al. 1995). The outcome of the interactions may be synergistic, antagonistic or additive depending on whether the combined effect on the target plants is greater, less than, or equal to the summed effect of the herbicides applied alone (Colby 1967; Nash 1981). A synergistic interaction occurs when the activity of two herbicides is more phytotoxic than either herbicide applied singly. A synergistic effect is beneficial in that it provides more effective weed control at lower herbicide concentrations; however it may also cause injury to sensitive rotational crops if the synergism of the two residual herbicides is not known (Zhang et al. 1995). In an additive interaction, also called "herbicide stacking" (Johnson et al. 2005), the injury observed in the target plants is the sum activity of the combined herbicides. With an antagonistic interaction, the efficacy of the combined herbicides is reduced and consequently results in decreased weed control but can also help to avoid unwanted crop injury (Zhang et al. 1995).

To examine interactions between soil-incorporated flucarbazone and sulfentrazone, we evaluated the combined effect of these two herbicides on sugar beet root and shoot inhibition. Root length inhibition was assessed in soil that was spiked with mixtures consisting of flucarbazone in the range from 0 to 15 ppb with sulfentrazone added at 50 ppb level, while shoot length inhibition was evaluated in soil that was amended with mixtures consisting of sulfentrazone in the range from 0 to 200 ppb with flucarbazone added at 6 ppb level. The expected inhibition was calculated using Colby's formula (Colby 1967):

$$E = X + Y - XY/100 \qquad (3)$$

where X is the plant growth inhibition (%) due to compound A and Y is the plant growth inhibition (%) due to compound B; comparing expected inhibition to the observed inhibition allows the nature of interactions to be revealed. The combined effect of flucarbazone and sulfentrazone was additive: the observed and expected root length inhibition of sugar beet in response to flucarbazone in combination with sulfentrazone were similar (Fig. 4a), as were the observed and the expected shoot length inhibition due to sulfentrazone in combination with flucarbazone (Fig. 4b). I_{50} values for observed and expected responses were not different at 0.05 level based on the asymptotic z-test. The additive effect of flucarbazone and sulfentrazone will help in weed control but may also increase risk of injury to rotational crops that are sensitive to both these herbicides.

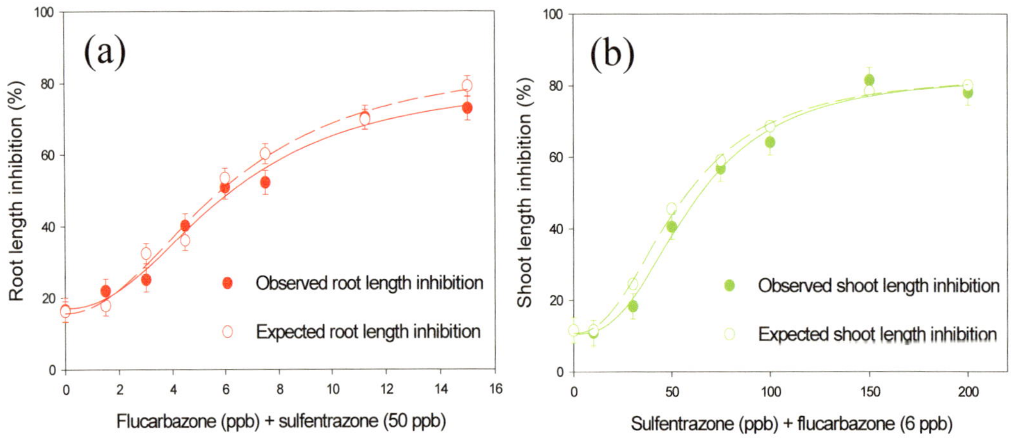

Fig. 4. (a) Root length inhibition of sugar beet in response to increasing concentration of flucarbazone in combination with 50 ppb sulfentrazone, and (b) shoot length inhibition of sugar beet in response to increasing concentration of sulfentrazone in combination with 6 ppb flucarbazone.

4. Effect of ammonium containing fertilizer on sugar beet bioassay

Typically plant response that is measured in a bioassay is not specific to one source. The lack of specificity may be desirable in that the presence of residues of all herbicides that detrimentally affect the same plant parameter are detected. However, other soil applied chemicals apart from herbicides may also alter the parameter measured in a bioassay and may change the outcome of the bioassay. We have reported that the detection of ALS-inhibiting herbicides in soil using a mustard root bioassay is influenced by N-fertilizer as mustard root length is shortened in response to ammonium ions (Szmigielski et al. 2011). Ammonium toxicity to plants is common and a change in root/shoot ratio is one of the symptoms of NH_4^+ toxicity (Britto & Kronzucker 2002).

To assess the effect of N-fertilizer on sugar beet roots and shoots, and consequently on flucarbazone and sulfentrazone detection in soil, ammonium nitrate was added to soil in the range from 0 to 200 ppm N, and root and shoot length was measured. Ammonium nitrate significantly reduced root length of sugar beet but the shoot length inhibition due to ammonium nitrate was very small and was less than 20% at the highest ammonium nitrate concentration tested (Fig. 5).

The combined response of sugar beet roots to flucarbazone and ammonium nitrate was examined by growing sugar beet plants in soil that was spiked with flucarbazone in the range of 0 to 15 ppb and mixed with ammonium nitrate added at 50 ppm N. The expected response due to flucarbazone in combination with ammonium nitrate was calculated using equation [3]. Since the expected root length inhibition was the same as the observed (Fig. 6), the combined effect of flucarbazone and N-fertilizer on sugar beet root length is additive.

Thus, root length reduction of sugar beet that is measured in a soil that received a recent application of ammonium containing or ammonium producing fertilizer may be misinterpreted as reduction due to herbicide residues and may yield false positive results. Because N-fertilizer interferes with the sugar beet root length bioassay, preferably soil sampling for the detection of residual herbicides should be completed preplant and before N-fertilizer field application, or at the end of the growing season.

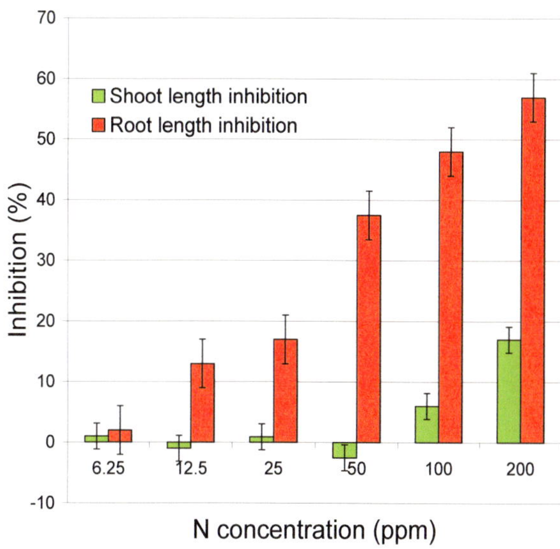

Fig. 5. Effect of increasing ammonium nitrate concentration in soil on shoot and root inhibition of sugar beet plants.

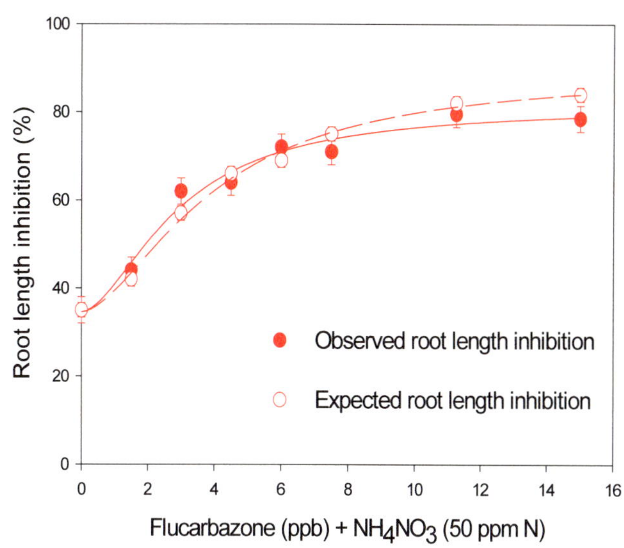

Fig. 6. Root length inhibition of sugar beet in response to increasing concentration of flucarbazone in combination with 50 ppm N added as ammonium nitrate.

5. Effect of landscape position on phytotoxicity and dissipation of flucarbazone and sulfentrazone

Farm fields with irregular rolling topography of low hills and shallow depressions are typical on the Canadian prairies (Fig. 7). Low-slope soils from depressions in the landscape typically have higher organic matter and clay contents and lower pH than up-slope soils from elevated parts of the terrain (Schoenau et al. 2005; Moyer et al. 2010). Furthermore, low-slope areas in the field generally have higher moisture content as a result of water accumulating in the depressions, while the up-slope areas are drier due to water runoff.

Fig. 7. Undulating landscape comprised of knolls and depressions in southwestern Saskatchewan (source: Geological Survey Canada).

Phytotoxicity of ALS- and protox-inhibiting herbicides is soil dependent, and the effect of organic matter, clay and soil pH on adsorption and bioavailability of these herbicides is well documented (Thirunarayanan et al. 1985; Renner et al. 1988; Che et al. 1992; Wang & Liu 1999; Wehtje et al. 1987; Grey et al. 1997; Szmigielski et al. 2009). Typically organic matter and clay decrease the concentration of bioavailable herbicide through adsorption of herbicide molecules to the reactive functional groups and colloidal surfaces. At alkaline soil pH, adsorption of weak acidic herbicides tends to decrease due to increased herbicide solubility in soil solution and due to repulsion of anionic herbicide molecules from negatively charged soil particles.

Dissipation of ALS- and protox-inhibiting herbicides in soil is governed by microbial and chemical processes. Microbial degradation is the primary mechanism as dissipation has been shown to be faster in non-sterile soil than in autoclaved soil (Joshi et al. 1985; Ohmes at al. 2000; Brown 1990). The dissipation rate of ALS- and protox-inhibiting herbicides varies with soil type and environmental conditions. Generally high organic matter content, high clay content and low soil pH decrease the dissipation rate by reducing the amount of herbicide available in soil solution for decomposition (Eliason et al. 2004; Goetz et al. 1990; Beckie & McKercher 1989; Ohmes et al. 2000; Grey et al. 2007; Main et al. 2004). Microbial and chemical decomposition both depend on soil water and temperature with faster dissipation occurring in moist and warm soils (Beckie & McKercher 1989; Joshi et al. 1985;

Walker & Brown 1983; Brown, 1990; Thirunarayanan et al. 1985). In flooded (saturated) soils decomposition may be reduced due to anaerobic conditions.

To examine the effect of landscape position on phytotoxicity and dissipation of flucarbazone and sulfentrazone, we used two soils that were collected from a farm field with varying topography in southern Saskatchewan, Canada. Soil from an up-slope position contained 0.9% organic carbon, 31% clay and had pH 7.9, while soil from a low-slope position contained 1.6% organic carbon, 51% clay and had pH 7.2. Flucarbazone phytotoxicity was assessed in the range from 0 to 15 ppb by measuring root length inhibition while sulfentrazone phytotoxicity was determined in the range from 0 to 200 ppb by measuring shoot length inhibition of sugar beet. Phytotoxicity of flucarbazone (Figure 8a) and of sulfentrazone (Figure 8b) was higher in the up-slope soil than in the low-slope soil. The I_{50} values determined from the dose-response curves were 3.5 and 5.7 ppb for flucarbazone, and 34.3 and 56.5 ppb for sulfentrazone in the up-slope and low-slope soil, respectively, and were different at 0.05 level of significance. Thus landscape position in a field has a considerable effect on bioavailability of flucarbazone and sulfentrazone, and different herbicide application rates may be required in fields of variable topography to achieve uniform weed control.

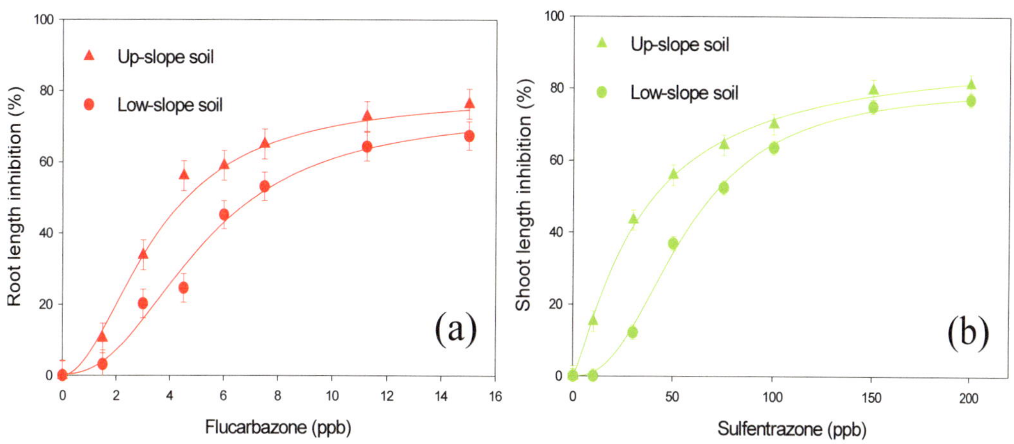

Fig. 8. Dose-response curves for (a) flucarbazone determined by root length, and (b) sulfentrazone determined by shoot length of sugar beet in soil from two landscape positions.

Flucarbazone and sulfentrazone dissipation in the two soils was examined under laboratory conditions of 25 C and moisture content of 85% field capacity. Soils were spiked with 15 ppb of flucarbazone and separately with 200 ppb of sulfentrazone, and at each sampling time the residual flucarbazone and sulfentrazone was determined using the sugar beet bioassay. Flucarbazone and sulfentrazone dissipation followed the bi-exponential decay model described in detail by Hill & Schaalje (1985):

$$C = a\,e^{-bt} + c\,e^{-dt} \quad (4)$$

where C is herbicide concentration remaining in soil after time t. In the bi-exponential decay model the dissipation rate is not constant and is fast initially and slow afterward, while in the first order decay model (when b = d in equation [4]) the dissipation rate does not change with time. Flucarbazone and sulfentrazone dissipation was more rapid in the up-slope soil than in the low-slope soil (Fig. 9a and 9b); flucarbazone half-life was 5 and 8 days, and sulfentrazone half-life was 21 and 90 days in the up-slope and the low-slope soil, respectively. Thus landscape positions in the field influence persistence of flucarbazone and sulfentrazone, and consequently may affect the potential for herbicide carry-over to the next growing season. However, because damage to sensitive rotational crops occurs when a herbicide is available to plants at harmful concentrations one year after application, risk of carry-over injury is controlled by the combined effect of herbicide dissipation and herbicide phytotoxicity, both of which are soil dependent; also the rotational crop must be susceptible to the residual herbicide concentration at the time of planting (Hartzler et al. 1989). Although flucarbazone and sulfentrazone persist longer in soil from depressions in the field, herbicide bioavailability is reduced in this soil, and thus residual flucarbazone or sulfentrazone may not pose a risk of injury to sensitive crops in low-slope areas. Predicting carry-over injury due to flucarbazone and sulfentrazone in farm fields with varying topography is a complex task and all factors that affect herbicide persistence and bioavailability have to be considered before choosing a rotational crop to grow.

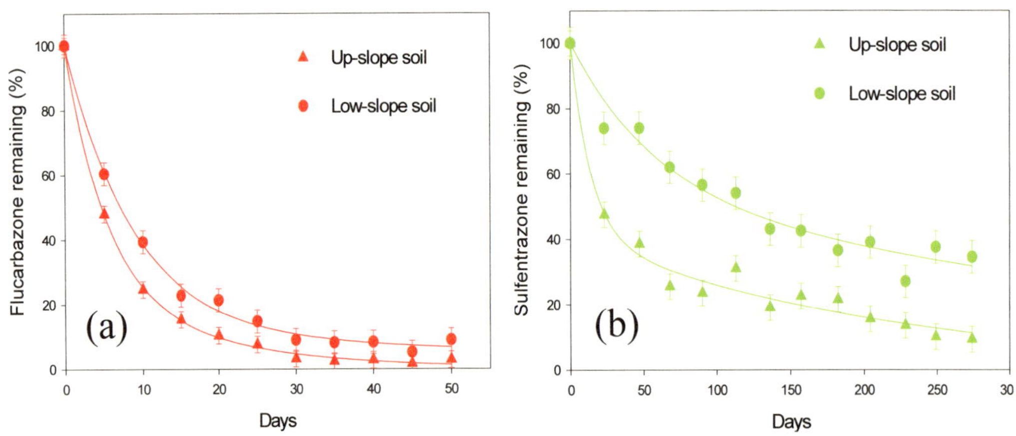

Fig. 9. Dissipation under laboratory conditions of (a) flucarbazone determined by root length, and (b) sulfentrazone determined by shoot length of sugar beet in soil from two landscape positions.

6. Practical considerations

Because sugar beet plants respond both to flucarbazone and sulfentrazone, a sugar beet bioassay allows for detection of these two herbicides in soil by evaluating both root and shoot inhibition. Growing sugar beet plants in Whirl-Pak™ bags is simple and provides a convenient method for assessing shoot and root length. Shoots are measured above the soil level and do not need to be harvested; this helps particularly with measuring shoots that are short and brittle at phytotoxic sulfentrazone concentrations. Roots are recovered from soil with water and consequently roots do not get broken or damaged before being measured.

Furthermore, as the bioassay is completed before roots grow to the bottom of the bag, root development in Whirl-Pak™ bags is not obstructed.

7. Conclusions

Using the sugar beet bioassay we determined: (1) that while flucarbazone primarily inhibits root length it also causes shoot reduction and while sulfentrazone primarily inhibits shoot length it also affects root development, (2) that the combined effect of soil-incorporated flucarbazone and sulfentrazone on root and shoot length inhibition of sugar beet is additive, (3) that N-fertilizer reduces root length of sugar beet but has little effect on shoot length and therefore the presence of freshly applied N-fertilizer may yield false positive results for flucarbazone residues, and (4) that flucarbazone and sulfentrazone phytotoxicity is higher and dissipation rate is faster in soils from up-slope than low-slope landscape positions under identical moisture and temperature conditions.

8. Acknowledgements

The financial support of FMC Corporation Canada, Arysta LifeScience Canada, and NSERC is gratefully acknowledged.

9. References

Anderson, R.L. & Barrett, M.R. (1985). Residual phytotoxicity of chlorsulfuron in two soils. *J. Environ. Qual.* Vol.14, pp.111-114, ISSN: 0047-2425.

Anderson, R.L. & Humburg, N.E. (1987). Field duration of chlorsulfuron bioactivity in the central Great Plains. *J. Environ. Qual.* Vol.16, pp.263-266, ISSN: 0047-2425.

Beckie, H.J. & McKercher, R.B. (1989). Soil residual properties of DPX-A7881 under laboratory conditions. *Weed Sci.* Vol.37, pp.412-418, ISSN: 0043-1745.

Blanco, F.M.G. & Velini, E.D. (2005). Sulfentrazone persistence in soybean-cultivated soil and effect on succession cultures. *Planta Daninha.* Vol.23, pp.693-700, ISSN: 0100-8358.

Bresnahan, G.A.; Koskinen, W.C.; Dexter, A.G. & Lueschen, W.E. (2000). Influence of soil pH– sorption interactions on imazethapyr carry-over. *J. Agric. Food Chem.* Vol.48, pp.1929-1934, ISSN: 0021-8561.

Britto, D.T. & Kronzucker, H.J. (2002). NH_4^+ toxicity in higher plants: a critical review. *J. Plant Physiol.* Vol.159, pp.567-584, ISSN: 0176-1617.

Brown, H.M. (1990). Mode of action, crop selectivity and soil relations of the sulfonylurea herbicides. *Pestic. Sci.* Vol.29, pp.263-281, ISSN:0031-613X.

Che, M.; Loux, M.M.; Traina, S.J. & Logan, T.J. (1992). Effect of pH on sorption and desorption of imazaquin and imazethapyr on clays and humic acid. *J. Environ. Qual.* Vol.21, pp.698-703, ISSN: 0047-2425.

Colborn, T. & Short, P. (1999). Pesticide use in the U.S. and policy implications: A focus on herbicides. *Toxicol. Ind. Health.* Vol.15, pp.241-276, ISSN: 0748-2377.

Colby, S.R. (1967). Calculating synergistic and antagonistic responses of herbicide combinations. *Weeds.* Vol.15, pp. 20-22, ISSN: 0043-1745.

Eliason, R.; Schoenau, J.J.; Szmigielski, A.M. & Laverty, W.M. (2004). Phytotoxicity and persistence of flucarbazone-sodium in soil. *Weed Sci.* Vol.52, pp.857-862, ISSN: 0043-1745.

FMC Corporation. (1999). Sulfentrazone, Product Profile. Philadelphhia: Agricultural Product Group.

Geological Survey Canada. Canadian Landscapes Fact Sheets. Moraines. Available from: http://gsc.nrcan.gc.ca/landscapes/pdf/moraines_e.pdf.

Goetz, A.J.; Lavy, T.L. & Gbur, E.E. (1990). Degradation and field persistence of imazethapyr. *Weed Sci.* Vol.38, pp.421-428, ISSN: 0043-1745.

Grey, T.L.; Vencill, W.K.; Mantrepagada, N. & Culpepper, A.S. (2007). Residual herbicide dissipation from soil covered with low-density polyethylene mulch or left bare. *Weed Sci.* Vol.55, pp.638-643, ISSN: 0043-1745.

Grey, T.L.; Walker, R.H.; Wehtje, G.R. & Hancock, H.G. (1997). Sulfentrazone adsorption and mobility as affected by soil and pH. *Weed Sci.* Vol.45, pp.51-56, ISSN: 0043-1745.

Günther, P.; Pestemer, W.; Rahman, A. & Nordmeyer, H. (1993). A bioassay technique to study the leaching behaviour of sulfonylurea herbicides in different soils. *Weed Res.* Vol.33, pp.177-185, ISSN: 0043-1737.

Hartlzler, R.G.; Fawcett, R.S. & Owen, M.D. (1989). Effects of tillage on trifluralin residue carryover injury to corn. *Weed Sci.* Vol.37, pp.609-615, ISSN: 0043-1745.

Hernández-Sevillano, E.; Villarroya, M.; Alonso-Prados, J.L. & García-Baudín, J.M. (2001). Bioassay to detect MON-37500 and triasulfuron residues in soil. *Weed Technol.* Vol.15, pp. 447–452, ISSN: 0890-037X .

Hill, B.D. & Schaalje, G.B. (1985). A two-compartment model for the dissipation of deltamethrin on soil. *J. Agric. Food Chem.* Vol.33, pp.1001-1006, ISSN: 0021-8561.

Hsiao, A.I. & Simth, A.E. (1983). A root bioassay procedure for the determination of chlorsulfuron, diclofop acid and sethoxydim residues in soils. *Weed Res.* 23:231-236. ISSN: 0043-1737

Johnson, E.N.; Moyer, J.R; Thomas, A.G.; Leeson, J.Y.; Holm, F.A.; Sapsford, K.L.; Schoenau, J.J.; Szmigielski, A.M.; Hall, L.M.; Kuchuran, M.E. & Hornford, R.G. (2005). Do repeated applications of residual herbicides result in herbicide stacking? In *Soil Residual Herbicides: Science and Management. Topics in Canadian Weed Science, ed.* R.C. Van Acker, 53-70, Volume 3. Sainte-Anne-de Bellevue, Québec: Canadian Weed Science Society - Société canadienne de malherbologie, ISBN: 0-9688970-3-7.

Joshi, M.M.; Brown, H.M. & Romesser, J.A. (1985). Degradation of chlorsulfuron by soil microorganisms. *Weed Sci.* Vol.33, pp.888-893, ISSN: 0043-1745.

Jourdan, S.W.; Majek, B.A. & Ayeni, A.O. (1998). Imazethapyr bioactivity and movement in soil. *Weed Sci.* Vol.46, pp.608-613, ISSN: 0043-1745.

Loux, M.M. & Reese, K.D. (1992). Effect of soil pH on adsorption and persistence of imazaquin. *Weed Sci.* Vol.40, pp.490-496, ISSN: 0043-1745.

Main, C. L.; Mueller, T. C.; Hayes, R. M.; Wilcut, J. W.; Peeper, T. F.; Talbert, R. E. & Witt, W.W. (2004). Sulfentrazone persistence in southern soils: bioavailable concentration and effect on a rotational cotton crop. *Weed Technol.* Vol.18, pp.346-352, ISSN: 0890-037X.

Martinez, C.O.; Silva, C.M.M.S.; Fay, E.F.; Maia, A.H.N.; Abakerli, R.B. & Durrant, L.R. (2008). Degradation of the herbicide sulfentrazone in a Brazilian Typic Hapludox soil. *Soil Biol. Biochem.* Vol.40, pp.879-886, ISSN: 0038-0717.

Mersie, W. & Foy, C.L. (1985). Phytotoxicity and adsorption of chlorsulfuron as affected by soil properties. *Weed Sci.* Vol.33, pp.564-568, ISSN: 0043-1745.

Moyer, J.R.; Coen, G.; Dunn, R. & Smith, A.M. (2010). Effects of landscape position, rainfall, and tillage on residual herbicides. *Weed Technol.* Vol.24, pp.361-368, ISSN: 0890-037X.

Moyer, J.R. & Esau, R. (1996). Imidazolinone herbicide effects on following rotational crops in southern Alberta. *Weed Technol.* Vol.10, pp.100-106, ISSN: 0890-037X.

Moyer, J.R.; Esau, R. & Kozub, G.C. (1990). Chlorsulfuron persistence and response of nine rotational crops in alkaline soils of southern Alberta. *Weed Technol.* Vol.4, pp.543-548, ISSN: 0890-037X.

Moyer, J.R. & Hamman, W.H. (2001). Factors affecting the toxicity of MON 37500 residues to following crops. *Weed Technol.* Vol.15, pp.42-47, ISSN: 0890-037X.

Nash, R.G. (1981). Phytotoxic interaction studies - techniques for evaluation and presentation of results. Weed Sci. Vol.29, pp.147-155, ISSN: 0043-1745.

Ohmes, G. A.; Hayes, R. M. & Mueller, T. C. (2000). Sulfentrazone dissipation in a Tennessee soil. *Weed Technol.* Vol.14, pp.100-105, ISSN: 0890-037X.

Pekarek, R.A.; Garvey, P.V.; Monks, D.W.; Jennings, K.M. & MacRae, A.W. (2010). Sulfentrazone carryover to vegetables and cotton. *Weed Technol.* Vol.24, pp.20-24, ISSN: 0890-037X.

Renner, K.A.; Meggitt, W.F. & Penner, D. (1988). Effect of soil pH on imazaquin and imazethapyr adsorption to soil and phytotoxicity to corn (*Zea mays*). *Weed Sci.* Vol.36, pp.78-83, ISSN: 0043-1745.

Schoenau, J.J.; Szmigielski, A.M. & Eliason, R.C. (2005). The effect of landscape position on residual herbicide activity in prairie soils. In *Soil Residual Herbicides: Science and Management. Topics in Canadian Weed Science, ed.* R.C. Van Acker, 45-52, Volume 3. Sainte-Anne-de Bellevue, Québec: Canadian Weed Science Society - Société canadienne de malherbologie, ISBN: 0-9688970-3-7.

Seefeldt, S. S.; Jensen, J.E. & Fuerst, E.P. (1995). Log-logistic analysis of herbicide dose-response relationships. *Weed. Technol.* Vol.9, pp.218-227, ISSN: 0890-037X.

Senseman, S.A. (2007). Herbicide Handbook, ninth ed. Weed Science Society of America, Lawerence, KS, ISBN: 0-911733-18-33.

Szmigielski, A.M.; Schoenau, J.J.; Irvine, A. & Schilling, B. (2008). Evaluating a mustard root-length bioassay for predicting crop injury from soil residual flucarbazone. *Commun. Soil Sci. Plant Anal.* Vol.39, pp.413-420, ISSN: 0010-3624.

Szmigielski, A.M.; Schoenau, J.J.; Johnson, E.N.; Holm, F.A.; Sapsford, K.L & Liu, J. (2009). Development of a laboratory bioassay and effect of soil properties on sulfentrazone phytotoxicity in soil. *Weed Technol.* Vol.23, pp.486-491, ISSN: 0890-037X.

Szmigielski, A.M.; Schoenau, J.J.; Johnson, E.N.; Holm, F.A. & Sapsford, K.L. (2011). Determination of thiencarbazone in soil by the mustard root length bioassay. *Weed Sci.* submitted, ISSN: 0043-1745.

Thirunarayanan, K.; Zimdahl, R.L. & Smika, S.E. (1985). Chlorulfuron adsorption and degradation in soil. *Weed Sci.* Vol.33, pp.558-563, ISSN: 0043-1745.

Vencill, W.; Grey, T. & Culpepper, S. (2011). Resistance of weeds to herbicides. Herbicides and Environment, Andreas Kortekamp (Ed.), ISBN: 978-953-307-476-4, InTech, Available from: http://www.intechopen.com/articles/show/title/resistance-of-weeds-to-herbicides

Walker, A. & Brown, P.A. (1983). Measurement and prediction of chlorsulfuron persistence in soil. *Bull. Environ. Contam. Toxicol.* Vol.30, pp.365-372, ISSN: 0007-4861.

Wang, Q. & Liu, W. (1999). Correlation of imazapyr adsorption and desorption with soil properties. *Soil Sci.* Vol.164, pp.411-416, ISSN: 0038-075X.

Wehtje, G.; Dickens, R.; Wilcut, J.W. & Hajek, B.F. (1987). Sorption and mobility of sulfometuron and imazapyr in five Alabama soils. *Weed Sci.* Vol.35, pp.858-864, ISSN: 0043-1745.

Whitcomb, C.E. (1999). An introduction to ALS-inhibiting herbicides. . *Toxicol. Ind. Health.* Vol.15, pp. 232-240, ISSN: 0748-2377.

Zhang, J.; Hamill, A.S. & Weaver, S.E. (1995). Antagonism and synergism between herbicides: trends from previous studies. *Weed Technol.* Vol.9, pp.86-90, ISSN: 0890-037X.

4

Investigation of Degradation of Pesticide Lontrel in Aqueous Solutions

E.A. Saratovskih
Institute of Problem of Chemical Physics, Russian Academia of Science Russia

1. Introduction

Development of new efficient methods for purifying water from industrial pollutants resistant to biodegradation is a challenge because of the current shortage of freshwater reserves in the world (Legrini et al., 1993; Skurlatov et al., 1994). It is known that application of pesticides (Ozelenenie, 1984; Bykorez, 1985; Burgelya & Myrlyan, 1985; Patel et al., 1991; Wan et al., 1994; Arantegui et al., 1995; Fliedner, 1997; Arkhipova et al., 1997), in particular, Lindane (Fliedner, 1997) and 2,4-D (Arkhipova et al., 1997), results in their long-term accumulation in soils and water reservoirs. Furthermore, pesticides' residual presence in foodstuffs and industrial crops causes serious diseases, including hereditary disorders (Eikhler, 1993; Calvert et al., 2004; Whyatt et al., 2004).

Numerous studies point to long-term preservation of many pesticides in natural waters: ground, river, or sea (Skurlatov et al., 1994; Yudanova, 1989; Yablokov, 1990). Organic chemicals, including the insecticide Lindane, were found in the fat of gray whales caught in the Arctic Ocean (Poliakova et al., 2005).

Lontrel (another commercial name is Clopyralid) is considered to be the herbicide of a wide range of action and, first, for the defense against weeds in grain crops and can be used in a mixture with 2M-4X or 2,4-D; $LD_{50} \approx 5000$mg/kg (Mel`nikov, 1987). Literature data on the mechanism of action of Lontrel are rather insignificant. There are indications (Hall et al., 1985) that Lontrel exhibits auxine-like activity. The most detailed studies are presented in our works. Metal complexes of Lontrel were not studied at all, except of our research group.

Formerly (Aliev et al., 1988; Saratovskikh et al., 1989b) we showed that 3,6-Dichloropicolinic acid (DCPA), the active principle of the herbicide Lontrel, readily formed complexes with metals, major environmental pollutants, and these complexes were stable under natural conditions. They are capable of participating in further complex formation with bioactive ligands due to the filling of the coordination sphere of the metal.

We have proved for the first time that pesticides themselves and their metal complexes interact with mono-, di-, and polynucleotides (Saratovskikh et al., 1988; 1989a). In all cases, two- or three-component complex systems are formed. It was shown that the pesticide complex with adenosine triphosphoric acid is formed due to the protonation of the N-7 nitrogen atom of the adenine heterocycle, and the nitrogen atom of the terminal NH_2 group can simultaneously be bound to the pesticide molecule due to the formation of a hydrogen bond. The formation of the pesticide complexes with ATP results in an energy deficient in the tissues of organisms. The effect of pesticides and their metal complexes induces the energy deficient of the cell, namely, inhibition of energy metabolism due to the formation of a complex with adenosine triphosphoric acid.

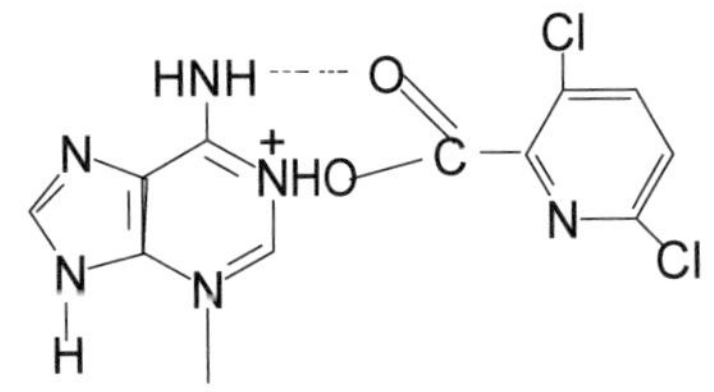

Structure of the [ATF-DCPA] complex.

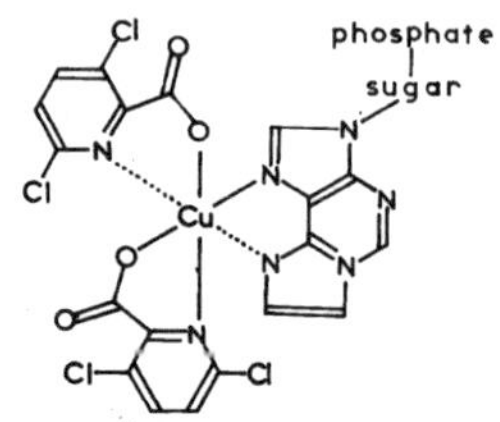

Structure of complex of ε-ATF with metal complex of DCPA - [ε-ATF-CuL_2].

We pioneered to show that Lontrel and its metal complexes simultaneously inhibit the activity of several enzymatic systems in organisms. These are oxidizing enzymatic systems, for example, NADH-oxidoreductase. In this case, the enzymatic activity is inhibited, first, due to the complex formation with dinucleotide (coenzyme) NADH. Second, this occurs due to the formation of the non-productive complex [enzyme-DCPA] in the active center of the enzyme. For DCPA comparison of inhibition of NADH-oxidoreductase $K_i = 10^4$ M, and competitive type of inhibition take place. The antireductase activity increases in the order: metal ion < DCPA < metal complex (Saratovskikh et al., 2005; 2007c).

We shown than DCPA and its metal complexes are bound into complex compounds with DNA and RNA (Saratovskikh et al., 1989a) and nativity of the DNA double helix is violated because of complex formation. The direct mutagenic effect was shown by us for the TA98 *Salmonella typhimurium* strain, mutations of the reading frame shift type are induced and promutagenity was revealed (Saratovskikh et al., 2007b).

Therefore, it is important to investigate its exposure to a microbial community of activated sludge (AS) and to sunlight, i.e., conditions mimicking those occurring in natural surface water bodies.

The possibility of using UV radiation for the decomposition of various chemical compounds was shown by several authors (Legrini et al., 1993; Guittonneau et al., 1988; Sundstrom et al., 1989; Castrantas & Gibilisco, 1990). Ultraviolet purification of water is superseding conservative chlorine treatment (Skurlatov et al., 1994; Skurlatov & Shtamm, 1997a, 2002). At present, over 1000 ultraviolet-purification devices of different capacities are in operation in 35 countries (Skurlatov et al., 1996; Kruithof et al., 1992). They finely replace the old method of treatment of drinking and waste waters based on the treatment with chlorine. A comparative estimation of the American specialists (Purus Inc., 1992) of the cost of UV quanta, ozone, and reagents used in processes of water preparation and water treatment is as follows: Cl_2 = 0,16$ per 1 mol; O_3 = 0,1$; photons 185, 254 nm (low-pressure Hg lamps, yield 40%) = 0,025$. The bactericide light quantum turned out to be the cheapest reagent (Skurlatov & Shtamm, 1997b). Sometimes the UV treatment is combined with the addition of oxygen or hydrogen peroxide. As shown (Shinkarenko & Aleskovskii, 1982), ozone formed upon the photochemical oxidation of oxygen dissociates, in turn, to the electron-excited oxygen atom and oxygen molecule in the singlet state. Having a high oxidizability, they attack a molecule of the contaminant and oxidize it (partially or completely). The UV treatment of waste waters produces no dibenzofurans, which are precursors of dioxins as it takes place for the treatment with ozone only (Beltran et al., 1993). The ultraviolet treatment of waste waters is well compatible with methods of biological purification from contaminants (Golubovskaya, 1978). Therefore, study of the products of UV-mediated DCPA degradation is of practical importance.

The objective of the present study was to investigate the possibility of DCPA degradation by biochemical and photochemical methods (taken alone or in combination with each other), in order to develop practical methods of degrading pesticides, the most challenging industrial pollutants.

Analysis of the composition of the reaction mixture and final products is a complicated task, which demands up-to-date highly sensitive methods, but it is of environmental importance. This study concerns the kinetics of DCPA degradation under natural conditions (deep-well, river, or sea water) and analysis of its products.

2. Materials and methods

2.1 Used substances, concentrations and replicates

3,6-Dichloropicolinic acid (DCPA) was studied, which is the main active principle of Lontrel and a copper–DCPA complex containing two DCPA molecules per copper atom - CuL_2. Experiments were performed with $1.22 \cdot 10^{-3}$ M DCPA solutions in distilled water, artificially made sea water, and river water from the Klyazma River. Solutions of CuL_2 with the concentration $0.4 \cdot 10^{-3}$ M. The artificial seawater was made from sea salt, containing the following anions (g/l): Na^+ (10.76), K^+ (0.39), Ca^{2+} (0.41), Mg^{2+} (1.30), SO_4^{2-} (2.70), Cl^- (19.35), Br^-(0.06), and CO_3^{2-} (0.07). The concentration of sea salt in the working solution was 35.5 g/l (Artobolevskii, 1977). River water was sampled near Biokombinat Village, Moscow Region, in the middle Klyazma River at the depth 1.5–2 m.

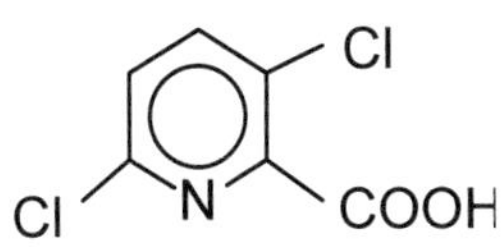

3,6-Dichloropicolinic acid - DCPA

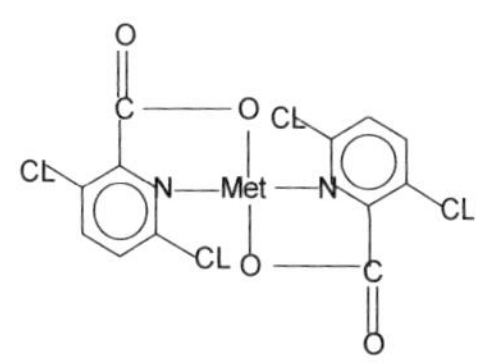

bis-(3,6-Dichloropicolinato)cuper(II)
CuL_2

All experiments were carried out three times. The measurement errors did not exceed 5-10%.

2.2 The biochemical oxidation of DCPA

Biochemical oxidation of DCPA was studied in 1 l laboratory aeration tanks under conditions of constant aeration and natural sunlight. Concentrated AS was sampled from an aeration tank of the Treatment plants in Chernogolovka, Moscow oblast. Five samples were studied, each containing 2 ml of AS. The samples were diluted in laboratory tanks to 1 l with a peptone medium. Sample 1 was taken as a control, not exposed to the mutagen (*N*-methyl-*N*-nitrosourea, nitrosomethylurea, NMU). NMU was added to samples 2, 3, and 4 to a concentration of 0.07% (Rapoport, 2010); samples 2 and 4 were exposed to NMU for 6 h, while sample 3 was exposed to NMU for 18 h. Samples 2–4 were treated with NMU once more for the same amount of time as in the initial treatment. Sample 4 was then treated the again after 28 days of the observation, while samples 2 and 3 were treated again after 44 days. Samples 1–3 were supplemented with DCPA at a starting concentration of $1.22 \cdot 10^{-3}$ M (0.23 g/l). Sample 4 was supplemented with a $Cu(L)_2$. This complex was synthesized according to (Aliev et al., 1988). We showed previously that this complex was a stronger herbicide than DCPA itself (Saratovskikh et al., 1990). It forms stable compounds with DNA (Saratovskikh et al., 1989a), mononucleotides, and dinucleotides (Saratovskikh et al., 1988). Such complexes can readily form in runoff from farmlands and in some industrial enterprises' wastewaters containing Lontrel and copper compounds. Sample 5 was supplemented with wastewater of a pilot Lontrel mixture, which contained approximately $\sim 10^{-3}$ M of the herbicide. Samples for the tests were taken four or five times per day for the first 3 days, once a day for the next 10 days, and then once every few days. The observation was performed for one year. The range of microbial species in the AS was determined according to (Belyaeva & Gyupter, 1969; Liperovskaya, 1977).

2.3 The photochemical oxidation of DCPA

Photochemical oxidation of DCPA under UV irradiation was performed in a quartz reaction vessel with internal diameter 2.2 cm and volume 25 ml at 4 cm from the ray emitter (in various sets of experiments, irradiation was performed with DRSh-1000, DRB-8, or BRA-15 lamps; 250–600 nm; BRA-15 lamp 1.3 mW/cm^2; Institute of Problems of Chemical Physics, Russia). The starting DCPA concentration was $5 \cdot 10^{-4}$ M. Oxygen, ozone, air, or argon were sparged (bubbled through the solution) with a capillary tube. Samples for measuring concentrations were taken at 10-min intervals. DCPA concentrations were measured with a Specord UV-VIS spectrometer (Karl Zeiss, Jena, Germany) according to light absorption at 283 nm.

2.4 Toxicological testings

The samples were subjected to biotests with two sentinel species: the infusorium *Tetrahymena pyriformis* and the luminescent bacterium *Beneckea harveyi*.

2.4.1 The biotest on *Tetrahymena pyriformis*

Tetrahymena pyriformis were cultivated in a peptone or carbohydrate–salt–yeast extract media (Yoshioka et al., 1985). Five milliliters of each sample were placed into three test tubes and one or two drops of a 3- to 4-day-old *Tetrahymena* culture were inoculated. Settled aquarium water was used as a control. The division period for *T. pyriformis* ranges within 4–6 h. The test lasted for 15 min. A change in the behavior or morphology of infusorian cells in the samples or their death was indicative of a strong toxic effect. Fifteen minutes after the mixing, three samples (one drop per sample) were taken from each test tube and fixed with iodine. Infusorian cells were counted under a microscope. In this method, toxicity coefficients *K* were calculated as follows:

$$K = [(Ac - Aex)/Ac]\ 100\%, \tag{1}$$

where Ac is the number of cells counted by microscopic examination in the control sample and Aex is the number of cells in an experimental sample. A sample considered toxic at $K >$ 50%. The statistical significance of values was determined by the Student's -test.

2.4.2 Biotest on *Beneckea harveyi*

In a chronic test, the toxicity of a sample was judged from the suppression of cell propagation after 24 h of incubation. A reduction in cell propagation by 50% or more was indicative of toxicity. A lyophilized preparation of *B. harveyi* was stored in a freezing chamber. Immediately before the experiments, the bacteria were suspended in 3% NaCl. For testing toxicity, 0.3–0.5 ml of the suspension was added to 0.5 ml of a water sample. The control experiment was performed with a 0.85% NaCl solution. Measurements were performed with a BLM-8801 luminometer (SKTB Nauka, USSR) with voltmeter detection. Toxicity was estimated from the decrease in bioluminescence of a sample relative to the control. The sample was considered toxic if its bioluminescence decreased by 50% or more. The level of bacterial luminescence is determined by the intensity of intracellular metabolism involving the enzyme luciferase. A decrease in luminescence may be related to either inhibition of the enzyme itself or to an effect of toxic substances on other links in the metabolic chain. The toxicity coefficient was calculated as:

$$T = [(Ic - Iex)/Ic]\ 100\%, \tag{2}$$

where Ic is the bioluminescence intensity in the control and Iex is the bioluminescence intensity in the sample tested. A sample is considered nontoxic at $T \le 19\%$, toxic at $19 < T \le 50\%$, and strongly toxic at $T > 50\%$.

2.5 Infrared spectra

Infrared spectra were recorded at 400–2200 cm^{-1} with a Specord 75IR spectrometer (Karl Zeiss, Jena, Germany) in KBr pellets (250 mg of KBr + 1.2 mg of a sample). Absorbance bands were identified according to established methods (Nakanisi, 1965; Sverdlov et al., 1970; Nakomoto, 1991).

2.6 Gas chromatography/mass spectrometry (GC/MS) analyses

Gas chromatomass spectrometry was performed with a Pegasus 4D chromatomass-spectrometer (LECO, Russia) under the following conditions: ionization energy 70 eV; 30 m RTX-5MS capillary silicon column; temperature program 40°C (5 min), 8°C/min, 300°C (10 min); scan range 28–450 Da. Qualitative identification was performed by reference to the WILEY mass spectrum library, including 270 000 compounds. Perdeuterated naphthalene was used as an internal reference for quantitative assay.

2.7 The elemental analysis

Analysis of photooxidation products was performed in solutions in distilled water after 13 and 38 h of UV irradiation with air bubbling. Aqueous solutions, tagged as "13" and "38," were extracted with dichloromethane 3 times for 10 min each. The extracts were combined, dried with anhydrous sodium sulfate, and concentrated in vacuum to 1 ml. Elemental analysis was performed after DCPA degradation. Samples were evaporated at 50–80°C, and C, N, Cl, and H were assayed in the completely dried residue (Klimova, 1975).

The contents of C, H and dry residue were determined by the modified Pregl method (Pregl, 1934) based on the ratio of molecular masses of the elements. A weighed sample of 3-5 mg was burned in a current of neat oxygen at t = 1040^{0}C. The amount of formed CO_2 and H_2O was determined by weighing the corresponding absorbers filled with ascarite (NaOH) and anhydrone (anhydrous Mg perchlorate - $Mg(ClO_4)_2$).

The content of N was determined by the Pregl-Dumas method (Steuermark, 1961). A weighed sample of 3-5 mg was burned in a current of CO_2 in the presence of CuO and then reduced by metallic copper. The volume of evolved gaseous nitrogen was determined with a gas meter.

The content of Cl was determined according to the Schöniger flask method (Schöniger, 1955). The analyzed substance (2-4 mg) was wrapped in filter paper placed in a Pt grid and hanged up to the cork of the flask. The flask was filled with oxygen. 2N KOH (1 ml), 0.5 мл H_2O_2 (0.5 ml), and bidistilled H_2O (5 ml) were placed on the flask bottom. The end of the paper (in which the weighed sample is wrapped) was ignited, and the flask was rapidly corked. Absorption was carried out for 0.5-1 hour. The contents was titrated with 0.01N $Hg(NO_3)_2$ in the presence of diphenylcarbazone to lilac-violet color. The content of Cl was determined by the titrant volume.

3. Results and discussion

Prior to the experiment, nine lower species were identified in the AS sample under study: algae, amoebae, sessile infusoria, and flagellates (Table 1). Addition of the pollutant reduced the number of species to seven. One group, blue-green algae, became predominant, apparently being the most resistant. Unicellular algae have also been reported as resistant to the herbicides Diuron and *o*-Phenanthroline (Laval-Martin et al., 1977).

Treatment with the mutagen caused a change in the species composition of AS and an increase in the range of the species. This increase was related to the fact that the populations of some species were too small to be detected before addition of NMU or DCPA. The presence of a certain pollutant may provide conditions for accelerated growth of those species that consume the pollutant as a preferable nutrient, thus allowing their identification. This phenomenon provides grounds for the purification of wastewaters from chemicals (Golubovskaya, 1978). In our experiment, the biocenosis that formed after NMU treatment included two phyla of lower plants (bacteria and algae) and five invertebrate classes (ciliates, sarcods, mastigiophores, nematodes, and rotifers). The number of species identified in AS sample 3, treated with NMU for 18 h, was greater than in sample 2, treated for 6 h (20 and 15 species, respectively). The population of rotifers notably increased, and sulfur bacteria and testaceous amoebae were identified (Table 1). This community successfully resisted the anthropogenic load. The toxicity of the sample treated for 18 h was lower than that after 6-hour treatment (Fig. 1). Long-term exposure (several months) reduced the range of species.

Organism	Control (without NMU treatment)	NMU treatment, 6 h	NMU treatment, 18 h	Repeated treatment, 6 h	Repeated treatment, 18 h
Ulothrix sp.	+	+	mass	+	+
Scenedesmus obliguus	+	+	+	+	
Chlorella vulgaris		+			
Flagellata sp.			+	mass	
Oicomonas socialis				+	+
Bodo globosus		+	+	+	+
Zooglea ramigera					occasional
Euglena viridis					+
Filamentous bacteria					occasional
Bacillus					+
Beggiatoa minima			+		+
Jromia neglecta sp.			+	+	
Arcella vulgaris			+		
Centropixis acullata	+		+		
Pamphagus hyalinus	+	+			+
Euglypha laevis				+	
Amoeba sp.				+	+
Aspidisca costata			+		+
Aspidisca lynceus		+	+		
Lacrimaria sp.		+		+	
Litonotus anser		+		+	
Chilodonella uncinata		+			
Vorticella alba		+	+		
Thuricola similes	+				
Telotpox		+	+		
Rotaria rotatoria	+	mass	+	+	+
Colurella sp.		+	+	+	+
Rotaria neptunia			+		+
Lelane Monostyla			+	+	+
Brachionus angularis		+			
Cephalodella gibba			+	+	+
Notommata sp.			+	+	+
Cephalodella forticula			+		+
Chaetonotus brevispinosus	+		+		

Table 1. Hydrobiological analysis of activated sludge.

The presence of DCPA, its copper complex, or industrial waste (samples 1-5) at concentrations used in the experiment exerted acute and chronic effects on the infusorium culture. Figure 1 shows that the toxicity of sample 1 remained high (~80%) after two months of monitoring. The toxicity of sample 2, treated with NMU for 6 h, remained high (~90%) for 36 days. Then, it decreased abruptly, and, after 56 days, the sample was virtually nontoxic. By the beginning of the third month, its toxicity reached 70%. As mentioned above, the decrease in toxicity after an 18-h treatment (as compared to the 6-h exposure to NMU) may be related to a change in the proportions of species in AS as a result of the mutagenic effect. The history of toxicity of sample 4 suggests that the CuL_2 had no pronounced toxic effect on the AS for 20 days after the onset of exposure to NMU. This result may be related to a decrease in the reactivity of DCPA in the copper complex. After 20 days, the complex appears to have been degraded and the toxicity of the sample abruptly increased. The history of toxicity of sample 5, containing industrial waste, was similar to that of sample 4. This finding suggests the presence of various complexes between DCPA and metals, whose decay after 20 days increases the toxicity of the sample dramatically.

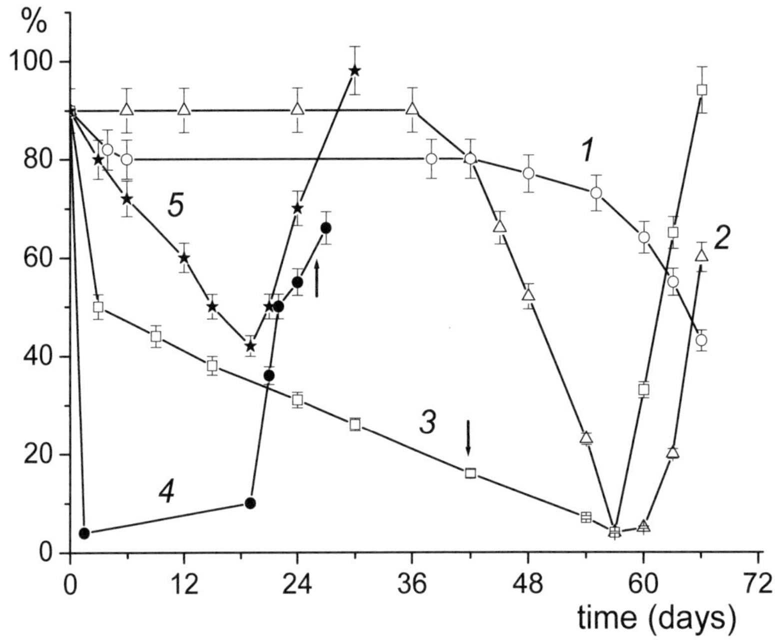

Fig. 1. Changes in the toxicity (with *B. harveyi*) of AS samples containing (*1, 2, 3*) DCPA, (*4*) the CuL_2 complex, and (*5*) industrial waste water: (*1*) control sample with intact AS, (*2*) AS treated with NMU for 6 h, and (*3*) AS treated with NMU for 18 h. Starting concentrations: DCPA, $1.22 \cdot 10^{-3}$ M; industrial wastewater, $\approx 1 \cdot 10^{-3}$ M; the CuL_2 complex, $0.4 \cdot 10^{-3}$ M. Temperature 25°C. Arrows indicate repeated NMU treatment.

As seen in Fig. 2, DCPA concentrations in samples 4 and 5 remained constant throughout the experiment. Most likely, it is only the proportions of various CuL_2 that changed, thereby altering the toxicity of these samples (Fig. 1). In samples 2–4, AS was treated with NMU to obtain mutations most resistant to the toxic substance under study. After the first 20 days (Fig. 2), DCPA concentrations changed neither in the control sample nor in samples exposed to the mutagen. After 18–20 days, DCPA concentrations in samples 1–3 began to decrease. The decrease in DCPA concentrations in samples 2 and 3 occurred faster than in the control sample. A steady state was established approximately 40 days after the beginning of the experiment. Another treatment with NMU was undertaken to further increase the oxidative potential of AS; however, this attempt did not cause a significant degradation of the substances under study. After 66 days, the content of DCPA decreased by 25%, while that in the sample not treated with NMU decreased by 20%.

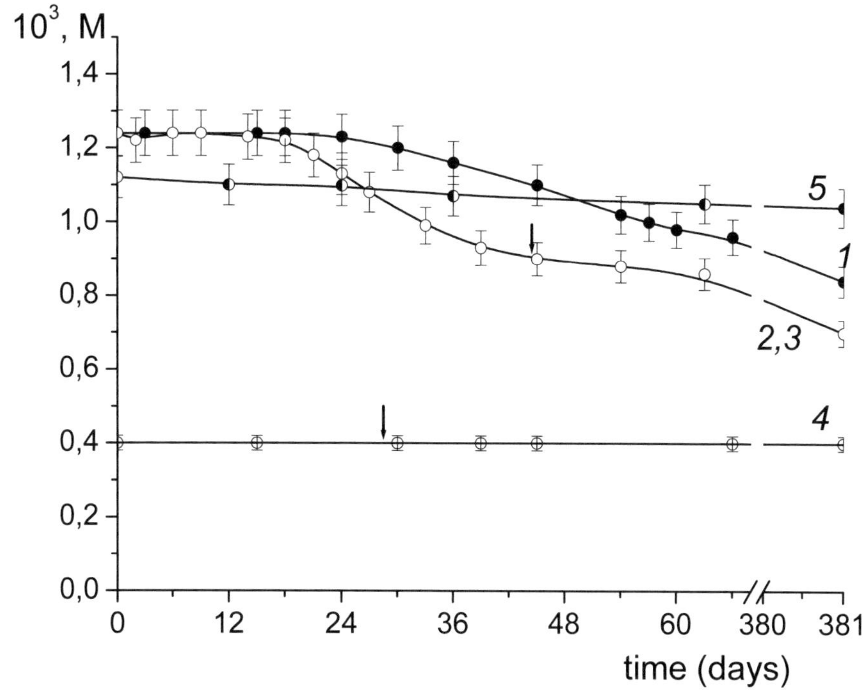

Fig. 2. Kinetic curves of the degradation of (*1, 2, 3*) DCPA, (*4*) the CuL_2 complex, and (*5*) industrial wastewater. Designations follow Fig. 1.

After one year of treatment of the pollutants with a solution of AS treated with the mutagen, DCPA concentration had decreased by 45% in total, i.e., by less than half; in the control (nonmutagenized) AS sample, the DCPA concentration had decreased by 30% in total. During the same one-year term, the concentration of toxic substances in sample 5, containing wastewater of the herbicide mixture, had decreased only by 7% in total.

Thus, our data indicate that DCPA the main active principle of herbicide Lontrel belongs to bioresistant organochlorine herbicides. Wastewater treatment in industrial treatment plants lasts only for 8 ÷ 11 h. During this treatment, Lontrel can form compounds with copper or other industrial pollutants. It is natural to assume that both Lontrel and its metal derivatives remain intact in treatment plants. Hence, the toxic substances can penetrate surface or subsurface water bodies and accumulate there, posing a hazard for microflora, plants, and fish. Because of Lontrel bioresistance, its application in agriculture for controlling weeds can exert selection pressure and give rise to mutant weeds, resistant to the herbicide. This effect has been shown for Simazine and Atrazine, which are resistant to biodegradation in soil (Fedtke, 1985). Bioresistance was the reason for prohibition of DDT, although it was nontoxic at recommended concentrations (Burgelya & Myrlyan, 1985).

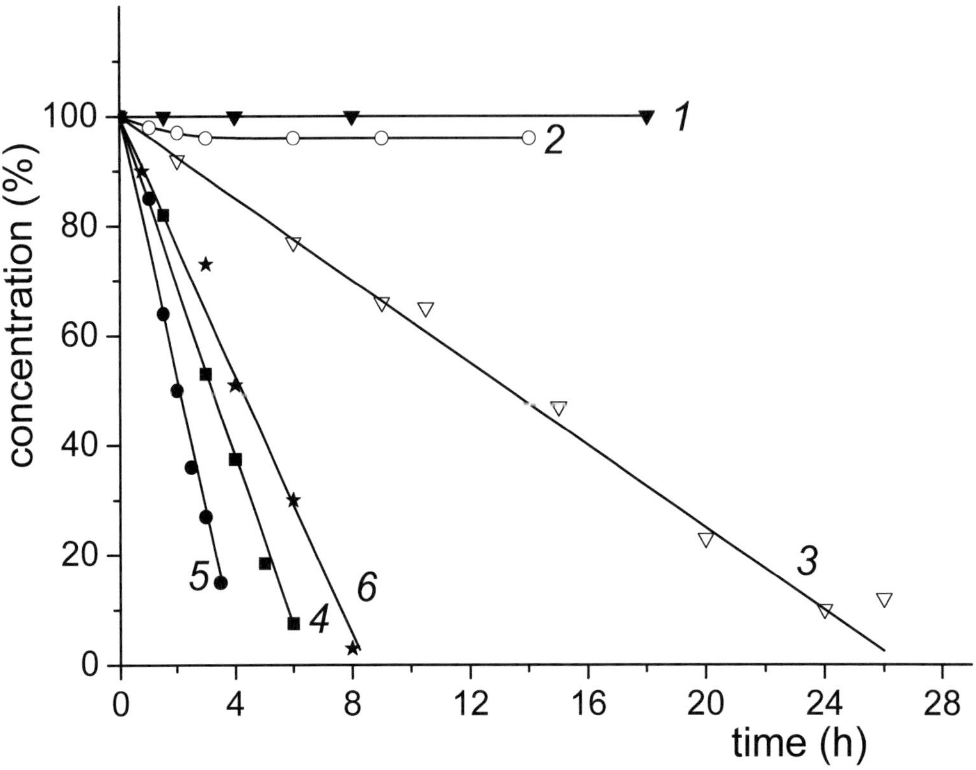

Fig. 3. Changes in DCPA concentrations (as percentages of the starting concentration), as determined by the time of UV irradiation: (*1*) without bubbling; (*2*) argon, DRB-8 lamp; (*3*) air, DRB-8 lamp; (*4*) ozone, DRB-8 lamp; (*5*) air, DRSh-1000 lamp; and (*6*) oxygen, hydrogen peroxide, BRA-15 lamp.

The high resistance of Lontrel to biodegradation under natural sunlight prompted us to perform an experiment on photochemical oxidation of the herbicide with the use of hardUV radiation, which is the only part of the solar UV spectrum (wavelength range, 400–360 nm) that reaches the surface of Earth, shorter waves being trapped by the atmosphere. Photochemical activity can be manifested by radiation within the absorption spectrum of a compound. In our case, for DCPA, the active wavelength is λ = 283 nm. Therefore, we used mercury lamps with emission lines within 254–579 nm. The energy of a quantum with a wavelength of 283 nm exceeds the energy of the C=N bond (~260 kJ/mol) by a factor of about 1.5, an amount of energy that may be sufficient for cleavage of the very stable pyridine ring, which is not oxidizable by AS microorganisms. The efficiency of UV irradiation in inducing redox degradation of various contaminants has been shown previously (Skurlatov & Shtamm, 1997b; Kruithof et al., 1992). It has been reported that Atrazine is degraded by UV irradiation alone within 20 min and by ozone alone within almost 3 h (Prado & Esplugas, 1999). Application of UV radiation, including its combination with the use of hydrogen peroxide, to the degradation of phenols has been reported (Legrini et al., 1993; Guittonneau et al., 1988; Sundstrom et al., 1989; Castrantas & Gibilisco, 1990).

No positive results were obtained in the first set of experiments, in which DCPA solutions were treated with UV radiation alone (Fig. 3, curve *1*). After 8 h of irradiation, optical density did not change; thus, the herbicide was not degraded at all.

To eliminate stagnation zones and to provide uniform illumination of the reaction-mixture layers distant from the UV lamp, the solution was bubbled with the inert gas argon. Figure 3, curve *2*, shows that the concentration of the starting substance decreased by 5% after 3 h of the reaction; after that, however, the reaction completely stopped.

To increase the oxidative effect of hard UV radiation, the reaction mixture was bubbled with air instead of argon. The cooperative action of UV, atmospheric oxygen, and mixing almost completely degraded the herbicide after 24–25 h of the reaction (Fig. 3, curve *3*). Simultaneous treatment with UV, oxygen, and hydrogen peroxide reduced the degradation time to 8 h (Fig. 3, curve *6*). Irradiation under the conditions of ozone bubbling accelerated DCPA oxidation four- to fivefold (as compared to air bubbling, other factors being the same). The herbicide was oxidized almost completely after 6–6.5 h of the photochemical reaction (Fig. 3, curve *4*).

Thus, bubbling of the reaction mixture with argon, i.e., mechanical mixing with inert gas, slightly accelerates DCPA degradation, but the use of air is much more efficient. In our experiments, agitation with air or oxygen flow (1) increased the reaction surface and (2) supplied the solution to the reaction zone. In addition, ozone is one of the products of photochemical oxidation of oxygen; when affected by UV irradiation, the ozone molecule, in turn, dissociates to an electronexcited oxygen atom and an oxygen molecule in the singlet state. These chemical species, having high oxidizing potentials, attack herbicide molecules and degrade them, either partially or completely (Shinkarenko & Aleskovskii, 1982).

Experiments were performed with DRB-8 and DRSh-1000 lamps. Complete DCPA degradation with a DRB-8 lamp and bubbling with air occurs within 24 h, whereas degradation with the use of a DRSh-1000 lamp takes 3–3.5 h, that is, a seven- to eightfold smaller amount of time. Thus, the rate of DCPA oxidation depends significantly on the power of the UV-radiation source.

The solution obtained by complete UV-induced degradation of DCPA was tested for toxicity according to changes in the enzymatic activity of luminescent bacteria. In the control experiment, the toxicity of DCPA was detected at concentrations of 10^{-7}–10^{-3} M in the absence of UV radiation.

The data presented in Table 2 indicate that no toxic effect was detected in the samples after either 5-min (effect on the cell membrane) or 30-min (effect on cellular metabolism) exposure throughout the whole range of the initial concentrations studied. Irradiated samples (containing products of complete herbicide oxidation) were toxic after 5-min exposure for all starting DCPA concentrations (Table 3). After 30-min exposure, toxicity was detected only in the sample with a starting concentration of 10^{-3} M, whereas samples with the starting concentrations of 10^{-5} and 10^{-7} M did not show any toxic effect. It may be inferred that a product of photochemical degradation of DCPA is inherently toxic to luminescent bacteria.

DCPA concent-ration, M	Toxicity coefficient (T), %											
	before irradiation				after irradiation				after irradiation and treated with AS for 24 h			
	exposure, min											
	5		30		5		30		5		30	
10^{-7}	0.90	non-toxic	11.80	non-toxic	44.0	toxic	1.80	non-toxic	4.00	non-toxic	3.90	non-toxic
10^{-5}	8.10		33.20		18.5		4.10		4.00		3.80	
10^{-3}	11.0		88.60		25.7		33.2		3.70		3.70	

Table 2. Toxicity coefficients of samples before and after UV irradiation as determined from the change of the luciferase activity of luminescent bacteria.

However, this substance, having a simpler structure than the pyridine ring, should be easily metabolized by AS microorganisms, as demonstrated by the experiment described below. After irradiation under the conditions of air bubbling for 24 h, samples with DCPA were placed into an aerated tank with AS for 24 h and the toxicity of the resulting solution was tested. According to the data obtained with the luminescent bacterium *B. harveyi*, the association of AS microorganisms successfully neutralized the contaminant formed after UV irradiation (Table 2). The solutions were nontoxic at all starting concentrations, even the highest one (10^{-3} M). The absence of toxicity in the samples was also shown with the infusorium *T. pyriformis* for all starting concentrations (Table 3).

DCPA concentration, M	Toxicity coefficient (*K*), %			
	exposure to AS for 24 h		exposure to AS for 48 h	
10^{-7}	0.81	nontoxic	15.8	nontoxic
10^{-5}	13.7	nontoxic	13.1	nontoxic
10^{-3}	10.5	nontoxic	16.9	nontoxic

Table 3. Toxicity coefficients of samples before and after UV irradiation and treatment with AS for 24 and 48 h as determined using the infusorium culture.

We studied the composition of products of the photochemical degradation of DCPA. The electronic spectra of the aqueous DCPA solutions show that the intensity of the band at 283 nm gradually decreased throughout the time of the experiment (Fig. 4). No additional bands or shift of the absorption maximum were observed in this region. Probably, the products of DCPA degradation had no intense bands in the UV or visible regions, or their concentrations were insignificant. The pH value of the solution decreased from 4.25 to 3.81.

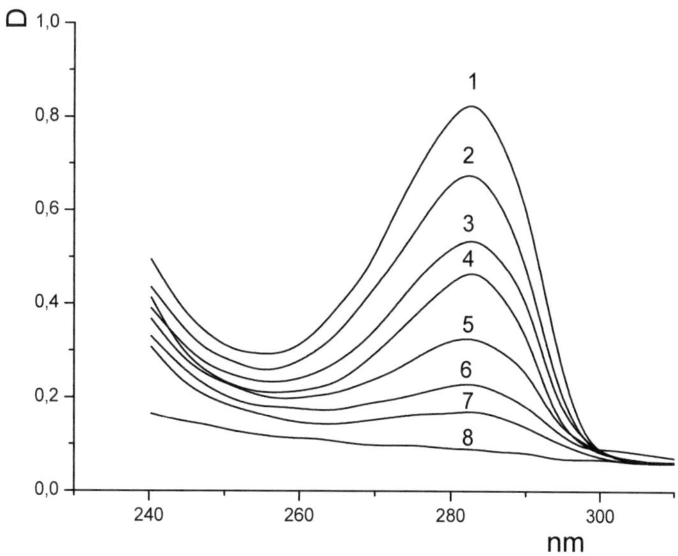

Fig. 4. Change of the DCPA electronic spectrum during irradiation. (*1*) Starting solution: $5 \cdot 10^{-4}$ M DCPA; (*2–8*), solution after UV irradiation for 6, 11, 13, 18.5, 23, 24, and 38 h, respectively.

Kinetic curves of photochemical DCPA degradation with oxygen bubbling are shown in Fig. 5. The figure reveals a difference among the shapes of the kinetic curves in distilled, sea, and river water. Curve *1* is linear: DCPA concentration decreases at a constant rate, reaching zero after 22 h. Curve *3* can be approximated by the log equation:

$$y=100\exp(-x/16), \tag{3}$$

whereas curve *2* can be approximated by the sum of 2 log curves:

$$y=30\exp(-x/3)+70\exp(-x/50). \tag{4}$$

The rates of DCPA degradation in both river and sea water were higher than in distilled water for the first 10 h.

However, the rate of DCPA degradation decreased significantly after irradiation for 7 h in seawater and 10 h in river water, being lower than in sample 1. After 45 h of observation, the DCPA concentrations in both samples were significantly different. The rate of DCPA photooxidation in river water proved to be higher than in seawater. According to our previous studies, this difference can be related to the fact that the herbicide forms UV-resistant complexes with metals occurring in natural media (Aliev et al., 1988; Saratovskikh et al., 1989b). At the beginning, 2 processes occur in the solution: DCPA degradation and binding to metals. As seawater is richer in metals, the concentrations of the metal complexes are higher, and the rate of DCPA degradation is lower than in river water (curve *3*).

Higher (Fig. 3), reported the kinetics of DCPA degradation by UV with air bubbling. According to electronic spectra, DCPA degraded by 50% after 13 h and by 90%, after 24 h. Complete DCPA degradation was achieved after 38 h irradiation.

Comparison of the IR spectra of the starting and UV-irradiated DCPA revealed substantial changes (Fig. 6, curves *1* and *2*). The intense absorbance band (AB) ν(C=O) at 1710 cm^{-1} was split into 4 bands: 1780, 1740, 1715, and 1690 cm^{-1}, which is attributable to DCPA degradation and the formation of other compounds. The AB at 1780 cm^{-1} can be related to ν(C=O) in the group 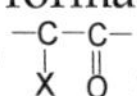 ; at 1740 cm^{-1}, to conjugated with an unsaturated C=C bond; at 1715 cm^{-1}, to asymmet-rical vibrations of two (C=O) bonds, and at 1690 cm^{-1}, to vibrations of C=O conjugated with the aromatic ring.

The ABs of fragment 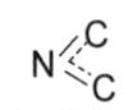 at 1560 and 1540 cm^{-1} (ν_{as} and ν_s of bonds C_{ar}.......N) in the starting DCPA shifted to higher frequencies (1620 and 1565 cm^{-1}). Apparently, the lone pair of nitrogen moved from the antibonding orbital to the complex-forming molecular orbital.

The sharp and intense ABs of the valence ν(C=C) and planar deformational δ_{II}(CO) vibrations at 1440, 1410, and 1305 cm^{-1} in the DCPA spectrum fused into a single very wide band with scarcely distinguishable maxima at 1450 and 1380 cm^{-1} in the spectrum of the degradation product. This may be related to the formation of a series of compounds derived from carboxylic groups COO^-, including those entering linear compounds.

Significant changes occurred in the range of low frequencies (800–500 cm^{-1}), including the ABs ν(Car -Cl) and ν(C-Cl). They involved band intensities and frequency shift. Very intense ABs appeared in the product spectrum at 605 and 530 cm^{-1}, probably related to C–Cl and Car–Cl vibrations, which is natural, because chlorine- containing compounds could be accumulated.

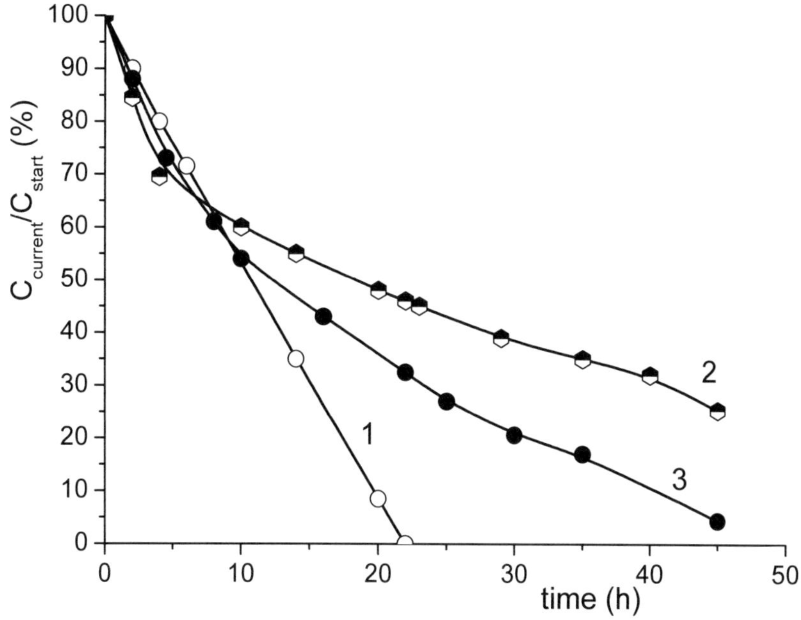

Fig. 5. Kinetics of DCPA concentration under the effect of UV irradiation with oxygen bubbling at 25°C. DCPA concentration 1.42·10^{-3} M. *1*, distilled water; *2*, artificial seawater; *3*, river water.

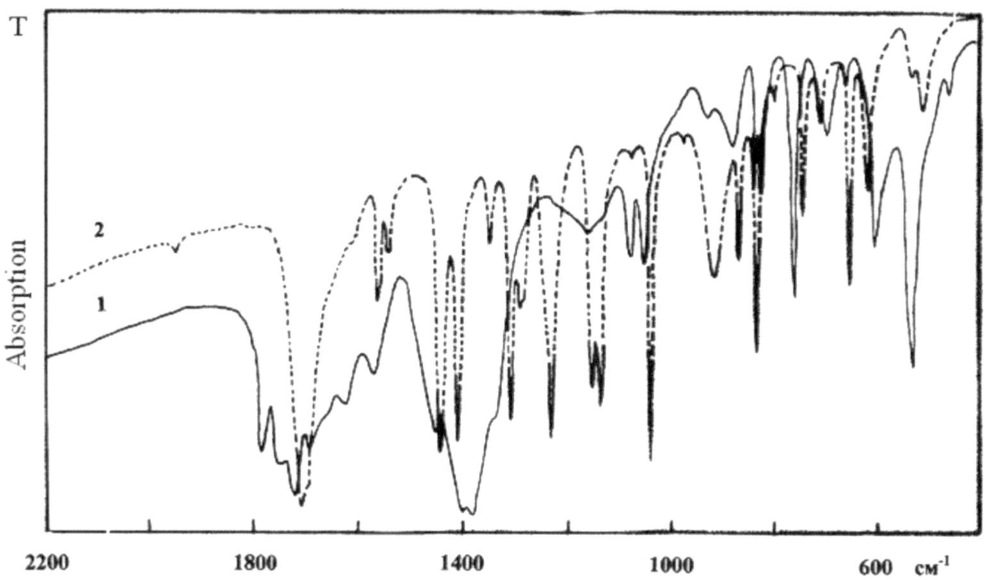

Fig. 6. Infrared spectra of (*1*) the product of DCPA photolysis; (*2*) intact DCPA.

It should be noted that the large widening of ABs in the regions 550 (large-amplitude proton vibration), 1400, and 1700 cm^{-1} can be related to associates with water. An intense and wide AB completely covered the region 3500–2700 cm^{-1}.

For final identification of compounds forming during UV degradation of DCPA, we performed GC/MS analysis of samples after UV irradiation. Ten compounds, including DCPA, were found in the sample taken after 13-h irradiation. The compounds and their ratios to DCPA are presented in Table 4. From these data, we deduced their molar concentrations.

No	Compound	13 h		38 h	
		percentage of DCPA	concentration, $\cdot 10^{-5}$ M	percentage of DCPA	concentration, $\cdot 10^{-6}$ M
I	4-Chloro-1,2-dimethylbenzene	0.05	1.71	0.17	1.16
II	Dichlorobutanol	0.06	2.00	0.26	1.75
III	3-Chlorobenzoyl chloride	0.06	1.65	0.28	1.54
IV	3,6-Dichloropicolinic acid	1.00	25.0	1.00	5.00
V	4-Chlorobenzoyl chloride	0.11	3.00	0.50	2.74
VI	Trichlorobutanol	0.18	4.87	1.32	7.14
VII	2,3,5-Trichloropyridine	0.12	3.16	-	-
VIII	Trichlorobutanol (isomer)	0.02	0.54	0.26	1.41
IX	Hexachlorocyclohexane	0.02	0.34	0.48	1.62
X	2,6-Dichloro-3-nitropyridine	0.11	2.74	-	-

Table 4. Areas of peaks of degradation products in percentage of DCPA after 13 and 38 h of UV irradiation and calculated concentrations (Note: –, not detected).

It is apparent from Table 4 that DCPA (compound IV) was predominant after 13 h of UV irradiation. However, 9 more compounds were identified in the mixture (I–X), which appeared as products of DCPA degradation. In addition to pyridine derivatives (primary degradation products), major components included substituted chlorobenzenes and chlorobutanols. The appearance of the latter can be explained by cleavage of the pyridine ring, as any aromatic ring, by secondary oxidation (Tretyakova et al., 1994). Further irradiation increased the concentrations of these products. Formation of chlorobenzenes can be explained by the Kost–Sagitulin rearrangement (Danagulyan, 2005), converting picoline derivatives to anilines. Subsequent oxidation of the amino group could give rise to the whole series of identified products of this kind.

It is more difficult to correlate the appearance of hexachlorocyclohexane (HCCH – compound IX) with the structure of the initial molecule, exposed to irradiation. It is reasonable to suggest that HCCH was a minor impurity in the starting pesticide.

It should be noted that some degradation products could be missed. In particular, salts of aniline derivatives, polyoxy compounds, or dicarboxylic acids could be too polar to be extracted from water and to get through to the chromatographic column (Lebedev et al., 1996).

After 13-h UV irradiation, the mixture contained $4.87 \cdot 10^{-5}$ M compound VI, 5 times less than DCPA. The amounts of compounds V, X, and VII were 10 times less than that of DCPA: $3.0 \cdot 10^{-5}$; $2.74 \cdot 10^{-5}$, and $3.16 \cdot 10^{-5}$ M, respectively. The amounts of compounds I, II, and III were 20 times less ($1.71 \cdot 10^{-5}$; $2.0 \cdot 10^{-5}$; and $1.65 \cdot 10^{-5}$ M, respectively), and the amounts of compounds VIII and IX, fifty times less ($0.54 \cdot 10^{-5}$ and $0.34 \cdot 10^{-5}$ M) than the amount of the herbicide to be degraded.

Continuation of UV irradiation changed the pattern. After 38-h irradiation, chloropyridine intermediates VII and X were completely degraded, apart from DCPA itself, whose concentration was $5 \cdot 10^{-6}$ M. The proportions of other degradation products increased considerably: three- to fivefold for compounds I-III and V (concentrations $\leq 1.16 \cdot 10^{-6}$; $1.75 \cdot 10^{-6}$; $1.54 \cdot 10^{-6}$; and $2.74 \cdot 10^{-6}$ M, respectively). The relative proportion of HCCH also increased significantly (IX; $\leq 1.62 \cdot 10^{-6}$ M), by a factor of virtually 25. The same was with trichlorobutanol (VIII $\leq 1.41 \cdot 10^{-6}$ M) and VI ($\leq 7.14 \cdot 10^{-6}$ M). Trichlorobutanol (VIII and VI) became the predominant component of the mixture. In the final sample, its concentration increased eightfold in comparison with 13-h irradiation and exceeded the concentration of DCPA.

The data of elemental analysis are shown in Table 5. The detected and predicted amounts were fairly close only for hydrogen. The contents of other elements differed significantly from the predicted values. This may be related to the fact that volatile oxides were released during evaporation; however, the humid sample released water with difficulty and began to melt.

Element	Found, %	Calculated, %
C	12.10-11.05	37.12
H	4.84-4.65	3.14
Cl	11.25-14.62	57.495
N	29.66-29.52	1.91
Ash	5.9-4.37	

Table 5. Data of the elemental analysis of the sample after 38 h of irradiation.

Our results indicate that DCPA is difficult to degrade. Photolysis resulted in cleavage of the pyridine cycle and formation of simpler compounds. However, the insecticide Lindane was identified among the photolysis products. Ultraviolet-mediated DCPA degradation in river and sea water was slower than in distilled water, probably because of formation of metal complexes. The rate of DCPA degradation in seawater at 25°C was lower than in river water. This fact is likely to hold true in natural ecosystems.

There is a firm opinion that the use of herbicides enhances the crop of the fields. However, many literature data (Skurlatov et al., 1994; Yablokov, 1990) indicate, most likely, an opposite fact: the use of "chemical remedies of plant protection" has stopped long ago to favor the crop and now gives an opposite effect (Fig. 7). This is because the microflora and humus matter of the fertile ground layer are annihilated (Golovleva & Fil`kenshtein, 1984; Saratovskikh & Bokova, 2007). The results of our studies showed the mechanisms of these negative processes (Saratovskikh et al., 1989a, 1988, 2005, 2007b, 2008b).

The surface areas of agricultural grounds decrease because of contamination with heavy metals and herbicides. This results in deep changes in the physicochemical, agrochemical, and biological properties of the arable land, an increase in the negative balance of humus (up to 1-3 t/hectare annually), and a decrease in the overall storage of biomass in soils. In

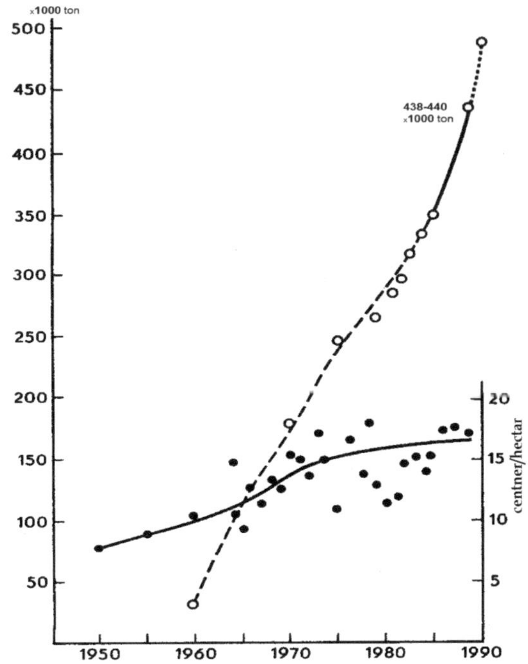

Fig. 7. Pesticides use in USSR and raising the level of crop yield. (Yablokov A.V., 1990). Scale of pesticide practice in USA has increased tenfold from the middle of 1970s to the early 1980s, weed losses increasing from 8 to 12 %.

the nearest 10-15 years the fertility of soils can decrease to a crop capacity of grain crops of 8-10 centner/hectare (Postanovlrenie, 2001).

The main soil-forming role belongs to the forest and grassy vegetation (Kusnetsov et al., 2005; Saprikin, 1984). Microorganisms (bacteria, microscopic fungi, and algae) play the most important role in the formation of soil fertility. Bacteria decompose organic residues to simple mineral compounds, perform processes of ammonification and denitrification, oxidize mineral compounds, and participate in nitrification (Jiller, 1988; Zviagintsev, 1987).

The organic component is presented by humus substances, which serve as a nutrient source for microorganisms and pedobionts. Soil fertility is determined by the content of humus substances in it. These substances are chemically and microbiologically stable. Due to the high content of ligand functional groups, they possess high complexation ability and are characterized by hydrophobic interactions. Black earths (chernozems) contain up to 15% humus, whereas the medium-humus soils contain up to 7% humus (Orlov, 1974). The change in the amount and qualitative composition of organic residues coming to the soil results in the situation that microorganisms use humus of the soil until its complete degradation (Mil`to et al., 1984).

We studied the change in the life cycles and population dynamics of soil-inhabitant collemboles of the species *Folsomia candida* and *Xenylla grisea (Hypogastruridae)* under the effect of various herbicides and metal complexes (Saratovskikh & Bokova, 2007). It was shown that under the action of DCPA the terms of appearance of the first sets of *F. Candida* increase from 10 to 48 days. The number of eggs in the sets decreases by 3 times. The duration of embryonic development elongates from 7 to 13 days. In a month the multiplicity of population decreases from 16 to 0.9 times. The biological activity of CuL_2 is multiply higher than that of DCPA. The action of CuL_2 results in a decrease in the population of adults. This effect is more negative for the population of posterity and an increase in whitebait even when using in low (10^{-7} M) concentrations. Evidently, the contact with herbicides is a reason for the violation of reproductive functions of the organism and decreases the population of the posterity of microarthropodes and also other microorganisms dwelling in the soil.

Pathology forms	Mean quantity on Russia	Environmental trouble area
an allergy to food in early childhood	70.0	400
bronchial (spasmodic) asthma	9.7	24
recurrent bronchitis	6.0	94
vascular dystonia	12.0	144
gastritis and gastroduodenitis	60.0	180
congenital malformation	11.0	140
encephalopathy	30.0	50
decrease intelligence quotient (IQ)>70%	30.0	138

Table. 6. Prevalence of pathology forms for children (per 1000 people) in environmental trouble area and on average over Russia.

The effect of herbicides and their metal complexes on hydrobionts is negative to the same extent (Saratovskikh et al., 2008a). So, DCPA and CuL_2 suppress the reproductive ability of hydrobionts, for example, infusorium *Tetrahymena pyriformis.* The effect is observed in a wide concentration rage from 10^{-1} to 10^{-7} M. Herbicides and their metal complexes decrease the activity of enzymes, for instance, luciferase of bacterium *Beneckea harveyi.* The inhibition of enzymes, for example, HADH-oxyreductase (Saratovskikh et al., 2005), ceases oxidation processes in polycellular organisms and results in the elimination of hydrobiont species and active silt and accumulation of contaminants in water ecosystems.

Larger inhabitants of flora and fauna disappear after representatives of the lowest trophic levels. The process gained the catastrophic character (Koptug, 1992).

Moreover, available published data show that the use of pesticides causes the most part of diseases of a modern human being (Table 6) (Gichev, 2003; Klyushnikov, 2005; Rakitsky et al., 2000) and is followed by lesions of the next generations of warm-blooded beings, including the man. This is the reason of many taken ill with cancroid (Popechitelev & Startseva, 2003). There is one more serious danger of using pesticides. It was indicated above that the pesticides have no selectivity of action. Gene-modified types of plants are developed to enhance the resistance of agricultural plants to the action of specific pesticides (Christoffers et al., 2002; Pyke et al., 2004; Sakagami et al., 2005). Based on the results presented, we may assert that the use of pesticides should drastically be reduced.

Nevertheless, this does not take place; on the contrary (Table 7), the absurdity of the situation is enhanced by the enlargement of application of gene-modified types of plants.

This cannot be explained from the scientific point of view, but economical reasons can be discussed. Diseases and death of people in all countries of the world, huge expenses of governments to (a) payment of sick leaves, (b) building of oncological and other medical centers, (c) payments of various medical insurances, and others, all these matters are profitable only for large chemical companies. Chemical companies produce: (a) pesticides, (b) gene-modified sorts of agricultural plants, and (c) drugs, which become more expensive and whose administration is accompanied by serious secondary effects.

USA	Economic loss from pollution of the air priced at a 20 billion US $ in year
Japan	damage bring of pollution of the environment averaged 5 trillion yen in year
Russia	damage bring of Chernobyl an accident averaged in 10 billion rouble (1990)
FRG	regulate application of pesticides on farms and on plot of land

Table 7. Economic loss from pollution of the environment.

4. Conclusion

Thus, 3,6-Dichloropicolinic acid (DCPA) the main active principle of herbicide Lontrel is poorly degradable by AS microbial association. Natural solar radiation does not affect its oxidation either, thus allowing the herbicide to accumulate in the environment. This results in dramatic changes in the composition of phytoplankton associations and decreases in the range of microbial species. These consequences may cause irreversible changes to the bioproduction of water bodies.

Application of chemical mutagenesis brings about AS with a broader range of species and, in turn, intensifies the oxidation of pollutants. Treatment of AS samples with NMU for 18 h resulted in more efficient detoxification as compared to the 6-hour treatment.

The rate of oxidation of DCPA by the action of UV radiation depends heavily on the source power. This rate increases three to fourfold when the reaction mixture is bubbled with oxygen or ozone as compared to air bubbling. Photochemical degradation of DCPA by UV radiation yields inherently toxic chemicals; however, they are successfully metabolized by the microbial association of the AS.

Ultraviolet irradiation (mimicking the natural sunlight action) did not degrade DCPA completely to environmentally safe products. The rate of DCPA degradation was notably lower when distilled water was replaced by river water and even lower in sea water. Chromatomass spectrometry revealed 9 compounds among the photolysis products, in addition to undergraded DCPA.

The use of "chemical remedies of plant protection" has stopped long ago to favor the crop and now gives an opposite effect. This is because the microflora and humus matter of the fertile ground layer are annihilated.

In the issue, it can be stated that both the pesticides and its decomposition products are high-toxicity substances. Therefore, the application of the pesticides must be reduced to minimum.

5. References

Aliev, Z.G.; Atovmyan, L.O.; Saratovskikh, E.A.; Krinichnyi, V.I. & Kartsev, V.G. (1988). Synthesis, structure, and spectral characteristics of copper-complexes with picolinic-acid derivatives. *Izv. Akad. Nauk SSSR, Ser.: Khim.*, No. 11, pp. 2495–2502.

Arantegui, J.; Prado, J.; Chamarro, E. & Esplugas, S.(1995). Kinetics of UV degradation of atrasine in a queous solution in the presence of H_2O_2. *J. Photochem. Photobiol.*, Vol. 88A, No. 1, pp. 65–74.

Arkhipova, M.B.; Tereshchenko, L.Ya. & Arkhipov, Yu.M. (1997). Photooxidative water purification from chlororganic pesticide 2,4D (2,4-di-chlor-phenoxyacetic acide). *Zh. Prikl. Khim.*, Vol. 70, No. 12, pp. 2016–2022.

Artobolevskii, I.I. (Ed.). (1977). *Politekhnicheskii slovar'* (Polytechnical Dictionary), Sov. entsikl., Moscow.

Beltran, F.J.; Rivas, J. & Acedo, B. (1993). Direct, radical and competitive reactions in the ozonation of water micropollutant. *J. Environ. Sci. Health. Part A. Environ. Sci. Engineering & toxic and hazardous substance control.* Vol. 28. No. 9. pp. 1947-1976.

Belyaeva, M.A. & Gyupter, L.I. (1969). Characterization of biocenosises of activated sludge in high-loaded aeration tanks and aeration tanks with a long period of aeration. *Biolog. Nauki*, No. 7, pp. 89–96.

Burgelya, N.K. & Myrlyan, N.F. (1985). *Geokhimiya I okruzhayushchaya sreda* (Geochemistry and Environment), Shtiintsa, Chisinau.

Bykorez, A.I. (Ed.). (1985). *Ekologiya i rak* (Ecology and Cancer), Naukova dumka, Kiev.

Calvert, G.M.; Plate, D.K.; Das, R.; Rosales R.; Shafetey, O.; Thomsen, C.; Male, D.; Beckman, J.; Arvizu, E. & Lackovic, M. (2004). Acute occupational pesticide-related illness in the US, 1998-1999: surveillance findings from the SENSOR-pesticide program. *Am. J. Ind. Med.*, Vol. 45, No. 1, pp. 14–23.

Castrantas, H.M & Gibilisco, R.D. (1990). UV destruction of phenolic-compounds under alkaline conditions. *Am. Chem. Soc.*, ACS Symp. ser. 422. Washington. pp. 77-85.

Christoffers, M.J.; Berg, M.L. & Messersmith C.G. (2002). An isoleucine to leucine mutation in acetyl-CoA carboxylase confers herbicide resistance in wild oat. *Genome,*. Vol. 45, No. 6, pp. 1049-1056.

Danagulyan, G.G. (2005). Kost-Sagitullin Rearrangement and Other Isomerization Recyclizations of Pyrimidines. (Review). *Khim. Geterotsikl. Soedin.*, No. 10, pp. 1445–1480.

Jiller, P. (1988). *The Structure of Co-societies and Ecological Niche.* Mir, Moscow.

Eikhler, V. (1993). *Yady v nashei pishche* (Poisons in Our LIfe), Mir, Moscow.

Fedtke, C. (1985). *Biochemistry and Physiology of Herbicide Action*, Agropromizdat, Moscow.

Fliedner, A. (1997). Ecotoxicity of poorly water-soluble substances. *Chemosphere*, Vol. 35, Nos. 1-2, pp. 295–302.

Gichev Yu.P. (2003). *Environmental Contamination and Ecological Stipulations of Human Pathology:* Analytical Review, (Ser. Ecology, Issue 68), GPNTB Siberian Division RAS, Novosibirsk.

Golovleva, L.A. & Fil'kenshtein, Z.I. (1984). Conditions of microbial degradaition of pesticides. *Agricultural Chemistry*. No. 3. pp. 105-119.

Golubovskaya, E.K. (1978). *Biologicheskie osnovy ochistki vody* (Biological Foundations of Water Purification), Vysshaya shkola, Moscow.

Government Regulation of the Russian Federation of November 8, 2001 no. 780 "On the Federal Target Program "Enhancement of Fertility of Soils in Russia for 2002-2005"."

Guittonneau, J.P.; de Laat, J.; Dore, M.; Duguet, J.P. & Bonnel, C. (1988). Comparative-study of the photodegradation of aromatic-compounds in water by UV and H_2O_2/UV. *Environ. Technol.*, Vol. 9, No. 10, pp. 1115–1128.

Hall, J.C.; Bassi, P. K.; Spencer, M. S. & Vanden Born, W. H. (1985). An evaluation of the role of ethylene in herbicidal injury induced by picloram and clopyralid in rapeseed and sunflower plants. *Plant Physiol.*, Vol. 79. pp. 18–23.

Klimova V.A. (1975). *The Main Micromethods for Analysis of Organic Compounds*. Khimiya, Moscow. 222 p.

Klyushnikov, V.Yu. (2005). Human organism as an indicator for environmental contamination. *Sensor*, No. 3, pp. 26-33.

Koptug V.A. (1992). *UNO Conference on the Environment and Development* (Rio de Janeiro, June 1992): Informational Review/ Siberian Division, RAS. Novosibirsk. 62 p.

Kruithof, J.C.; van der Leer; R.Chr.; Hajnen, W.A.M. & Kruithof, P. (1992). Practical experiences with UV disinfection in the Netherlands. *J. Water SRT-Aqua*, Vol. 41, pp. 25 - 37.

Kusnetsov, V.I.; Kozlov, N.I. & Homiakov, P.M. (2005). *Mathematical Simulation of Forest Evolution for Purposes of Controlling Forestry*. LENAND, Moscow.

Laval-Martin, D.; Dubertret, D. & Calvayrac, R. (1977). Effect of atrazine and methabenzthiazuron on oxygen evolution and photosynthetic properties of a *Chlorella pyrenoidosa* and *Euglena gracilis*. *Plant Sci. Lett.*, Vol. 10, No. 2, pp. 185–195.

Lebedev, A.; Moshkarina, N.; Buryak, A. & Petrosyan, V. (1996). Water chlorination of nitrogen containing fragments of humic material. In: *Technological civilization impact on the environment.* Karlsruhe. p. 86.

Legrini, O.; Oliveros, E. & Braun, F.V. (1993). Photochemical processes for water treatment. *Chem. Rev.*, Vol. 93, No. 2, pp. 671–677.

Liperovskaya, E.S. (1977). In *Obshchaya ekologiya. Biotsenologiya. Gidrobiologiya* (General Ecology. BiocenoBiocenology. Hydrobiology), *Itogi Nauki Tech.*, Vol. 4, VINITI, Moscow. pp. 25–29.

Mel`nikov, N.N. (1987). *Pesticides. Chemistry, Technology, and Use*. Khimiya. Moscow.

Mil`to, N.I.; Karbanovich, A.I. & Vorochasova, B.T. (1984). *The Role of Microflora in Soil Protection from Agroindustrial Contaminations.* Nauka i Tekhnika, Minsk.

Nakanisi, K. (1965). *Infrakrasnye Spektry i Stroenie Organicheskikh Soedinenii*, Mir, Moscow.

Nakomoto, K. (1991). *IK spectry i spectry KR neorganicheskikh I koordinatsionnykh soedinenii*, Mir, Moscow.

Orlov, L. S. (1974). *Humic Acids of Soils,* Mos. State Univ., Moscow.

Ozelenenie, problemy fitogigieny i okhrany gorodskoi sredy (Planting of Greenery: Problems of Phytohygiene and Protection of Urban Environment), (1984). Nauka, Leningrad.

Patel, B.M.; Moye, H.A. & Weinberger, R. (1991). Post column formation of fluorophores from nitrogenous pesticides by UV-photolysis. *Talanta*, Vol. 38, No. 8, pp. 913–924.

Poliakova, O.V.; Mndgojan, K.K.; Khruchsheva, M.L.; Ilyashenko, V.Yu. & Lebedev, A.T. (2005). Environmental applications of GC-GC-MS for the Analyses of Super

Complex Mixtures of Organic Pollutants, *Proc. VI Congr. Intern.* GRUTTEE. Aix-les-Bains, 2005. Berlin. p. 32.

Popechitelev, E.P. & Startseva, O.N. (2003). *Analytical Studies in Medicine, Biology, and Ecology.* Vysshaya Shkola, Moscow. 280 p.

Prado, J. & Esplugas, S. (1999). Comparison of different advanced oxidation processes involving ozone to eliminate atrazine. *Ozone Sci. Eng.*, Vol. 21, pp. 39–52.

Pregl F. (1934). *Quantitative Organic Microanalysis.* Goskhimizdat, Moscow. 203 p.

Pyke, F.M.; Bogwitz, M.R.; Perry, T.; Monk, A.; Batterham, P.& McKenzie, J.A. (2004). The genetic basis of resistance to diazinon in natural populations of *Drosophila melanogaster. Genetica* (Dordrecht), Vol. 121, No. 1, pp. 13-24.

Rakitsky, V.N.; Koblyakov, V.A. & Turusov V.S. (2005). Nongenotoxic (epigenetic) carcinogens: pesticides as an example. A critical review. *Teratog.carcinog. Mutagen.*, Vol. 20, No. 4, pp. 229-240.

Rapoport, J.A. (2010). *Microgenetics*, Reprint, Moscow.

Sakagami, H.; Hashimoto, K.; Suzuki, F.; Ogiwara, T. & Fujisawa, S. (2005). Molecular requirements of lignin-carbohydrate complexes for expression of unique biological activities. *Phytochem.*, Vol. 66, No. 17, pp. 2108-2120.

Saprikin, F.Ia. (1984). *Geochemistry of Soils and Nature Conservancy,* Nedra, Leningrad.

Saratovskikh, E.A.; Kondrat'eva, T.A.; Psikha, B.L.; Gvozdev, R.I. & Kartsev, V.G. (1988). Complex-formation of some pesticides with adenosine triphosphoric acid. *Izv. Akad. Nauk SSSR, Ser.: Khim.*, No. 11, pp. 2501–2506.

Saratovskikh, E.A.; Lichina, L.V.; Psikha, B.L. & Gvozdev, R.I. (1989a). Character of the reaction of dinucleotides and polynucleotides with some pesticides. *Izv. Akad. Nauk SSSR, Ser. Khim.*, No. 9, pp. 1984–1989.

Saratovskikh, E.A.; Orlov, V.S. & Krinichnyi, V.I. (1989b). EPR spectroscopic study of metallocomplexes of 3,6-dichloropicolinic acid. *Izv. Akad. Nauk SSSR, Ser. Khim.*, No. 11, pp. 2477–2481.

Saratovskikh, E.A.; Papina, R.I. & Kartsev, V.G. (1990). Binding of ATP by pesticides as a possible primary mechanism of inhibition of seed germination and plants growth. *Sel'skokhoz. Biol.*, No. 5, pp. 152–159.

Saratovskikh, E.A.; Korshunova, L.A.; Gvozdev, R.I. & Kulikov, A.V. (2005). Inhibition of the nicotinamide adenine dinucleotide-oxidoreductase reaction by herbicides and fungicides of various structures. *Izv. Akad. Nauk SSSR, Ser. Khim.*, No. 5, pp. 1284–1289.

Saratovskikh, E.A. & Bokova, A.I. (2007). Influence of herbicides on the population of soil-dwelling collembola. *Toxicological Review,* No. 5, pp. 17-23.

Saratovskikh, E.A.; Glaser, V.M.; Kostromina, N.Yu. & Kotelevtsev, S.V. (2007a). Genotoxicity of the pestiside in Ames test and the possibility to formate the complexeses with DNA. *Ecological genetics,* Vol. 5, No. 3, pp. 46- 55.

Saratovskikh, E.A.; Korshunova, L.A.; Roschupkina, O.S. & Skurlatov, Yu.I. (2007b). Kinetics and Mechanism of Inhibition of Enzymatic Processes by Metal Compounds. *Khim. Fiz.*, Vol. 26. No. 8. pp. 46-53.

Saratovskikh, E.A.; Kozlova, N.B.; Baikova, I.S. & Shtamm, E.V. (2008a). Correlation between the toxic properties of contaminants and their constants of complexation with ATP. *Khim. Fiz.*, Vol. 27. No. 11. pp. 87-92.

Saratovskikh, E.A.; Psikha, B.L.; Gvozdev, R.I. & Skurlatov, Yu.I. (2008b). Kinetic model of transfer process of Technogenic contaminants through liposomal membranes. *Khim. Fiz.*, Vol. 27, No. 7, pp. 59–65.

Schöniger, W. (1955). *Microchim. Acta.* pp. 123-129.

Shinkarenko, N.V. & Aleskovskii, V.B. (1982). The chemical properties of singlet molecular oxygen and its significance in biological systems. *Usp. Khim.*, Vol. 51, No. 5, pp. 713–718.

Skurlatov, Yu.I.; Duka, G.G. & Miziti, A. (1994). *Vvedenie vekologicheskuyu khimiyu* (Introduction in Ecological Chemistry), Vysshaya shkola, Moscow.

Skurlatov, Yu.I.; Shtamm, E.V.; Pavlovskaya, N.N.; Lukov, S.A. & Mironova, T.A. (1996). Problemy pit'evogo vodosnabzheniya i puti ikh resheniya. *Rossiisko-skandinavskii nauchno-tekhnicheskii seminar* (Problems of Drinking Water Supply and Ways of Their Solution. Russian-Scandinavian Science and Technology Seminar), Rakhmanin, Yu. A. (Ed.). Nauka, Moscow. pp. 82–85.

Skurlatov, Yu.I. & Shtamm, E.V. (1997a). Role of the redox, free-radical and photochemical processes in natural waters, upon waste-water treatment and water preparation. *Khim. Fiz.*, Vol. 16, No. 12, pp. 55–68.

Skurlatov, Yu.I. & Shtamm, E.V. (1997b). Ultraviolet Radiation in Processes of Water Preparation and Water Treatment. *Water Supply and Sanitary Technology.* No. 9. pp. 14-19.

Skurlatov, Yu.I. & Shtamm, E.V. (2002). Some sun in tap water. *Ekologiya i Zhizn'*, Vol. 25, No. 2, pp. 24–27.

Steuermark Al. (1961). *Quantitative organic microanalysis.* Academic Press. New York, London. 665 p.

Sverdlov, L.M.; Kovner, M.A. & Krainov, E.P. (1970). *Kolebatel'nye Spektry Molekul* (Vibrational Spectra of Molecules), Nauka, Moscow.

Sundstrom, D.W.; Weir, F.A. & Klei, H.E. (1989). Destruction of aromatic pollutants by UV-light catalyzed oxidation with hydrogen-peroxide. *Environ. Progr.*, Vol. 8, No. 1, pp. 6–12.

Tretyakova, N.Yu.; Lebedev, A.T. & Petrosyan, V.S. (1994). Degradative pathways for aqueous chlorination of orcinol. *Environ. Sci. Engineer.*, Vol. 28, pp. 606–613.

Wan, H.B.; Wong, M.K. & Mok, C.Y. (1994). Comparatue study on the guantin yields of direct photolysis of organophosphorus pesticides in aguences solution. *J. Agric. Food Chem.*, Vol. 42, No. 11, pp. 2625–2630.

Whyatt, R.M.; Rauh, V.; Barr, D.B.; Camann, D.E.; Andrews, H.F.; Garfinkel, R.; Hoepner, L. A.; Diaz, D.; Dietrich, J.; Reyes, A.; Tang, D.; Kinney, P. L. & Perera, F.P. (2004). Prenatal insecticide exposures and birth weight and length among an urban minority cohort. *Environ. Health Perspect.*, Vol. 112, No. 10, pp. 1125–1132.

Yablokov, A.V. (1990). *Yadovitaya priprava* (Poisonous Relish), Mysl', Moscow.

Yoshioka, Y.; Ose, Y. & Sato, T. (1985). Testing for the tixicity of chemicals with *Tetrahymena pyriformis. Sci. Total Environ.*, Vol. 43, Nos. 1-2, pp. 149–157.

Yudanova, L.A. (1989). *Pestitsidy v okruzhayushchei srede* (Pesticides in the Environment), GPNTB Siberian Division AN SSSR, Novosibirsk.

Zviagintsev, D.G. (1987). *Soils and Microorganisms,* Mos. State Univ., Moscow.

5

Distribution and Potential Effects of Novel Antifouling Herbicide Diuron on Coral Reefs

M.A. Sheikh[1,5], T. Oomori[1], H. Fujimura[1], T. Higuchi[1], T. Imo[1,6], A. Akamatsu[1], T. Miyagi[2], T. Yokota[3] and S. Yasumura[4]

[1]*Department of Chemistry, University of the Ryukyus, Okinawa,*
[2]*Okinawa Prefectural Institute of Health and Environment, Okinawa*
[3]*Water Quality Control Office, Okinawa Prefectural Bureau, Okinawa*
[4]*WWF Japan, Minato-ku, Tokyo*
[5]*Research Unit, The State University of Zanzibar,*
[6]*Faculty of Science, Samoa National University,*
[1,2,3,4]*Japan*
[5]*Tanzania*
[6]*Samoa*

1. Introduction

1.1 Characteristics of diuron

N′-(3,4-dichlorophenyl)-N, N-dimethylurea (diuron) is a herbicide belonging to the phenylamide family and the subclass of phenylurea (Fig. 1). It is a colourless crystalline compound in its pure form, non-ionic, with a moderate water solubility of 42 mg /L at 20 °C. It remains a solid at ambient temperature (25 °C) with a melting point of 158–159 °C. Its vapour pressure is 0.009 mPa at 25 °C and has a calculated Henry's law constant of 0.000051 Pam^3 /mol suggesting that diuron is not volatile from water or soil (Giacomazzi and Cochet, 2004).

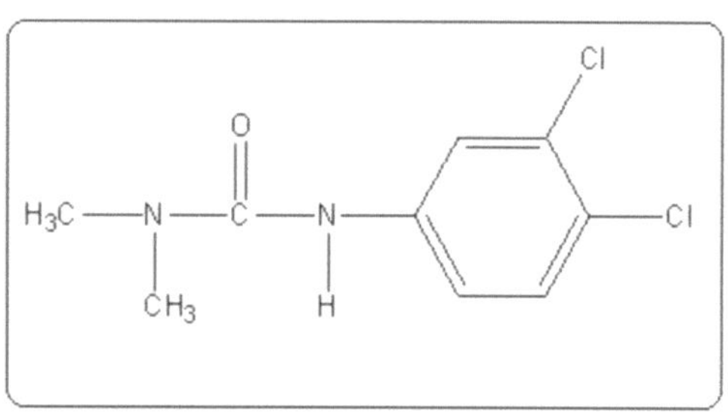

Fig. 1. Chemical structure of diuron.

Diuron has been used to control weeds on hard surfaces such as roads, railway tracks, and paths. It is also used to control weeds in crops such as pear and apple trees, forests, ornamental trees and shrubs, pineapples, sugar cane, cotton, alfalfa and wheat. Furthermore, diuron is widely used as a marine antifouling compound.

1.2 Contamination status and potential effects of diuron to coral reefs

Coral reefs are widely distributed in tropical and subtropical shallow waters (Smith and Kinsey, 1978; Suzuki and Kawahata, 2003; Inoue et al., 2005). They are characterized as highly productive carbon systems for both organic and inorganic carbon (Smith and Kinsey, 1978). Photosynthesis and calcification are the main biogeochemical processes in the coral reef ecosystems (Smith 1973; Suzuki and Kawahata, 2003).

In recent decades, coral reefs have begun to face many threats caused by both natural and anthropogenic sources. The sustainability of these ecosystems have been thrown into doubt by a number of challenges including: marine pollution, global environmental changes, and outbreaks of the crown-of-thorns starfish and coral disease. It has been estimated that 27% of the world's coral reefs have already been lost and 31% have been projected to be degraded by 2030 (Wilkinson, 2000). More integrated efforts and new conservation approaches are necessary to minimize further catastrophe in the future of coral reef ecosystems.

Diuron is considered a priority hazardous substance by the European Commission (Malato et al., 2002). Countries including the UK, Sweden, Denmark and France have restricted the use of diuron in antifouling paints (Konstantinou and Albanis, 2004; Giacomazzi and Cochet, 2004).

Diuron inhibits photosynthesis in plants by binding site of photosystem II (PSII), which limits the electron transfer (Vandermeulen, et al., 1972). Eco-toxicological studies have shown that diuron induces significant impacts on corals (Jones, 2005), as shown by the reduction of ^{14}C incorporation in *Madracis mirabilis* (Owen et al., 2002), the reduction of $\Delta F/Fm'$ in *Stylophora pistillata, Seriotopora hystrix* and *Acropora formosa* (Jones, 2005), the loss of symbiotic algae in *Montipora digitata* and *S. hystrix* (Jones, 2004), and the detachment of soft tissue in *Acropora tenuis* juveniles (Watanabe et al., 2006). In addition, the herbicide has been associated with serious impacts on other marine ecosystems such as mangrove diebacks (Bell and Duke, 2005).

Diuron is very persistent in the environment and can remain from one month to up to one year in a given ecosystem (Giacomazzi and Cochet, 2004). Diuron has been detected in marine environments from various regions such as western Japan (Okamura et al, 2003), the UK (Boxall et al., 2000), Spain (Ferrer and Barcelo, 1999; Martinez et al., 2000), The Nertherlands (Lamoree et al., 2002), and Sweden (Dahl and Blanck, 1996). Diuron can undergo abiotic degradation such as hydrolysis, photodegradation, as well as biotic degradation (Giacomazzi and Cochet, 2004).

In Japan, diuron has been extensively used in antifouling paints in shipping and agricultural activities (Okamura et al., 2003). In 2004 alone, ~11 tons of diuron was used for sugar cane crops in the Okinawa Prefecture, which was the highest amount of diuron used in Japan outside of the Tokyo metropolitan. In addition, the urban areas of Okinawa mainland, apply a significant amount of diuron as a weed control (Kitada, 2007).

Despite the extensive usage of diuron in Ryukyu Archipelago, and the associated toxicological implications in coral reefs, very little is known about the baseline levels and potential physiological effects of diuron in coral reef waters around the Ryukyu Archipelago. So far, only one study has reported the diuron contents in river sediments around mainland Okinawa (Kitada, 2007). Yet, the risks posed by pollution of coral reef waters with the soluble fraction of toxic chemicals need to be given priority.

Therefore, this study provides a combination of the results of a systematic monitoring and behavior of diuron in coral reef ecosystems around the Ryukyu Archipelago as well as the ecotoxicological impacts of the herbicide diuron on coral reefs.

2. Materials and methods

2.1 Study area

The study was conducted around the Shiraho coral reefs, adjacent areas and main Naha Bay located in the Ryukyu Archipelago, South-western Japan. The Ryukyu Archipelago is located between 24 and 30 °N, constituting the southern part of the Nansei Islands (Fig. 2). The Archipelago is a chain of more than 100 Islands lying in between Kyushu (mainland Japan) and Taiwan, separating the East China Sea from the Pacific Ocean.

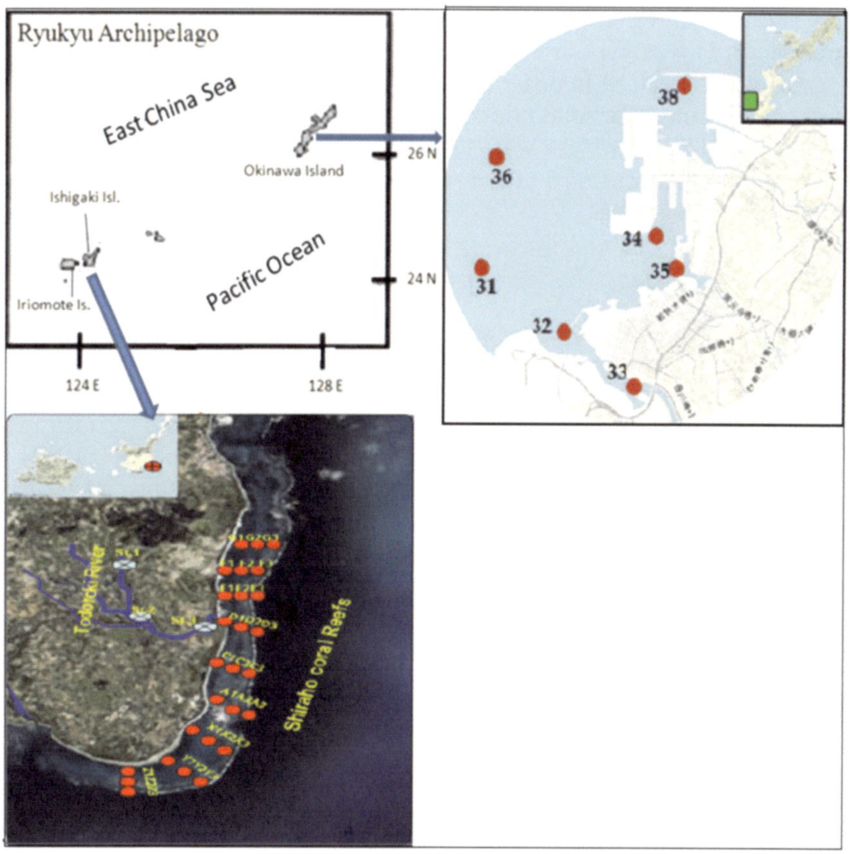

Fig. 2. Sampling locations.

During the main sampling event, the Shiraho reef was divided into nine transects, where three sampling points were established in each transect. Ten sampling points were selected along the Todoroki River. An extensive survey of diuron in the waters around the Shiraho coral reefs and inflow from the Todoroki River was carried out during various seasons in 2007, 2008 and 2009. A total of 22 and 191 water samples were analyzed for the Todoroki River and Shiraho coral reefs, respectively. In Naha Bay, 42 samples were collected between Sept., 2007 and Feb., 2008).

At each location, a 1 L sample of water was collected in an acetone-washed amber bottle. Samples were returned to the laboratory and stored at < 4°C in a cold dark room and extracted within 10 days.

2.1.2 Sample pre-treatment

Diuron in water was analyzed following the solid phase extraction LC/MS method of analysis of pesticides in drinking water as recommended by the Ministry of Health, Labor and Welfare, Japan (Okinawa Prefectural Enterprise Bureau, 2006). Water samples were pre-concentrated in the solid phase extraction cartridges (PLS-3, GL sciences, Japan). Prior to the extraction of the diuron, the columns were conditioned with 10 mL of acetonitrile, followed by methanol and milli-Q water, respectively. 10 mL of 0.2 M EDTA was added to 1 L of the water sample and pH was kept at 3.5. 1 mL of 1mg/L diuron D-6 ($C_9H_4Cl_2D_6N_2O$) was spiked as a surrogate standard in order to monitor the recovery of diuron. Water samples were eluted using an automatic solid phase extraction controller (Shimadzu, Japan) at a flow rate of 20 mL/min. PLS-3 cartridges were then dried under nitrogen gas for 5 min. Diuron was eluted from the column using 5 mL of acetonitrile. Finally, acetonitrile was evaporated to 0.2 mL with pure nitrogen gas.

2.1.3 Instrumental analysis

The analysis of diuron was achieved using LC-MS (Agilent 1100LC/MSD SL System) under the following operating conditions: Column; Agilent ZORBAX Eclipse XDB-C18 4.6 × 30 mm,1.8 µm, Column temperature, 40 °C. Mobile phase; A, HCO_2 H/Water (0.1%), B, acetonitrile, Gradient 95% A-(liner gradient 5min)-80 % B (7 min). Flow rate; 0.5 mL min^{-1}. Injection volume 10 µl. MS conditions; Ionization mode, negative ion-ESI SIM/Scan mix mode, Desolvation gas, nitrogen 12 L min^{-1}, Desolvation temperature, 350 °C, Capillary voltage, 2.5 kV, SIM monitor ion; m/z 231 and 237. The detection limit was 0.02 µg/L. The recovery of Diuron D-6 was > 90 % in the spiked samples. Data for the environmental samples were corrected for recovery values.

2.2 Laboratory incubation experiment

2.2.1 Coral sample preparation

A colony of coral *Galaxea fascicularis*(Fig.3) was collected from the shallow zones in front of the University of the Ryukyus Tropical Biosphere Research Center (TBRC), Sesoko Island (127°25′E26°39′N) with permission from the Okinawa Prefectural Government (# 17-04). The colony was then transported in a bucket with approximately 5 L of seawater and then transferred immediately into an open-circuit, fresh seawater aquaria exposed to sunlight through a black mesh roof. The coral colony was tagged and cut into 1.5-2.0 cm pieces that were anchored in PVC tubes on acryl resin screws. The fragments were acclimatized for ~4 weeks in the aquarium before being brought to the laboratory for the experiment.

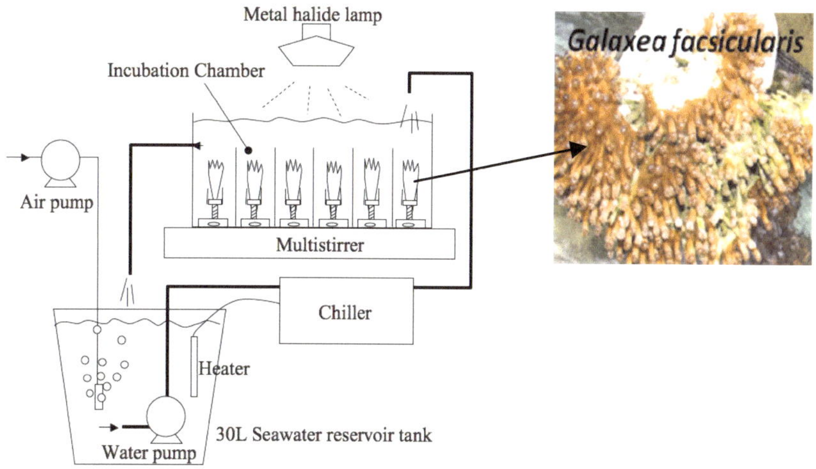

Fig. 3. Schematized diagram showing the experimental set-up (seawater circulation system).

The study was conducted in a continuous-flow seawater aquarium (15 x 30 x 20 cm^3) (Fig 3.). Seawater temperature (27 ^oC) was carefully controlled by Chiller (G x C-200, China) while light intensity was controlled by a metal halide lamp (Neo Beam Light 24W, KAMIHATA, Japan) which provided illumination at the coral surface at a Photosynthetic Available Radiation (PAR) of 300 µmol m^{-2} s^{-1} during a 12:12-h light/dark cycle , respectively.

Stock solutions of diuron (Sigma-Aldric, Germany) were prepared in filtered seawater using acetone (PCB and pesticide analysis grade) to improve dissolution. Another stock solution containing only acetone in filtered seawater was prepared for control treatment.

2.2.2 Exposure experiment

Corals were exposed to various treatments (0 ng L^{-1} (control), 1000 ng L^{-1}, and 10,000 ng L^{-1}) of diuron for 96 h. Six replicates of the coral *Galaxea fascicularis* were used for each treatment. Each coral were incubated in an acryl chamber (0.18 L) for a duration of 2 h. Water circulation was stopped during the incubation period. The seawater samples were collected at the beginning and at the end of incubation. The pH of seawater was recorded *in-situ* using pH meter (Orion 290 A+Thermo, USA). The total alkalinity (A_T) was measured using an Auto Titration System (TIM 860 Radiometer, France) within 10 days after sampling.

Changes in inorganic carbon production (IP) and organic carbon production (OP) were then determined by the alkalinity and total inorganic carbon depletion method (Smith 1973; Smith and Kinsey 1978; Fujimura et al., 2001) as follows:

$$IP = -0.5 \cdot \Delta A_T \cdot \rho \cdot V/\Delta t \cdot A \quad (1)$$

$$OP = \Delta C_T \cdot \rho \cdot V/\Delta t \cdot A - IP \quad (2)$$

IP= inorganic production, OP=organic production, ΔA_T= Change of total alkalinity, ρ= density of seawater, Δt= Change of incubation time, A=surface area of coral, V=volume of sea water used for coral incubation, ΔC_T=Change of total inorganic carbon. C_T was obtained from pH and alkalinity according to carbonate equilibrium as described by Fujimura et al., (2001).

2.3 Statistical analysis

Statistical analyses were performed using SPSS 16 to test for significant differences between the results of the individual treatments. Differences between doses (treatments) were tested for significance using a one-way analysis of variance (ANOVA), with an α of 0.05.

3. Results

3.1 Distribution of diuron

The levels of diuron in the samples were ranged between ND (not detected)-753.8 ng L^{-1}. The maximum concentration (753 ng L^{-1}) of diuron was found in the portion of the Todoroki river draining to Shiraho coral reefs. The level of contamination of diuron in the waters around the Ryukyu Archipelago is not as high in comparison to western Japan and other bodies of water around the world. For example, concentrations of up to 3,050 ng L^{-1} were detected in the Seto Inland Sea, (Okamura et al., 2003); 42,000 ng L^{-1}in lagoon water in Italy (Gennaro et al., 1995); 768 ng L^{-1} in the marina UK, (Boxall et al., 2000); 6,742 ng L^{-1} estuaries UK, (Thomas et al., 2001), 2,000 ng L^{-1} in Mediterranean coast, Spain, (Martinez et al., 2000), and 1,130 ng L^{-1} in Marinas Netherlands, (Lamoree et al., 2002). However, the maximum level detected in this study has exceeded the permitted maximum levels of 430 ng L^{-1} as set by the Dutch National Institute of Public Health and the Environment (Giacomazzi and Cochet, 2004). Thus, diuron could pose significant risks and general ecological health concerns.

Diuron residues showed significant spatial variations in the sampling areas (Fig. 4, Table 1). Based on the results, it is evident that the distribution reflects anthropogenic activities and possible diuron sources. The detection frequency of diuron at Naha Bay was comparable to the Shiraho reef. In addition, diuron was frequently detected at station 35 (Fig. 4), indicating an active source of diuron in Naha Bay. The frequency of diuron detection of diuron in Okinawa mainland is comparative to the coastal waters in southern California (93%) (Sapozhnikova et al., 2007).

The concentration of diuron in the Todoroki River was relatively high at upstream, ST1, 753. 8 ng L^{-1}, compared to the rest of the points towards the lagoon (Table 1). Also, relatively high concentrations of diuron were found at transect F and E at the Todoroki River mouth (Table 1). These results suggest that the Todoroki River could be a potential source of diuron from farms to the Shiraho coral reefs. In Ishigaki Island, diuron is extensively used for agricultural activities.

This research also showed seasonal variations. The concentrations of diuron in the Shiraho coral reefs were relatively higher during November (Winter) as opposed to August-September (Summer) and Spring (May-June), 2007, 2008 and 2009 (Table 1). Thus, there is a possibility that diuron is retained in the soils in summer during the maximum usage

	Spring (May 2007)	Summer (Aug,2007)	Winter (Nov., 2007)	Spring (May, 2008)	Winter (Nov, 2008)	Winter (Jan, 2009)
A1	ND	0.62	2.07	0.23	0.26	ND
A2	ND	0.9	ND	0.36	5.43	ND
A3	ND	ND	2.18	0.31	5.8	ND
C1	ND	ND	ND	1.45	ND	ND
C2	ND	ND	ND	ND	ND	ND
C3	ND	ND	ND	0.83	19.3	ND
D1	0.88	0.87	2.25	ND	7.64	ND
D2	0.7	ND	ND	ND	ND	ND
D3	ND	ND	ND	ND	7.34	ND
E1	6.27	1.15	12.9	6.21	15.9	6.4
E2	ND	ND	0.23	ND	ND	ND
E3	ND	0.59	3.25	ND	10.2	ND
F1	1.28	ND	0.73	9.23	0.18	12.9
F2	1.61	ND	ND	ND	0.82	ND
F3	1.24	ND	15.61	ND	ND	ND
G1	0.92	0.93	ND	9.83	ND	ND
G2	0.78	1.5	ND	0.06	10.8	ND
G3	0.67	ND	ND	2.97	0.06	ND
X1	ND	ND	ND	ND	ND	ND
X2	ND	1.34	ND	1.61	ND	ND
X3	ND	0.72	ND	ND	ND	ND
Y1	ND	ND	24.2	3.97	2.81	ND
Y2	ND	ND	ND	ND	ND	ND
Y3	ND	ND	**90.02**	ND	0.92	ND
Z1	ND	ND	39.07	ND	ND	ND
Z2	ND	ND	8.48	1.48	ND	ND
Z3	ND	ND	ND	ND	7..22	ND
	Todoroki River					
ST1	12.5	NS	NS	68.6	**753.8**	NS
ST2	1.2	NS	NS	3.06	37.6	NS
ST3	1.7	NS	NS	7.64	50.1	NS
ST4	NS	NS	NS	NS	42.8	NS
ST5	NS	NS	NS	NS	11.2	NS
ST6	NS	NS	NS	NS	25.2	NS
ST7	NS	NS	NS	NS	21	NS
ST8	NS	NS	NS	NS	10.3	NS
ST9	NS	NS	NS	NS	77.7	NS
ST10	NS	NS	NS	NS	67.8	NS

NS means No Sample; ND means Not Detected

Table 1. Spatial and temporal variation of diuron around Shiraho coral reefs (ng L^{-1}).

(summer) in the region and released to the aquatic systems during high rainfall during December (Winter). The retention of diuron in the soils is may due to the fact that it has a long half-life 43-2180 days (Sapozhnikova et al., 2007). In Naha Bay, relatively high levels were detected in September compared to other months (Fig. 4). This season coincides with the high boating season in Okinawa Island, thus suggesting that the main source of diuron is from antifouling paints released from ships. Okamura et al., (2003) found considerable levels of diuron inside the fishing ports in Western Japan.

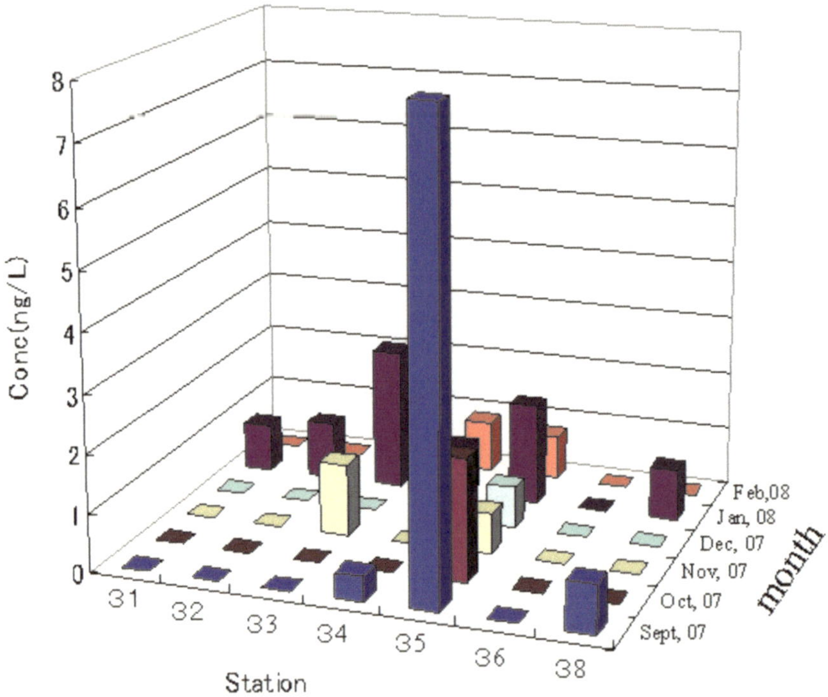

Fig. 4. Spatial and temporal distribution of diuron in Naha Bay (Sep. 2007-Feb. 2008).

3.2 Effect of diuron on coral metabolism

3.2.1 Effects of diuron on photosynthesis rate

Photosynthesis was significantly reduced (ANOVA $p < 0.05$) when the coral was exposed to 10,000 ng L^{-1}. Diuron concentrations of 10,000 ng L^{-1} caused rapid decrease in photosynthesis after 96 h exposure (Fig. 5). The results show that 1,000 ng L^{-1} reduced photosynthesis 6.5 % relative to the control but was not significant (ANOVA $p > 0.05$). The photosynthesis was gradually decreased from control, 1000 ng L^{-1} to 10,000 ng L^{-1} (Fig. 5).

3.2.2 Effects of diuron on calcification rate

Diuron had a significant impact on calcification rate only at the highest concentration of diuron (10,000 ng L^{-1}) (ANOVA $p < 0.05$), the calcification rate dropped to 67.3 % less than the control (Fig. 5).

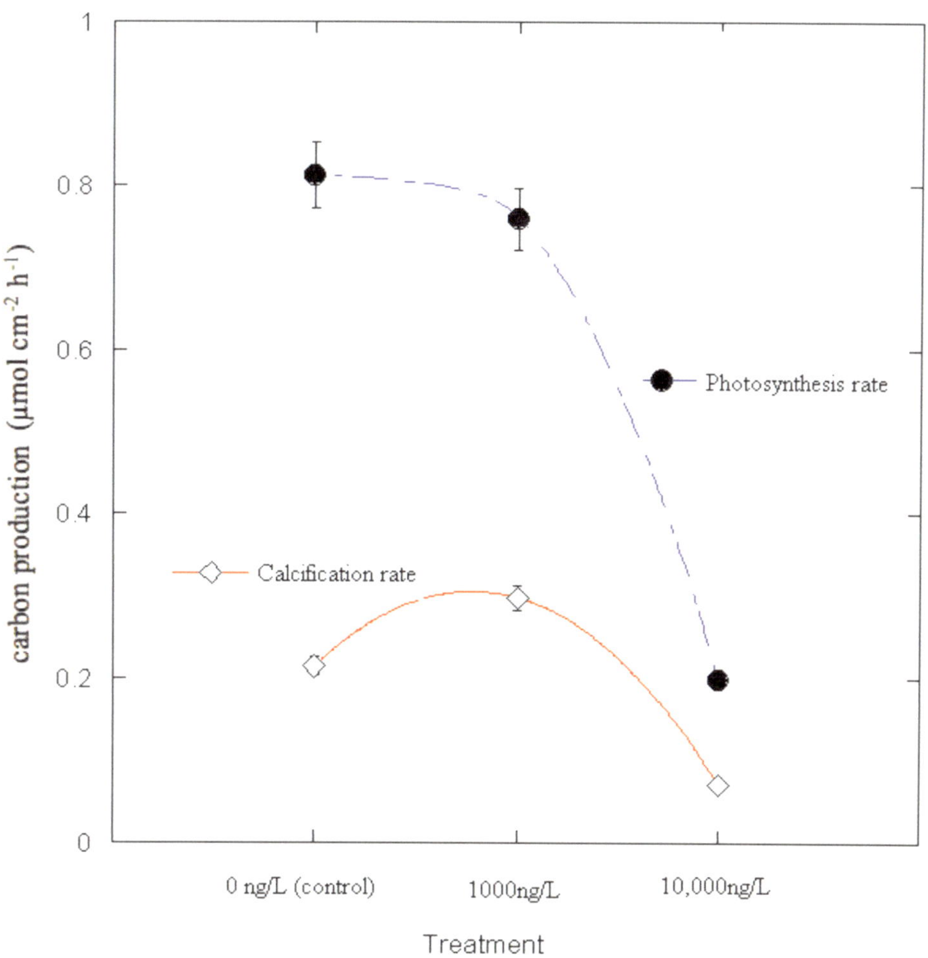

Fig. 5. Effects of diuron on calcification and photosynthesis.

Diuron (DCMU) inhibits photosynthesis by blocking electron transport in phosystem II (PSII), eventually causing immediate disruption in the symbiosis between zooxanthellae and host coral (Råberg et al., 2003). When temporarily bound, they can disrupt photosynthetic electron flow, and can eventually lead to a loss of excitation energy at photosystem reaction center (Jones and Kerswell, 2003).

Our results clearly show that the photosynthesis rate gradually decreased when corals were exposed to treatments of 1,000 and 10,000 ng L^{-1} of diuron. These results are supported by previous studies by Jones & Heyward (2003) and Owen et al. (2002) which also showed that diuron inhibit the photosynthesis of corals by blocking the conversion of excitation energy into chemical energy (Jones, 2005).

The calcification of corals is controlled by a mutualistic relationship with zooxanthellae. The zooxanthellae utilize the waste products of a host such as CO_2 from respiration (Liat et al., 2005). Our results demonstrate that a 10,000 ng L^{-1} treatment of diuron caused a sharp decrease in the calcification of *Galaxea fascicularis* after 96 h exposure. This may be caused by the disruption of symbiosis between zooxanthellae and the host coral. Diuron might also block the energy which may be needed to trigger calcium uptake from the seawater coeleteron of coral *Galaxea fascicularis* (Al-Horani et al., 2003). These findings suggest that the deterioration of calcification rate might be caused by the blockage of energy transfer by the PSII compounds. This is attributed to a study conducted by Chalker and Taylor (1975), which showed that energy is needed for the calcification process to transport ions and for the formation of organic matrix.

4. Potential eco-toxicological effects of diuron on coral reefs

The average levels of diuron detected around the Ryukyu Islands including the Shiraho coral reef waters were more than three-fold less than the lowest observable effects concentration (LOEC) to corals as shown by laboratory eco-toxicological studies for example Reduction of ΔF/F` in *Stylophora pistillata* 250 µg L^{-1} (Jones & Heyward., 2003); Loss of algae in *Pocillopora damicornis* 30,000 ng L^{-1} (Negri et al., 2004); reduction of respiration in *Porites cylindrica* 10,000 ng L^{-1} (Råberg et al., 2003); Loss of algae in *Seriotopora hystrix* 10,000 ng L^{-1} (Jones , 2004).

These results suggest that the present contamination of diuron in the coastal waters around the coral reefs and adjacent areas in the Ryukyu Archipelago are not at an alarming stage, and does not seem to pose a serious threat to the health of corals during acute exposure. However, it is important to highlight that the maximum value of diuron concentration (753 ng L^{-1}) detected has already approached the threshold limit, and thus could have a significant risk on corals in the future, solely or synergistically. It remains uncertain in the field of marine science how chronic exposure to the environmentally relevant concentrations of diuron coupled with other environmental stressors such as acidification, nutrients, temperature and sedimentation will affect corals.

5. Conclusions

Based on the findings of our investigation, we may conclude the following;

5.1 The results show that, presently, low background levels of diuron contamination (<1, 000 ng L^{-1}) have no detectable impacts on the survival of coral reefs at present. However, in order to maintain the coral health, the risk studies for hazardous chemicals, including diuron, should remain a matter of concern.

5.2 Agricultural activities, such as those involved with sugarcane plantations, significantly contribute to diuron contamination in coral reefs around this region rather than antifouling paints from ships.

5.3 The concentration of diuron of 1,000 ng L^{-1} effectively reduces photosynthesis rate of coral under short-term exposure but does not seem to affect calcification rate.

5.4 According to environmental concentrations of diuron reported in elsewhere, it is possible that the rate of photosynthesis in corals will be inhibited in the current study sites and alike.

5.5 Further scientific investigations regarding for long-term exposure to novel antifouling chemicals, including diuron, to corals are still needed in order to implement appropriate monitoring and guideline levels of these in coastal uses.

6. References

Al-Horan, F. A., Al-Moghrabi, S. M., de Beer, D. (2003). The mechanism of calcification and its relation to photosynthesis and respiration in the scleractinian coral Galaxea fascicularis. *Mar. Biol.* 142, 419-426.

Bell, A. M., Duke, N. C., 2005. Effects of photosystem II inhibiting herbicides on mangroves-preliminary toxicological trials. Mar. Pollut. Bull. 51, 297-307.

Boxall A.B.A., Comber, SD., Conrad, A, U., Howcroft, J., Zaman, N. (2000). Inputs, monitoring and fate modeling of antifouling biocides in UK estuaries. *Mar. Pollut. Bull.*, 40, 898-905.

Chalker, D. J., Taylor, D. L. (1975). Light enhanced calcification, and the role of oxidative phosphorylation in calcification of coral Acropora cervicornis. *Proc. Royal Soc. Lond. Biol.* Sci. 190, 323-331.

Dahl, B and Blanck, H. (1996). Toxic effects of the antifouling agent Irgarol 1051 on periphyton communities in coastal water microcosms. *Mar. Pollut. Bull.*, 32:342-50.

Ferrer, I and Barcelo´, D. (1999). Simultaneous determination of antifouling herbicides in marina water samples by on-line solid-phase extraction followed by liquid chromatography-mass spectroscopy. *J. Chromat. A*, 854, 197- 206.

Fujimura, H., Oomori, T., Maehira, T., Miyahira, K. (2001). Change of coral carbon metabolism influenced by coral bleaching. *Galaxea*, 3, 41-50.

Gennaro, M. C., Abrigo, C., Giacosa, D., Rigotti, L., Liberatori, A. (1995). Separation of phenyl urea pesticides by ion-interaction reversed phase high performance liquid chromatography: diuron determination in lagoon water. *J. Chromat. A.* 718, 81-8.

Giacomazzi S., Cochet, N. (2004). Environmental impact of diuron transformation. *Chemosphere*, 56, 1021-1032.

Inoue, M., Suzuki, A., Nohara, M., Kan, H., Edward, A., Kawahata, H. (2005). Coral skeletal tin and copper concentrations at Pohnpei, Micronesia: possible index for marine pollution by toxic anti-biofouling paints. *Environ. pollut.*, 129, 399-407

Jones, R. (2005). The ecotoxicological effects of Photosystem II herbicides on corals. *Mar. Pollut. Bull.*, 51, 495–506.

Jones, R. J. (2004). Testing the Photoinhibition model of coral bleaching using chemical inhibitors. *Mar. Ecol. Prog. Ser.* 284, 133–145.

Jones, R. J., Heyward, A. (2003). The effects of Produced Formation Water (PFW), an effluent from the offshore oil and gas industry, on coral and isolated symbiotic dinoflagellates. *Mar. Freshwater Res.* 54, 1–10.

Jones, R.J. and Kerswell, A. (2003). Phytotoxicity evaluation of Photosystem II (PSII) herbicides on scleractinian coral. *Mar. Ecol. Prog. Ser.*, 261, 149-159.

Kitada, Y. (2007). Distribution and behavior of hazardous chemical substances in water and sediments collected from rivers and adjacent to coral reefs in Okinawa Island, Japan. PhD dissertation, Tohoku University, Japan, 90-94.

Konstantinou, I. K., Albanis, T. A. (2004). Worldwide occurance and effects of antifouling paint booster biocides in aquatic environment. *Env. Int.* 30, 235-248.

Lamoree, M. H., Swart, S. P., van der Horst., van Hattum B. (2002). Determination of diuron and antifouling paint biocide Irgarol 1051 in Dutch Marinas and coastal waters. *J. Chromatogr. A.* 970, 183-190.

Liat, A., Dahan, D., Golan, Y., Vago, R. (2005). Effect of light regimes on the microstructure of the reef-building coral Fungia simplex. *Mat. Sci. and Eng.* C, 25, 81-85.

Malato, S., Blanco, J., Fernandez-Alba, R., Aguera, A., Rodrigues, A. (2002). Photocatalytic treatment of water-soluble pesticides by photo Fenton and TiO2 using solar Energy. *Catal. Today* 76, 209-220.

Martinez, K., Ferrer, I., Barcelo, D. (2000). Part-per-trillion level determinations of antifouling pesticides and their byproducts in seawater samples by solid -phase extraction followed by high-perfomance liquid chromatography atmospheric pressure chemical ionization mass spectrometry. *J. Chromatogr. A.* 879, 27-37.

Negri, A., Vollhardt, C., Humphrey, C., Heyward, A., Jones, R.J., Eaglesham, E., Fabricius, K., 2004. Effects of the herbicide diuron on the early life history stages of coral. Mar. Poll. Bull. 51, 370–383.

Okamura, H., Aoyama, I., Ono, Y., Nishida, T. (2003). Antifouling herbicides in the coastal waters of western Japan. *Mar. Pollut. Bull.,* 47, 59–67.

Okinawa Prefectural Enterprise Bureau Annual report, 2006. (In Japanese).

Owen, R., Knap , A., Toaspern, M., Carbery, K. (2002). Inhibition of coral photosynthesis by the antifouling herbicide Irgarol 1051. *Mar. Pollut. Bull.,* 44, 623–632.

Råberg, S., Nyström, M., Erös, M., Plantman, P. (2003). Impact of the herbicides 2, 4-D and diuron on the metabolism of coral *porites cylindrical. Mar. Env. Res.* 56, 503-514.

Sapozhnikova,Y., Wirth, E., Schiff, K., Brown, J., Fulton, M. (2007).Antifouling pesticides in the coastal waters of Southern California. *Mar. Pollut. Bull.,*54,1962-1989.

Smith S V. (1973). Carbon dioxide dynamics: a record of organic production, respiration, and calcification in the Eniwetok reef flat community. *Limnol. Oceanogr.* 18 106-120.

Smith S V, Kinsey D W. (1978). Calcification and organic carbon metabolism as indicated by carbon dioxide. In; Stoddart DR, Johannes RE (Eds.) *Coral reefs: research methods,* pp 469-484. Unesco, Paris.

Suzuki, A., Kawahata, H. (2003). Carbon budget of coral reef ecosystems: an overview of observations in fringing reefs, barrier reefs and atolls in the Indo pacific regions, *Tellus.* 55B, 428-444.

Thomas, K. V., Fileman, T.W., Readman , J. W., Waldock, M. (2001). Antifouling paint booster biocides in the UK coastal environment and potential risks of biological effects. *Mar. Pollut. Bull.,* .42, 677-688.

Vandermeulen, J. H., Davis, N. D., Muscatine, L. (1972). The effect of Inhibitors of Photosynthesis on Zooxanthellae in Corals and other Marine Invertebrates, *Mar. Bio.* 16, 185-191.

Watanabe, T., Yuyama, I., Yasumura, S. (2006). Toxicological effects of biocides on symbiotic and aposymbiotic juveniles of the hermatypic coral *Acropora tenuis. J. Exp. Mar. Biol. Ecol.,* 339, 177–188.

Wilkinson, C. R. (Ed.), 2000. Status of Coral Reefs of the World: 2000. Australian Institute of Marine Science, Western Australia, p. 363.

6

Herbicide Reduction Methods

Martin Weis, Martina Keller and Victor Rueda Ayala
University of Hohenheim
Germany

1. Introduction

1.1 The necessity for weed management

Weed control is crucial to prevent high yield losses. Oerke estimated the potential yield loss of weeds rising up to 34% worldwide (Oerke, 2006). Chemical weed control with herbicides has been the major control tool for the last decades in developed countries.

These management practices have proven to be useful, effective and worth the efforts for many farmers, but chemical weed control in the long term has to face changing conditions and circumstances. They are used in a constantly changing and adaptive environment, and therefore their effectiveness can vary over time and location. However, the extensive and unsound use of herbicide has resulted in the development of herbicide resistance in several weeds. In addition the launch of herbicides with new modes of action has slowed down considerably (Rueegg et al., 2007). This can partly be ascribed to the consolidation process of the agrochemical industry resulting in less overall research and development infrastructure as in the increased regulatory requirements for registration. In Europe, the review process of pesticides under the Directive 91/414/EEC has additionally narrowed the spectrum of approved herbicidal active ingredients.

Apart from the positive effect by eliminating weed competition, herbicides can have a negative impact on the environment. Today, traces of herbicides can easily be found in surface and ground water. Public concerns due to these negative side-effects have led to political action plans to reduce herbicide use, especially in Europe these considerations are taken to a large geographic scale.

Thus strategical needs for resistance management and the smart use of the existing and remaining herbicides in combination with other weed management tools will be crucial to keep these weed control tools viable for further use. Also political pressure requires to rethink the use of herbicides and to promote integrated weed management systems. Therefore, the aim of this book chapter is to present ways and options to reduce, complement and replace herbicides.

1.2 Political framework

In the European Union new guidelines are being defined for integrated pest management (IPM) (European Parliament & Council of the EU, 2009, Article 14(5)). As part of the new regulations, the weed management and proper herbicide use are to be reconsidered and put into practice from the year 2014 onwards. The European guideline will be transcribed into national law to that date. The general policy of IPM is defined by the following steps (Zornbach, 2011)

I. preventive management of pests
II. monitoring
III. control decision
IV. preference for non-chemical methods
V. pesticide selection
VI. reduction of pesticides to the necessary amount
VII. resistance management
VIII. monitoring of treatment success

This procedure will be also mandatory for the application of herbicides. The explicit statement to prefer non-chemical management methods and to reduce the amount of herbicides demands for new methods in weed management approaches. The possible improvements range from the development of information technologies for decision support to new implements for site-specific herbicide usage or alternative weeding methods. On the European and national level research activities are funded to develop technologies supporting this policy, e.g. within the ENDURE and PURE projects (Pesticide Use-and-risk Reduction in European farming systems with Integrated Pest Management) and on national level there are initiatives to develop new practises, e.g. in Germany the innovation funding program of the German Federal Agency for Agriculture and Food (BLE - Bundesanstalt für Landwirtschaft und Ernährung). Some of the research efforts, leading to technology supporting the new policy, will be outlined in this chapter.

2. Integrated weed management

Integrated weed management (IWM) is one part of integrated pest management (IPM), the latter targeting at all pests occuring in crops. The European research network ENDURE defines IPM as:

> IPM is a sustainable approach to managing pests by combining biological, cultural and chemical tools in a way that minimises economic, environmental and health risks.

Integrated weed management combines all applicable methods to reduce the effect of weeds in the cropping systems. The general goal of IWM is to enhance the weed management by using different methods to reduce the weed pressure. Swanton et al. (2008) illustrate the approach as "use of many little hammers". Synergistic effects are expected, leading to higher overall success rates in weed management than using only one of the weed suppressing measures. The combination of different methods has the advantage, that the overall effect can lead to a better weed suppression not only in one growing period, but also over larger time scales. Since many of the methods are 'orthogonal', in the sense that they are not depending on each other and work with different modes of action, a combination of these can add up the effects of the treatments. Weed management happens to be carried out in constantly changing conditions: depending on the weather conditions, concurrence situation (e.g. according to the crop type and rotation), weed pressure, seed bank, the weeds biological properties and herbicide resistance status, some measures may be more effective than others. A long-term strategy has to take into account the changes of the weed population and of the influencing environmental factors. Weeds, like any biological entity, can adapt to changes in their environment. Such changes can be human-induced or general, environmental ones (climatic shifts, newly introduced species).

Crop rotation is a classical technique with a weed suppressing effect, widely adopted in the agricultural practise. It can disturb the regeneration cycle and accumulation of weeds, leading to lower infestation densities of the crop relevant weed species over the years. Weed management depends on the cropping system to which it is applied, since the cropping system can have a significant influence on the weed pressure. Recent precision farming technology may lead to a wider adoption of different cropping systems: Corre-Hellou et al. (2011) compared weed suppression for cereal–pea intercropping systems and found a significant reduction of weeds compared to sole pea crops. The wider application of intercropping is supported by precisely working equipment, supporting the management starting from the sowing until harvest with improved methods for the planning, treatment, monitoring and harvest.

An reintroduction of cover crops, which have not been used intensively in the last decades, also exhibits unused weed-supressing potential (Blackshaw et al., 2001). Undersown species can suppress the weeds due to their allelopathical effects or by early coverage, preventing germination and providing competition.

Sowing time, sowing rates and row spacing are closely related to the crop type and management practises, and all have an influence on the weed development and should be optimised to reach low weed infestation levels. A long-term strategy should include the prevention of seed set and distribution of weed seeds. The measures range from tillage and harvesting operations within the field to contamination prevention of seeds for sowing. An accumulation of weed seeds in seed banks needs to be addressed, at least certain levels of seed and resulting weed density should be avoided.

Swanton et al. (2008) identifies four key components for IWM management: i) tillage (depth, soil type) and its influence on vertical weed seed distribution. With reduced tillage practices like no-tillage and strip-tillage seeds are close to the surface and are subject to predation and a more uniform emergence, and rotational tillage is considerable. ii) time of weed emergence, relative to the crop and the iii) critical period for weed control, both relevant for the yield effect of the weeds on the crop have to be taken into account for optimised management decisions. iv) Weed germination during the harvest window, after the critical period, can affect yield quality and the harvesting operation, but are not regarded to be an issue for the seed bank development.

The adoption of IWM practices by the farmers has two aspects: one the one hand most farmers are aware of the implications and use one or more of the techniques affecting the weed distribution. On the other hand such measures are usually not practised as conceptual long-term strategy, and the benefits and problems are not perceived as levelling out over time. Often the most effective, chemical solution is preferred to achieve a "clean acre", as long as there are no obstacles like herbicide resistance.

The introduction of IWM into the practise therefore requires training efforts on the farmers side. Risks for the several options should be assessed and modelled in expert systems, which then can aid consultants and farmers in their decision-making.

Llewellyn & Pannell (2009) studied the perception and adoption of knowledge about IWM techniques and resistance development by farmers through workshops. They found significant changes in workshop participants behaviour and their intention to extend practices to different weed management methods. In Australia training and information resources were created to help farmers and advisors implement integrated weed management according to a manual (Storrie et al., 2006).

Jones et al. (2006) created a risk model for alternative weed management and found seasonal variability to have a substantive influence on the accuracy of the risk assessment and the estimation of the benefits. According to case studies conducted by the ENDURE network in maize crops (ENDURE, 2008), comparable weed control can be achieved with different levels of IWM, concluding that a reduction of herbicides apparently has a proportional manner: the more non-chemical measures are added the more can herbicide input be lowered. Models have been created to tackle complex interaction of the components for a sustainability analysis of management systems, including measures for weed management (ENDURE, 2009).

3. Alternative weed management without herbicides

There are several methods to control weeds without herbicides. It is well known that conventional farming heavily depends on herbicides for weed control, therefore most alternative measures are developed and used in ecological farming. Perhaps, only preventive methods such as soil tillage and seed bed preparation are used in both, conventional and organic farming (Bond et al., 2003; Rueda et al., 2010). In the previous section it was stated that the necessary weed management should be a balanced approach by including many different measures to successfully reduce weed pressure, towards an integral management plan. This procedure is applicable to both, conventional and organic, farming systems. Combination of many measures has the advantage that weed control can be extended throughout the whole crop growing period and also in large scales, and due to many modes of action the risk of resistant species will severely decrease. Several modes of action can be distinguished to control weeds without herbicides: burial of weeds in soil, cutting plants into pieces, uprooting, rupture of the plant cell and desiccation, etc. which are described below.

If plants are uprooted, they loose proper soil contact, disrupting their supply of water and nutritions. Destroying or disturbing the function of leafs or stems lead to a delayed development or death. Some methods target the seeds in the soil, disturbing their ability to develop into plants and thereby reducing the germination probability. Tools to achieve this are mechanical or use heat, applied either through flame, electrocution (Blasco et al., 2002) or laser light (Heisel et al., 2001; Nadimi et al., 2009).

3.1 Mechanical weeding (harrowing)

Mechanical weed control is referred to as cultivating tillage, because it mostly comprises a shallow soil cultivation after sowing or planting with tools such as harrows and hoes (Rueda et al., 2010). These tools can be used to perform a whole crop cultivation (crop and weed plants), inter-row and intra-row cultivation. Cultivating tillage mainly controls annual weeds through uprooting, tearing plants into pieces and burial of weeds into soil; however perennial weeds are little affected (Jensen et al., 2004; Kurstjens & Kropff, 2001; Kurstjens & Perdok, 2000; Rueda et al., 2010). Since controlling weeds with mechanical tools generally is a trade-off between weed control and crop damage due to cultivation, post-emergence cultivation must be combined with pre-emergence methods to overcome the poor selectivity (Melander et al., 2005). Cultivating tools like harrows and hoes may be used for cultivating row crops, and also small grain cereals or legumes (Jensen et al., 2004; van der Weide et al., 2008). Weed harrowing is carried out on the whole crop area and hoeing is applied inter-row, however, crop plants may also be affected (Melander et al., 2005).

Weed control by mechanical means has generally a lower efficacy than herbicides, therefore the challenge is to improve mechanical weeding to make it competitive with chemical control. Five experiments on weed harrowing in winter wheat (*Triticum aestivum* L.) were carried out in Denmark and Germany, aiming to increase the cultivation selectivity and weed control efficacy while reducing the crop damage by soil coverage, as a result of harrowing (Rueda-Ayala et al., 2011). Three of the experiments were kept weed free either with herbicides or due to lack of natural weed germination. These experiments were also harrowed to measure crop tolerance to soil cultivation and thus burial in soil. Two experiments were only harrowed, in order to control weeds and calculate the highest achievable yield as a consequence of weed control by harrowing. At each location, there were at least one weed free experiment and one experiment with high weed competition in the same field, to compare weed control and crop tolerance to harrowing. Four intensity levels in Denmark and five in Germany were created by increasing the number of passes with the spring-tine harrow on the same day of cultivation, an untreated control was included in any case. The most dominant weed species were *Veronica persica* (47%), *Viola arvensis* (31%), *Poa annua* (8%), and *Chenopodium album* (7%), in Denmark, and *Matricaria inodora* (65%), *Cirsium arvense* (16%), *Alopecurus myosuroides* (10%), and *Galium aparine* (6%) in Germany.

The selectivity of harrowing was a bit higher in Denmark than in Germany, although the crop growth stage at the time of harrowing was irrelevant for the selectivity. The first pass with the harrow controlled 93% and 68% of the weeds in Denmark and Germany, respectively. Likewise, four passes controlled 97% and 98% in Denmark and Germany, respectively. In order to achieve 80% weed control, it was required to achieve about 2 to 12% soil cover in Denmark and about 22 to 30% soil cover in Germany. The crop yield loss by 25% soil covering as a result of harrowing was about 0 to 5% in Denmark and 0 to 1% in Germany. When comparing the uncultivated plots with the herbicide-treated plots, it was found that the effectiveness of weed harrowing was comparable to the herbicide treatment. The average crop yield gain by harrowing with an optimal intensity in Denmark was 13%, which was not different from the 14% yield gain obtained by weed control with herbicides. In Germany it was not possible to measure the effect of herbicides, due to lack of natural weed germination, however, the average yield gain obtained by controlling weeds due to harrowing was as high as 27%. These results are a proof that weed harrowing is a very effective tool to control weeds, and under favourable soil conditions, applying the optimal intensity and at the right crop growth stage, these benefits are comparable with those obtained with herbicides.

3.2 Thermal weeding, flame weeding, manual weeding

While mechanical weed control is the most widely used non-chemical weed control method in arable crops, there are also other means to reduce the amount of weeds. Although the latter are often applied to control weeds in non-arable areas, e.g. on roads or rails, these can nevertheless be used for farming, too. Especially in high-value crops these can be considered, even though the measures may be accompanied by a higher energy, material or labour input.

3.2.1 Soil solarisation, soil steaming, and hot water

Soil solarisation achieves dis-infestation of soil due to the heat resulting from trapped soil radiation by covering the moist soil with a plastic film during several weeks, which has the consequence of killing soil-borne pathogens and to avoid weed seed germination by thermal inactivation (Candido et al., 2011; Cimen et al., 2010). This technique, alone or combined with other techniques, has been found to effectively control weeds like green bristlegrass (*Setaria viridis*), common purslane (*Portulaca oleracea*), redroot amaranth (*Amaranthus retroflexus*), bindweed (*Convolvulus spp*), black night-shade (*Solanum nigrum*), canadian thistle (*Cirsium arvense*), and even weed species unsusceptible to selective herbicides (Candido et al., 2011). The same author found that the total weed density and biomass were still significantly lower in solarised soil than in the untreated plots also after the harvest of the second crop under greenhouse and field conditions, therefore being a reliable possibility to replace the few chemicals available to manage weeds in conventional lettuce production.

Bàrberi et al. (2009) highlighted that soil solarisation can be combined with hot steaming, as a measure to reduce the use of methyl bromide for soil disinfection. Most of the weed seeds are killed with steam heating the soil from 70–100°C to a depth of at least 10 cm, but weeds below this layer may be unaffected and germinate when the soil is disturbed to that depth (Bond et al., 2003). Bàrberi et al. (2009) also mention that although steaming may also kill the beneficial soil microflora, it helps to overcome soil solarisation's limits: viable alternative in Mediterranean and tropical areas, limited use in summer months and occupation of vast areas with plastic films up to 3 months. *Alopecurus myosuroides*, *Fallopia convolvulus* and *Setaria viridis* were the most sensitive species to steaming alone, with about 77% control efficacy. Additionally, the effect of hot water (85°–95°C) and hot foam on weeds has also been reported as a good control method without affecting crops; the foam is intended to remain on the weeds and prolong the effect of the heat (Bond et al., 2003).

3.2.2 Flame weeding

Weeds are killed by flaming due to the intense wave of heat that ruptures the plant cells and desiccates them, thus a foliar contact treatment is required, and any long-term effects depend on the plant recovery and the subsequent weed emergence (Bond et al., 2003). Flaming equipment uses liquefied petroleum gas (propane) and the burners can produce up to 1900°C (Knezevic et al., 2011). Flaming can be applied on the whole vegetation or directed to unwanted weeds, with the advantage that burners can be used when soil conditions do not allow mechanical weeding or herbicide resistant species are to be controlled (Bond et al., 2003). Leaf relative water content, growth stage of the weeds and timing of treatment application are influential factors for the efficacy of flaming, according to greenhouse experiments by Knezevic et al. (2011). These authors found that flaming was more effective when conducted on the afternoon between 3 P.M. and 6 P.M. at early growth stages of the plants. However, since flaming does not disturb the soil, some weeds (especially grasses) may escape weed control because their growing point during early growth stages is below the soil surface, thus is protected from the flame (Bond et al., 2003; Knezevic et al., 2011). Furthermore, some crop plants were also affected by the flame, and therefore it is important to apply the treatment as late as possible before the crop emerges (Hansson & Svensson, 2011).

3.2.3 Manual weeding

Pulling the weeds by hand or weeding with hand tools reaches the desired level of absolute site-specific weed control (Rueda et al., 2010). The major constraints are the low capacity in terms of hectares per hour and the high labour costs per hectare, especially if the weed density is extremely high and other measures fail. For instance, Hansson & Svensson (2011) found that if flame weeding was used one day too early, it may increase the hand weeding costs by approximately 280 euros per hectare. In a review of hand weeding tools used in Europe some implements are still very effective, e.g. to control deep-rooted grass weeds (Bond et al., 2007). Other tools are available to cut or move soil to cover small weeds, while the operator is standing or kneeling on the surface, such as the draw hoe, swan-necked hoe, onion hoe, collinear hoe, Dutch hoe, Swiss oscillating hoe, stirrup hoe. In developing countries, hand weeding tools may be combined with animal drawn tools to reduce the human labour, increase the area cultivated per hour and become a cheaper option than herbicides (Benzing, 2001).

3.2.4 Further alternatives to control weeds

Bond et al. (2003) made a review of other ways to control weeds. Electrocution, an electrical discharge of about 15,000 V, can kill weeds (Blasco et al., 2002), but even at a low weed density of 15 plants m^{-2}, a big amount of energy is required, and therefore it would not be suitable as primary weed control tool. Microwave radiation uses ultra high frequency (UHF) electromagnetic energy at a larger wavelength than light, and this radiation can kill weeds, however this technique is very slow and expensive.

Lasers emit hight amounts of light that can inhibit the plant growth due to the heat generated during absorption, but often do not kill weeds. Heisel et al. (2001); Mathiassen et al. (2006); Nadimi et al. (2009) have demonstrated in experiments with CO_2 lasers the possibility to regulate the weed growth centre. However, more research is required to validate the viability for this technology to be usable. Another alternative is biological weed control, through its classical approach (introducing a biocontrol agent, e.g. micro-organism or insects), by inundating a region with a large numbers of biocontrol agent or through a conservative approach where a detailed ecological knowledge is required about the weed and the biocontrol agent. An example of the latter is allelopathy, which is the release of allelochemicals from a plant or a fungal species that is toxic to other plant species.

4. Site-specific weed management

Site-specific weed management (SSWM) provides opportunities to reduce the amount of herbicides to a minimum of the required amount (Christensen et al., 2009). Several aspects of the policy in section 1.2 have to be integrated for a successful use of SSWM: items III, V, VI and VIII are addressed directly, and the resulting data sets can further be used for monitoring tasks (II).

Nowadays the application of herbicides is usually done on the field level: the herbicides are selected according to the crop in the field, then a mixture is prepared and uniformly applied to the whole field. A common practice is to estimate the mean weed infestation and use thresholds for the weed species, which are to be considered. The weed density thresholds and occurrence of weed species are used to compose the necessary herbicide mixture. Afterwards this mixture is applied to the field, usually with a sprayer that can regulate the amount of herbicide according to the velocity, resulting in a uniform application on the field in terms of a constant amount per area, measured in litres per hectare.

This approach does not take into account the variability of the weed density within the field. Therefore herbicides are used even in weed free parts of the field, where this would not be necessary. Weeds are often found to have an irregular distribution within a field, with high densities in small-scale patches (Dieleman & Mortensen, 1999; Gerhards & Christensen, 2003). These irregularities arise from varying numbers of seeds in the soil (the seed bank), and germination probabilities can be different according to the local soil conditions, water availability and several other factors. Some influencing factors are crop-related: the general ability of the crop to suppress weeds and compete for the resources. Local variations of coverage and time of emergence can also arise from irregular sowing densities and depths.

Site-specific weed management (SSWM) addresses the within-field variability explicitly. To be able to change management practices on the sub-field level several prerequisites are necessary. The procedure to implement SSWM includes the following steps:

1. The actual weed infestation/weed pressure has to be assessed within the field. Christensen et al. (2009) identify an automation of the weed mapping component as a missing prerequisite for SSWM. Approaches have been described in the literature to build up such systems, and they will be outlined in section 4.1.
2. Decisions have to be made for the (local) treatment. Decision components for weed management have been developed to aid farmers and researchers, an overview is given in section 4.2.
3. The field must be subdivided into management zones, each zone delineating areas of similar treatment.
4. Documentation of the management for further use (section 4.3).
5. Application technology (sprayers) with the ability to vary the treatment is needed (section 4.4) to carry out the treatment plan according to the management zones.

These steps resemble the general workflow cycle in precision agriculture, which have often been visualised as a circle (Srinivasan, 2006, p. 23): starting with data sampling the following steps are data processing, decision making, application and a subsequent control of the effectiveness of the taken actions (e.g. by yield/weed mapping). The introduction of precision farming technology by the farmers still faces some challenges: new technology has to provide benefits, either economically or in terms of a reduced manual work load. The interest to invest in such implements and to change the necessary management decisions, varies between farmers. Reichardt et al. (2009) interviewed farmers (in Germany) regularly about their knowledge and adoption status and found the need for better information, e.g. in education, and that only parts of the farmers are introducing precision technology. The basic instrumentation like GPS and steering aids have the highest uptake rate among farmers. More work has to be done to implement such technology in a user-friendly way and deal with the complexity of the task itself and the many different components, which all have to work together.

4.1 Sensing weeds

A prerequisite for a broad introduction of SSWM is the possibility to automatically detect weeds. Manual weed sampling is too expensive in terms of labour, the costs for the working hours in the field reduce the economic benefit of the possible savings. The variation of the weed infestation within a field has to be assessed prior to the application (Wiles, 2005).

There are two different cases for the treatment which have to be considered separately: pre-emergence and post-emergence application of herbicides. In the first case the plants are not yet germinated, and herbicides are applied to the soil to prevent germination. It is clear, that in this case there is nothing specific which could be sensed and analysed within the field. An analysis of the seed bank is very labour intensive (Wiles & Brodahl, 2004; Wiles & Schweizer, 2002) and involves taking soil samples and analysis of the samples in a laboratory. To achieve this, a lot of manual work has to be carried out, and therefore this approach does not scale to larger areas. On the other hand it is possible to estimate the size and location of seed banks from germination rates, such that historical maps of the weed infestation could be used for a precision application (Christensen & Heisel, 1998; Williams & Mortensen, 2000).

For the second case (post-emergence application), weed detection automation has been successfully applied. Brown & Noble (2005); Singh et al. (2011); Weis & Sökefeld (2010) review the methods and approaches for an automated weed sampling. There are only a few commercially available sensor systems on the market, which can be used for weed density estimation. WeedSeeker (NTech Industries Inc., Ukiah, CA, USA), Detectspray (North American Pty, Ltd., NSW, Australia, 1995) and Crop Circle (Holland Scientific, Inc., Lincoln, NE) are products, which have been used in research for this task (Andújar et al., 2011; Biller, 1998; Biller et al., 1997; Sui et al., 2008). An additional requirement for the sensing, which is not yet fulfilled by the sensors on the market, is the identification of different weed species to be able to use selective herbicides (Gutjahr & Gerhards, 2010). This cannot be achieved with the optical technology in the named products, since they only measure values correlating with the general coverage. In crop-free areas, e.g. between the rows or during pre-emergence of the crop, these systems can be used, since from the coverage the biomass of the weed and therefore the infestation can be estimated. With additional, a priori information about the occurring weed species this can be used to generate spraying decisions. Also, if broadband herbicides are to be applied, these sensors provide valuable data. A similar approach is possible with remote sensing data (Thorp & Tian, 2004).

The full benefit of site-specific weed management can be achieved, if selective herbicides are used on the sub-field level. A reason for this can be found in the heterogeneous distribution of each species: each species can show a different distribution pattern within the field (Gutjahr et al., 2009; Sökefeld, 2010).

Two general approaches for site-specific weed sensing and herbicide application can be distinguished: an 'offline' and 'online' procedure. For the offline approach the two steps —sensing and application— are sequential, which means that the weed infestation is assessed first, then assembled offline to a weed infestation map and an application map and the latter is used in a second step in the field to apply the herbicides (Oebel & Gerhards, 2006). An example of the offline approach is given in Figures 1 and 2. In a first step the weed infestation situation was acquired manually and with a sensor-system in a maize crop located at the research station of the University of Hohenheim, Ihinger Hof, in Renningen near Stuttgart, Germany (May 2008). Plant coverage measurements are shown as grey values in a grid. From the coverage the patchiness of the weed distribution can already be seen. A more detailed image analysis led to a species discrimination. The latter are depicted as coloured points, measurements without detected weeds are white. The manually identified weed patches for

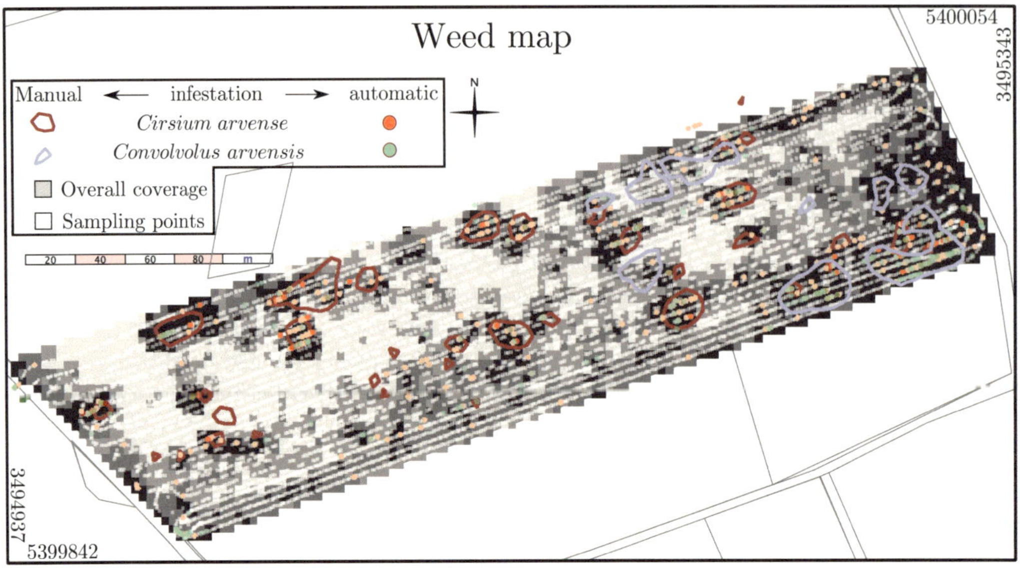

Fig. 1. Weed map, the dots are the results of the automatic weed sampling, the polygons were measured manually. Additionally the overall coverage and sampling points are shown. The coordinates are given in Gauß-Krüger projection (DHDN Zone 3).

two perennial weed species (*Cirsium arvense* and *Convolvulus arvensis*) are outlined as polygons and overlaid. Further details of the analysis system for the measurement of these maps can be found in Weis (2010). The information about the weed infestation was then used to create an application map for a site-specific treatment of *Cirsium arvense* (Fig. 2). The application map contains the management zones, composed as a regular grid of quadratic polygons. For each zone a treatment decision was derived: green zones are to be sprayed, white zones do not need treatment. The comparison with the manually acquired patches shows that the overall decision based on the sensor measurements is suitable for the treatment. Only small patches were missed by the system, mostly due to the sampling distance, since the patches were either between the measured tracks or the measurements did not 'hit' plants in low-infested patches. Due to the zones' structure and singular false positives, a slightly larger area compared to the manual measurements was marked for treatment. Nevertheless only 30% of the total field area were treated, leading to herbicide savings of 70%.

An online approach combines both steps, integrating the sensing, decision making and application during a single pass (Blasco et al., 2002; Tian, 2002), e.g. with robotic approaches (as described in section 5.1). This can be achieved with smart sensors for the weed detection in combination with a decision and control component on the tractor. Such systems are in development for commercialisation and will give an opportunity for sensor guided applications on a large scale.

4.2 Decision rules and decision support systems

The economic weed density threshold is the density at which the costs of an application equal the monetary loss due to the yield decrease caused by the weed infestation. Economic thresholds provide simple rules to decide whether weed control is (economically) justified or not on the field level and thus reduce the herbicide use compared to prophylactic herbicide

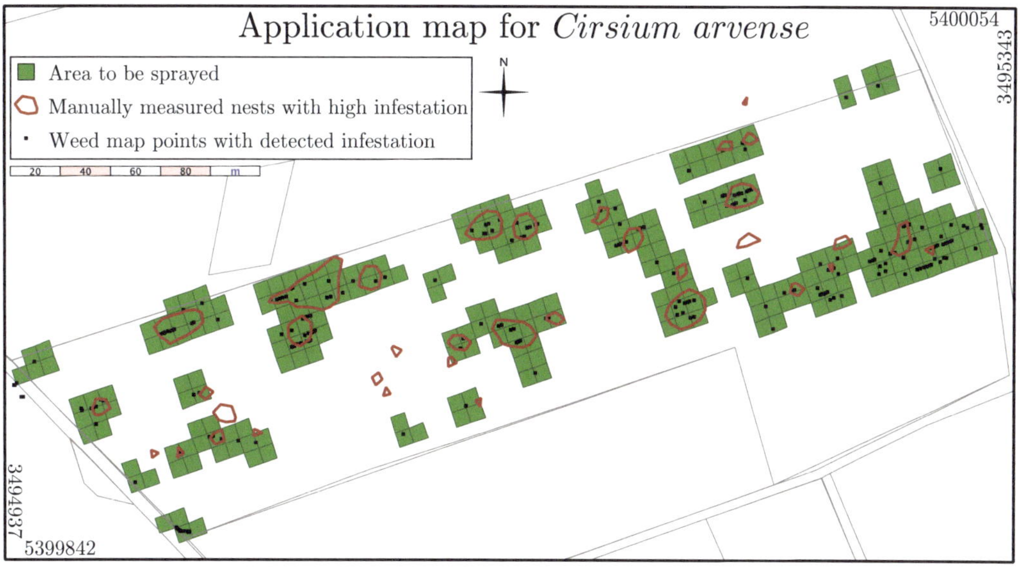

Fig. 2. Application map for *Cirsium arvense*, generated from the automatically measured weed map in Fig. 1. The weed map values were interpolated to a grid of 6×6 m with a 3 m buffer.

applications (Coble & Mortensen, 1992; Gerowitt & Heitefuss, 1990). The decision whether to spray or not depends on the average weed infestation of a field, assessed early in the season. The determined thresholds vary in the literature e.g. Gerowitt & Heitefuss (1990) used threshold values of 20-30, 40-50, 0.1-0.5 m^{-2} for grass weeds, broadleaved weeds and for cleavers (*Galium aparine*) respectively in small annual grains. For broadleaved weeds also a threshold of 5-10% of ground cover was tested (Gerowitt & Heitefuss, 1990). Economic thresholds have generally not been adjusted to changing prices of grain and herbicides and thus serve merely as a guideline (Gerhards et al., 2011; Gutjahr & Gerhards, 2010).

In contrast to economic thresholds, Decision Support Systems (DSSs) are more sophisticated decision aids. They are valuable in situations in which the amount of available information (e.g. crop growth stage, weed species, weather conditions, herbicide spectrum, their economic and ecological properties and implications) is too large for intuitive decision-making (Mir & Quadri, 2009). For uniform field herbicide applications, many DSSs have been presented in research journals and several are also implemented and available as web-based, desktop and/or pocket (on-site usage) solution (Bennett et al., 2003; Rydahl, 2003). Frequently required inputs for DSSs are the grown crop, its competitive ability (e.g. growth stage, vigour), weed species, weed densities, weed growth stages, herbicide and application costs, the expected price for the crop and the expected yield without weed competition (Gutjahr & Gerhards, 2010).

Wiles et al. (1996) distinguished between efficacy-based and population-based DSSs for weed control. The former provide herbicide recommendations and in some cases recommendations for the optimal herbicide dose for the actual weed infestation of a field. An integral part of these DSS are databases storing information about herbicide performance under various conditions (crop, weed species, growth stage etc.), as reported by Christensen et al. (2009). Examples of this type of DSSs are SELOMA (Stigliani & Cosimo, 1993) or Crop Protection

Online (Rydahl, 2003). In addition, population-based DSS incorporate information about weed biology, ecology and management through deterministic models (Christensen et al., 2009; Wiles et al., 1996). Examples are HERB (Wilkerson et al., 1991) or GESTINF (Berti & Zanin, 1997). Gutjahr & Gerhards (2010) provide a review of various DSSs in weed control. DSSs for weed control have considerably evolved and are still evolving from quite simple models using only few aspects relevant to weed control to DSSs that integrate more aspects as optimum dose rates, weather conditions, considerations regarding the environment and the problem of herbicide resistance development (Rydahl et al., 2008). For example, the DSS Weed Manager integrates two time scales, a within single season and an overall scale for several years in rotation perspectives. A multi-stage heuristic decision model for the in season decision model and for the over several years rotation perspective they used stochastic dynamic programming in this DSS (Parsons et al., 2009). Furthermore, DSSs for weed control vary in the number of crops and weed species included, some are only designed for a few crops and a limited number of weeds. Others are functional for the main crops and weeds at national or regional level (Rydahl et al., 2008). One drawback of DSSs is that they are seldom transparent to the users and that the models do not allow the user to interact or change neither the algorithms nor the model parameters. Furthermore, in countries, in which DSSs have already been implemented and are available to farmers, less than 3% of the farmers were using them (Rydahl et al., 2008).

As stated before weeds occur in patches and therefore patch spraying, i.e. spraying only restricted to areas with weed densities high enough to cause economic yield loss, should allow considerable herbicide savings. Gerhards & Oebel (2006) could achieve herbicide savings from 6-81% for herbicides against broad-leaved weeds and 20-79% for herbicides against grass weeds in cereals. In trials carried out by Wiles (2009), 34% of the field areas could be left untreated in average. It is important to notice that these potential savings depend on the actual weed distribution and the employed decision algorithms.

Only few DSSs for patch spraying have been presented so far (Gutjahr & Gerhards, 2010). The system DAPS (Decision Algorithm for Patch Spraying) was developed for barley (*Hordeum vulgare* L.) and winter wheat (*Triticum aestivum* L.) and was tested in a 5-year field trial by Christensen et al. (2003). The model consists of three main components: i) a competition model, which uses the density equivalent model approach presented by Berti & Zanin (1994). ii) a herbicide dose-response model and iii) an algorithm which derives the economically optimal herbicide dose for a given weed infestation (Christensen et al., 2003). In the field trial this model for patch spraying was compared with PC-Plant Protection, a control and a so called mean DAPS in a completely randomized design with 10 replicates. For the DAPS each plot consisted of 6 management zones of 12×12 m. In each managment zone weed density was determined in circular sample of $0.25\,m^2$. The field trial indicated no significant differences in yield, however significant herbicide savings of DAPS compared with PC-Plant Protection (Christensen et al., 2003). HPS-online has been suggested by Gutjahr & Gerhards (2010), but it has not been implemented yet. Weed cover of single weed species or weed groups, determined by bi-spectral cameras and classification algorithms, and economical basis data serve as input. Parameters of yield loss functions for single weed species, and parameters of dose-response curves for the respective weed and herbicide combinations are used to derive a yield gain function in dependence of the applied dose. Using the costs of weed control and the expected price for the crop, the optimal dose can be determined for each management zone. In addition, it is planned to allow the farmer to adjust the system according to current conditions, e.g. crop status or weather conditions (Gutjahr & Gerhards, 2010). For this suggested model robustness/predictability of yield loss functions has to be researched, since they can differ between locations and years. In addition most yield loss functions have been determined with weed density as explanatory variable. Furthermore dose-response parameters are not all available, and dose-response curves have been determined mainly in the greenhouse and the transferability of such results to field conditions is difficult.

Generally, for site-specific spraying the term DSS can be misleading, as the algorithms should not suggest or recommend the herbicide or the control needed, instead they take the decision whether to spray or not, or to spray more or less for the respective management zone. This is especially the case for an on-line approach, when the herbicides or the herbicide mixture have been chosen in advance, the sprayer is already loaded and the decision to spray or not to spray has to be taken in real-time at the sprayer's regular speed. The use of GPS and the geo-referenced documentation of spraying decisions would open the possibility to provide the farmer with a log file or a map highlighting locations in the field, where the system had identified patches of weeds which could not be fully controlled by the loaded herbicide, rising the awareness of the farmers to potential problems and to allow him to take actions. In addition, the decision algorithms should only require little computing time.

Concluding, DSSs support farmers' decision making in weed control and can optimize weed control economically and minimize the negative environmental effects of weed control. However, the use and adoption of DSSs by farmers and consultants have to be brought forward. In addition, more field data has to be gathered to support and calibrate decision algorithms and calculation models of DSSs (Rydahl et al., 2008). For site-specific weed control, the two presented examples and the above mentioned results of field trials show the potential for considerable herbicide savings if site-specific spraying is adopted by farmers. Ecological benefits of patch spraying can be considerable. On one hand the amount of applied herbicide is reduced as herbicides are only applied in field areas, where herbicides are needed. On the other hand, in field areas where no herbicides are applied, flora and fauna can establish without disturbance, enhancing the overall biodiversity (Timmermann et al., 2003). Neither DAPS nor HPS-online is available to farmers (yet), but launches of commercial products integrating automated weed sensing, decision algorithms and application technology are to be expected soon.

4.3 Data handling

As seen before, a lot of data is needed for precise treatments: measurement data acquired in the field or by remote sensing, application maps, yield maps and soil maps, all of them with a spatial component. For decision support not only such spatial data, but also the weather conditions, active ingredients of pesticides, timing information, regulation rules, equipment information, farm data and a lot more have their role and need to be integrated for the final decision about the optimal management.

In precision farming, geographic positions have to be acquired during the treatments to identify the areas which have already been treated. This way the documentation of the taken actions helps to avoid overlaps directly during the treatment. If prescriptions maps were prepared, then a terminal software needs to look up the actual position in the map and use its information additionally for the control decision. Therefore geographic information systems (GIS), GPS positioning and terminals to control the equipment play an important role in precision agriculture. Some farm management information systems already include the possiblity to store certain geodata, although most of them do not target and handle precision farming data explicitly. Partial solutions exist to create application maps that can be transferred to the farm equipment for applications in the field.

Storage of geodata itself is only one aspect of the necessary data handling, additionally metadata is needed to describe the data sets itself, answering at least the following questions: who acquired the data, how, when and which equipment was used, what was the goal and usage perspective and how accurate is the data, which preprocessing steps have been applied? This way the data sets can be found and understood later, which is necessary for further analysis steps, if they are based on this data. If for example an economic analysis of the long-term strategy is performed, then all data sets acquired during the time frame considered can be relevant inputs.

By coupling locally acquired data sets and external sources, e.g. weather data and pest prediction service data, new application fields are supported. The complexity of DSSs for integrated pest/weed management can be tackled by a suitable modularisation into components. These components do not necessarily need to be run on one system or created and implemented by one manufacturer, if suitable interfaces allow the exchange of data between them. By interconnecting components the expertise of different fields and sources can be joined. A service-oriented architecture based on web-enabled services provides the technical framework for the implementation of system components that can be joined together as needed. Services like recommendations based on the data and functionality of such components are an option for the portfolio of consultants and as such should be developed in cooperation with the experts in the field.

The standardisation of data and services may find technological solutions soon, as they are the topic of recent research efforts (Nash et al., 2010). Difficulties arise mainly from semantical barriers, which prevent the correct interpretation of data and its meaning in different contexts. Farm manangement information systems (FMIS) should include such standards and thus can add functionality in a modular manner. As FMIS are used by the end-users (farmers), they provide the tool for integration of expert systems, which can be created and maintained as modular solutions. Such standardisation may require an important contribution from researchers and the willingness to implement them in commercial products, and therefore might need considerable time before becoming usable. The introduction of the ISOBUS standard, defining interfaces for interoperation of machinery components, into agriculture needed roughly a decade until all relevant manufacturers were providing functional interfaces. Such a standard on the other hand allows the development of new applications with less effort (Iftikhar & Pedersen, 2011). AgroXML is another proposed standard for farm management data exchange (Schmitz et al., 2009), and has successfully been applied to farming tasks supporting a distributed system structure (Steinberger et al., 2009). If the regulations were available in a formal, machine-readable rule format, then operations can be planned and checked according to these regulations. Authorities and researchers as well as FMIS providers need to work together to reach this goal and provide useful tools for the end-users. The necessary prerequisites have been defined (Nash et al., 2011; Nikkilä et al., 2010) and proof-of-concept solutions were developed to show the feasibility of the approach in the European project FutureFarm (Nikkilä et al., 2010). Especially for the weed population modeling the WeedML standard has been proposed (Holst, 2010), fostering coooperation and exchange between researchers in this field. Systems integrating this technology can become an integral part for the decision support of farmers.

4.4 Application technology for weed management

The application technology for chemical weed management has seen advances in the last decade, leading to more precise application of herbicides in the field and thus reducing the amount of herbicides applied. The equipment to apply herbicides to the field plays an important role for an optimized treatment.

One concern for an optimum treatment quality is the reduction of drift. In windy weather conditions the drift effect can lead to an uneven treatment, because the spray liquid moves from the envisaged position and can stack up in neighbouring areas. The resulting, unwanted variation within the field can on the one hand lead to poor weed control due to lower amounts, on the other hand damage the crop in vulnerable growth stages and also the environment in areas with higher amounts. It can also lead to pollution of non-target areas outside of the field, often in shelter-belts where the wind velocity is reduced. The drift can especially be a problem for targeted omission of sensitive areas, e.g near water or biotopes. To comply with restrictions, optimal drift reduction is one crucial prerequisite. It can be achieved by selection and calibration of the equipment, and naturally by applying under good weather conditions (no wind). One way to reduce the drift is the selection of nozzles with larger orifice size producing larger droplets or special drift-reducing nozzles, which for example incorporate air into the spray droplets. The droplet size is also dependent on the spray pressure and additives that increase spray viscosity. Bigger droplets are not as susceptible to wind as smaller ones. The selection of the right nozzle is not only dependent on the drift effect, but also relying on other circumstances. Smaller droplets can have advantages for the uptake efficiency by the plant, since the more homogeneous wetting raises the probability for absorption into the leaf. Adjuvants additionally can be used to intensify the contact of the droplets to the leaf surface and aid the uptake through the epidermis.

Nowadays most sprayers are able to control the amount of herbicides to a uniform level by feedback control systems. By pressure variation they control the amount according to the driving speed, assuring constant amounts of spray liquid per area unit.

4.4.1 Variable rate technology

For a precise treatment and variation of the herbicide application within a field, sprayer technology has to be able to adapt the rates according to a spraying plan. Variable rate technology (VRT) became available in the last decade and entered the market for precision applications (Sökefeld, 2010).

A basic variation of the amounts can be realised by switching on and off the whole boom or parts of it. In the latter case the whole boom width is divided into parts which can be controlled independently of each other. The parts can be sections of fixed length or down to the single nozzle with an individual nozzle control. With such systems it is possible to avoid overlaps, since the nozzles or sections can be switched off in areas which have already been treated. They can also be used to leave out no-treatment zones and fulfil distance requirements (e.g. near running waters).

Technically the flow control and thereby the amount of a herbicide mixture can be achieved by pressure variation. If the pressure is lowered centrally, then the amounts on the whole boom width are reduced. There are upper and lower limits for flow rate, depending on the pressure operation interval of the nozzles. Pressures outside this interval lead to insufficient droplet sizes. Other systems use solenoid valves, which are directly integrated at each nozzle and allow to control the flow based on an electromechanical principle. Mixing the fluid with air in the nozzles can reduce the flow down to the half. Varying orifices in the nozzles are another way to control the output, this can be achieved either by a moving, steerable component within each nozzle or by combining several nozzles into one holder and switching between them. The presented technology can vary the amount of a prepared herbicide mixture.

If the herbicide mixture itself should be varied within the field, additional techniques have to be used. Either each herbicide gets mixed beforehand into several tanks and sprayed independently of each other, or the mixing takes place on the sprayer. A late mixing has the advantage to lower the amount of mixture within the whole system, which is favourable for the cleaning procedure and the minimised amount of remainders. In the extreme case herbicides are mixed near/in the nozzles into fresh water by direct injection systems (Schulze-Lammers & Vondricka, 2010). Because in this case the mixing takes place under pressure, the resulting problems have to be addressed: small amounts of liquid and varying viscosity have to be mixed into relatively large amounts of water, such that the resulting fluid is homogeneous before reaching the nozzle (Vondřička, 2007).

There are sprayers appearing on the market explicitly targeting precision farming applications, implementing such techniques. The Pre-Mix-System (Amazone) has a water tank and an additional tank with a preliminary mixture and can therefore vary the concentration down to zero during the operation by mixing these two components. The VarioInject system (Lechler) is a direct injection system, which can be mounted in the rear of the sprayer and mix the raw herbicide ingredients on demand with water. This way mixture remaining can be reduced to a minimum and only the herbicide actually applied to the field is used.

5. Herbicide-tolerant crops

Since their introduction in 1996 herbicide-resistant crops have been planted on a rapidly increasing areas, amounting worldwide to 83.6 Mha in 2009 and even more if crops with stacked traits are considered (Gianessi, 2008; James, 2009). In general, herbicide-resistance has been the dominant trait in biotech crops. In the process, glyphosate [N-(phosphonomethyl)glycine]-resistant soybean (*Glycine max* (L.) Merr.), maize (*Zea mays* L.), canola (*Brassica napus* L.) and cotton (*Gossypium hirsutum* L.) were most important (Duke & Powles, 2009; James, 2009; Owen, 2008). The major herbicide-resistant crop growing countries are USA, Brazil, Argentina, India and Canada (James, 2009). In Europe, the cultivation of herbicide-resistant crops has mainly been restricted to field trials dudue to public concerns and opposition (Davison & Bertheau, 2007; Kleter et al., 2008).

Despite the controversial debate in Europe, herbicide-resistant crops have several advantages. The use of herbicide-resistant crops, such as glyphosate- and glufosinate-resistant ones, broadens the spectrum of controlled weeds and provides new mode of actions to be used in-crop. This is especially important to control weed population resistant against other herbicides. In addition these herbicides are rather environmentally friendly and are easily degraded in soil (Knezevic & Cassman, 2003) and due to their broad spectrum they can replace several herbicides which would be used alternatively (Duke, 2005).

Gianessi (2005) calculated considerable savings in the amount of applied herbicide in the US agriculture due to glyphosate-resistant crops, whereas Benbrook (2001) found an increase in herbicide use in herbicide-resistant crops compared to conventional crops. Duke (2005) stated that more studies suggested a decrease in herbicide use in herbicide-resistant crops or a comparable amount of herbicide use than an increase. However, if farmers rely merely and consequently on this tool of herbicide-resistant crops, increased tolerance and resistance of weeds can spread rapidly and shifts within weed communities will occur readily (Knezevic & Cassman, 2003). In glyphosate-resistant soybean for example, *Ipomoea* and *Commelina* species as winter annuals are becoming much more common and problematic. The easiest way to control these more frequently occurring weeds, is to add tank-mix partners to glyphosate, which again results in higher use of herbicides (Culpepper, 2006). In addition there is the risk of gene escape i.e. transfer of resistant genes to other plant species, which can result in very difficultly controllable weeds and high herbicide inputs to control them (Knezevic & Cassman, 2003). One trend is to combine several tolerance genes in herbicide-resistant crops, this will decrease the single selection pressure of a distinct herbicide (Green, 2009), but also increase again the use of herbicides.

The sound use of herbicide-resistant crops can provide a tool to reduce herbicide use and allowing the use of more environmentally friendly herbicides. However, a smart combination with other IWM management tools is a prerequisite to sustain these opportunities.

5.1 Robotic weeding

Robots were introduced into production systems a long time ago and have found their place for tasks, which are repetitive and therefore error-prone or are carried out in dangerous environmental conditions. A robot can be defined as a machine, which is able to sense its environment, analyse the situation and decide for an action according to a task specification. Actuation is then initiated with a control component ensuring the correct operation. A certain degree of 'intelligence' is needed to react on the changing surrounding and act accordingly. Therefore often artificial intelligence techniques are implemented in this field. Such technology found its place mainly in controlled environments (e.g. industrial production lines) and has proven to conduct repetitive tasks in an efficient manner. The extension of the operation to agricultural fields is on the way, and some machinery already implement part of the robotic properties (Blackmore et al., 2007). The security of the operation of unmanned vehicles is one of the obstacles, which has to be addressed. Human supervision and interaction nowadays is still necessary, the automation of subtasks on the other hand steadily develops. Many implements for field operation already include sensing, steering and control systems for their unguided operation. In agriculture, these implements can be modular: tractors implement parts of robotic navigation, sensors can be mounted to sense the status of the crop or soil and terminals are used to make decisions and control implements according to their abilities (Blasco et al., 2002). Robots integrate all of the aforementioned technologies (sensing, decision support, actuators), but also require additional techniques for the navigation. Combinations of such technology therefore can be regarded as robots, e.g. the proposed weed sensing and technology already works to a large extent without human intervention, since the decision can be based on sensor data, and the decision and actuation (spraying) are automated and do not require human interaction. Tractors with auto-steering guided by GPS already reduce the amount of work for the driver, such that the driver can focus on other tasks. The future of robots in agricultural production systems can either advance in the automation and control of large machines or the development of smaller machines for special local operations. Robotic weeding is an approach to automate the labour intensive task of manual weed scouting and/or weeding. It has the potential to be carried out not only on the canopy or local (row) scale, but operate on the plant level. Autonomous machines could take over parts of the task, either for the autonomous creation of weed maps or the weed management on small scales. Operation times of robots are an argument for their introduction: tedious and time-consuming tasks can be done by robots in a 24/7 manner. If implements are available that target single plants, like micro sprayers (Midtiby et al., 2011) or hoes (Melander, 2006), then the operation of these can be carried out on a robot. The treatment of single plants limits the driving speed, as opposed to the development towards faster and larger implements with higher field area capacity. This can be counteracted by the use of multiple, smaller robots, which in turn are more flexible in their use (Blackmore et al., 2007). It is likely that parts of the machinery undergo development with robotic technology and the final solution will be a combination of task specific implements, which can be combined individually, creating task specific robotic automation as needed. The sensor developments and decision components researched lead the way and their integration will lead to new possibilities for the management.

Some problems still need to be tackled, before an introduction into wider practice will take place: the security of operation, energy constraints on smaller machines. Support and supervision of such technology on the other hand open new fields for businesses.

6. Conclusion

Weeds still are the cause of high yield losses, and alternative measures for weed control are required, because of the rising problems with herbicide residues in the environment and food. The alternative weeding methods without herbicides described in this chapter present a high potential to successfully compete with herbicide treatments. For instance, weed harrowing or a combination of flaming with mechanical tools, has shown an increase in crop yields due to the achieved weed control, up to a similar or even higher level than that obtained with chemical control. Considering these methods within a balanced approach such as a integrated weed management plan, there is a good chance to fulfil the political framework, at least in Europe, to prefer non-chemical weed control methods and to move towards the integrated pest management. However, it requires some risk acceptance and training efforts by the farmers to accomplish a good decision making plan. Existing sensors to assess the complex crop- weed- and soil variability contribute to reduce the use of herbicides towards a site-specific weed management approach, because then they could be only used on a sub-field level. Site-specific weeding also profits from the opportunities of information systems, data handling and decision support systems. Especially the latter is relevant, as DSS can optimize weed control economically and from an environmental point of view. In addition, this technology will allow monitoring the management success over a larger time-scale. In Europe, herbicide-resistant crops may gain some attention in the future, at least on a research level, for their potential to reduce herbicide application or to use only active ingredients which harm the environment less. However, public concern and opposition will still be a big barrier to overcome. More research is necessary to validate the performance and risks of such crops, and then training and public information is needed, as not only the farmers need to know about the pros and cons, but also the consumers. Finally, robotic weeding seems a promising technology to become successful in industrialized countries to reduce chemical weed control, once accurate and robust methods for automatic and real-time weed discrimination are developed. Nevertheless, once again expert knowledge is the most essential part for decision making technology, and there is still much to investigate, in order to tackle the constraints like security of the operator, energy consumption, time of operation and purchase cost of a robot weeding system. But even without highly engineered equipment considerable amounts of herbicides can be saved. The right management decisions have to be taken and multiple measures for weed control should be introduced into the existing production systems and their well-established practices.

7. References

Andújar, D., Ángela Ribeiro, Fernández-Quintanilla, C. & Dorado, J. (2011). Accuracy and feasibility of optoelectronic sensors for weed mapping in wide row crops, *Sensors* 11(3): 2304–2318.
URL: *http://www.mdpi.com/1424-8220/11/3/2304/*

Benbrook, C. (2001). Do GM crops mean less pesticide use?, *Pesticide outlook* 12(5): 204–207.

Bennett, A. C., Price, A. J., Sturgill, M. C., Buol, G. S. & G., W. G. (2003). HADSS, pocket HERB, and webHADSS: Decision aids for field crops, *Weed Technology* 17: 412–420.

Benzing, A. (2001). *Agricultura Orgánica: Fundamentos para la Región Andina [Organic agriculture: principles for the Andean regions]*, Neckar-Verlag, Villingen-Schwenningen.

Berti, A. & Zanin, G. (1994). Density equivalent: a method for forecasting yield loss caused by mixed weed populations, *Weed Research* 34: 327–332.

Berti, A. & Zanin, G. (1997). Gestinf: a decision model for post-emergence weed management in soybean (*Glycine max* (l.) merr.), *Crop Protection* 16: 109–116.

Biller, R. H. (1998). Reduced input of herbicides by use of optoelectronic sensors, *Journal of Agricultural Engineering Research* 71: 357–362.
URL: *http://www.sciencedirect.com/science/article/B6WH1-45J55FB-7/2/43cd9c341a6ae320d4ee5b88c65966fb*

Biller, R., Hollstein, A. & Sommer, C. (1997). Precision application of herbicides by use of optoelectronic sensors, *Precision Agriculture* pp. 451–.

Blackmore, B., Griepentrog, H., Fountas, S. & Gemtos, T. (2007). A specification for an autonomous crop production mechanization system, *Agricultural Engineering International: the CIGR Ejournal* IX(PM 06 032): 1–24.
URL: *http://www.cigrjournal.org/index.php/Ejounral/article/view/900/894*

Blackshaw, R. E., Moyer, J. R., Doram, R. C., & Boswell, A. L. (2001). Yellow sweetclover, green manure, and its residues effectively suppress weeds during fallow, *Weed Science* 49(3): 406–413.

Blasco, J., Aleixos, N., Roger, J. M., Rabatel, G. & Moltó, E. (2002). Ae–automation and emerging technologies: Robotic weed control using machine vision, *Biosystems Engineering* 83: 149–157.
URL: *http://www.sciencedirect.com/science/article/B6WXV-474G449-2/1/77aae51c283a3e8e193f2f0a7f9fd8aa*

Bond, W., Turner, R. & Davies, G. (2007). A review of mechanical weed control, *Technical report*, HDRA, Ryton Organic Gardens, Coventry, CV8 3LG, UK.
URL: *http://www.gardenorganic.org.uk/organicweeds*

Bond, W., Turner, R. & Grundy, A. (2003). A review of non-chemical weed management, *Technical report*, HDRA, Ryton Organic Gardens, Coventry, CV8 3LG, UK HRI, Wellesbourne, Warwick, CV35 9EF, UK.
URL: *http://www.organicweeds.org.uk*

Bàrberi, P., Moonen, A. C., Peruzzi, A., Fontanelli, M. & Raffaelli, M. (2009). Weed suppression by soil steaming in combination with activating compounds, *Weed Research* 49(1): 55–66.
URL: *http://onlinelibrary.wiley.com/doi/10.1111/j.1365-3180.2008.00653.x/abstract*

Brown, R. & Noble, S. (2005). Site-specific weed management: sensing requirements – what do we need to see?, *Weed Science* 53(2): 252–258.
URL: *http://dx.doi.org/10.1614%2FWS-04-068R1*

Candido, V., D'Addabbo, T., Miccolis, V. & Castronuovo, D. (2011). Weed control and yield response of soil solarization with different plastic films in lettuce, *Scientia Horticulturae* 130(3): 491–497.
URL: *http://www.sciencedirect.com/science/article/pii/S0304423811004079*

Christensen, S. & Heisel, T. (1998). Patch spraying using historical, manual and real-time monitoring of weeds in cereals, *Zeitschrift für Pflanzenkrankheiten und Pflanzenschutz* Sonderheft XVI: 257–263.

Christensen, S., Heisel, T., Walter, A. M. & Graglia, E. (2003). A decision algorithm for patch spraying, *Weed Research* 43: 276–284.

Christensen, S., Søgaard, H., Kudsk, P., Nørremark, M., Lund, I., Nadimi, E. & Jørgensen, R. (2009). Site-specific weed control technologies, *Weed Research* 49(3): 233–241.

Cimen, I., Turgay, B. & Pirin¸ V. (2010). Effect of solarization and vesicular arbuscular mychorrizal on weed density and yield of lettuce (*Lactuca sativa* l.) in autumn season, *African Journal of Biotechnology* 9(24): 3520–3526.
URL: *http://www.scopus.com/inward/record.url?eid=2-s2.0-77953939652&partnerID=40&md5=a41ade3c7f567239aa3c65fd4e6cf9cc*

Coble, H. D. & Mortensen, D. A. (1992). The trheshold concept and its application to weed science, *Weed Technology* 6: 191–195.

Corre-Hellou, G., Dibet, A., Hauggaard-Nielsen, H., Crozat, Y., Gooding, M., Ambus, P., Dahlmann, C., von Fragstein, P., Pristeri, A., Monti, M. & Jensen, E. (2011). The competitive ability of pea–barley intercrops against weeds and the interactions with crop productivity and soil n availability, *Field Crops Research* 122(3): 264–272.
URL: *http://www.sciencedirect.com/science/article/pii/S037842901100116X*

Culpepper, A. S. (2006). Glyphosate-induced weed shifts, *weed technology* 20(277–281).

Davison, J. & Bertheau, Y. (2007). EU regulations on the traceability and detection of GMOs: difficulties in interpretation, implementation and compliance, *CAB reviews: Perspectives in agriculture, veterinary science, nutrition and natural resources* 77.

Dieleman, J. & Mortensen, D. (1999). Characterizing the spatial pattern of *Abutilon theophrasti* seedling patches, *Weed Research* 39: 455–467.

Duke, S. O. (2005). Taking stock of herbicide-resistant crops ten years after introduction, *Pest management science* 61: 211–218.

Duke, S. & Powles, S. B. (2009). Glyphosate-resistant crops and weeds: now and in the future, *Agbioforum* 12: 346–357.

ENDURE (2008). Integrated weed management (iwm) case study – report on field studies, literature review, general conclusions and recommendations and future iwm research, *Deliverable DR1.6*, ENDURE - European Network for Durable

Exploitation of crop protection strategies. Project number: 031499. URL: *http://www.endure-network.eu /content/download/5616/43701/file/ENDURE_DR1.6.pdf*

ENDURE (2009). Presentation of DEXiPM arable crops A qualitative multi-criteria model for the assessment of the sustainability of pest management systems, *Deliverable DR2.14a*, ENDURE - European Network for Durable Exploitation of crop protection strategies. Project number: 031499. URL: *http://www.endure-network.eu/content/download/5616/43701/file/ENDURE_DR1.6.pdf*

European Parliament & Council of the EU (2009). Directive 2009/128/EC of the European Parliament and of the Council of 21st October 2009 establishing a framework for Community action to achieve the sustainable use of pesticides (Text with EEA relevance), *Official Journal of the European Union* L 309: 71–86. 24.11.2009. URL: *http://eur-lex.europa.eu/LexUriServ/LexUriServ.do?uri=OJ:L:2009:309:0071:0086:EN:PDF*

Gerhards, R. & Christensen, S. (2003). Real-time weed detection, decision making and patch spraying in maize, sugar beet, winter wheat and winter barley, *Weed Research* 43(6): 385–392.
URL: *http://dx.doi.org/10.1046/j.1365-3180.2003.00349.x*

Gerhards, R., Gutjahr, C., Weis, M., Keller, M., Sökefeld, M., Möhring, J. & Piepho, H. (2011). Using precision farming technology to quantify yield effects due to weed competition and herbicide application, *Weed Research* . to appear.

Gerhards, R. & Oebel, H. (2006). Practical experiences with a system for site-specific weed control in arable crops using real-time image analysis and GPS-controlled patch spraying, *Weed Research* 46(3): 185–193.

Gerowitt, B. & Heitefuss, R. (1990). Weed economic thresholds in the F. R. Germany, *Crop Protection* 9: 323–331.

Gianessi, L. P. (2005). Economic and herbicide use impacts of glyphosate-resistant crops, *Pest management science* pp. 241–245.

Gianessi, L. P. (2008). Review economic impacts of glyphosate-resistant crops, *Pest management science* 64: 346–352.

Green, J. M. (2009). Evolution of glyphosate-resistant crop technology, *Weed Science* (57): 108–117.

Gutjahr, C. & Gerhards, R. (2010). *Precision Crop Protection - the Challenge and Use of Heterogeneity*, in Oerke et al. (2010), 1 edn, chapter Decision rules for site-specific weed management, pp. 223–239.
URL: *http://springer.com/978-90-481-9276-2*

Gutjahr, C., Weis, M., Sökefeld, M. & Gerhards, R. (2009). Influence of site specific herbicide application on the economic threshold of the chemical weed control, *in* A. Bregt, S. Wolfert, J. Wien & C. Lokhorst (eds), *EFITA conference '09*, EFITA (European Federation for Information Technology in Agriculture), Wageningen Academic Publishers, Wageningen, Netherlands, pp. 557–565. Only available on CD-rom (PDF-file) - Papers presented at the 7th EFITA conference, Joint International Agricultural Conference, 6.-8. July.
URL: *http://www.efita.net/apps/accesbase/bindocload.asp?d=6549&t=0&identobj=4K9sKV1z&uid=57305290&sid=57&idk=1*

Hansson, D. & Svensson, S.-E. (2011). Effect of flame weeding at different time intervals before crop emergence, *in* D. C. Cloutier (ed.), *Proceedings 9th EWRS Workshop on Physical and Cultural Weed Control, 28–30 March 2011*, Samsun, Turkey, p. 65.

Heisel, T., Schou, J., Christensen, S. & Andreasen, C. (2001). Cutting weeds with a CO2 laser, *Weed Research* 41: 19–29.
URL: *http://dx.doi.org/10.1111/j.1365-3180.2001.00212.x*

Holst, N. (2010). WeedML: a tool for collaborative weed demographic modeling, *Weed Science* 58(4): 497–502.
URL: *http://www.bioone.org/doi/abs/10.1614/WS-D-09-00013.1*

Iftikhar, N. & Pedersen, T. B. (2011). Flexible exchange of farming device data, *Computers and Electronics in Agriculture* 75(1): 52–63.
URL: *http://www.sciencedirect.com/science/article/pii/S0168169910001894*

James, C. (2009). Global status of commercialized Biotech/GM crops: 2009, *Technical Report 41*, International service for the acquisition of agri-biotech applications, ISAAA: Ithaca, NY. URL: *http://www.isaaa.org/resources/publications/briefs/41/download/isaaa-brief-41-2009.pdf*

Jensen, R. K., Rasmussen, J. & Melander, B. (2004). Selectivity of weed harrowing in lupin, *Weed Research* 44(4): 245–253.
URL: *http://www.blackwell-synergy.com/doi/abs/10.1111/j.1365-3180.2004.00396.x*

Jones, R., Cacho, O. & Sinden, J. (2006). The importance of seasonal variability and tactical responses to risk on estimating the economic benefits of integrated weed management, *Agricultural Economics* 35(3): 245–256.
URL: *http://onlinelibrary.wiley.com/doi/10.1111/j.1574-0862.2006.00159.x/abstract*

Kleter, G. A., Harris, C., Stephenson, G. & Unsworth, J. (2008). Review comparison of herbicide regimes and the associated potential environmental effects of glyphosate-resistant crops versus what they replace in Europe, *Pest management science* 64: 479–488.

Knezevic, S. Z. & Cassman, K. G. (2003). Use of herbicide-tolerant crops as a component of an integrated weed management program, *Crop management* . URL: *http://www.plantmanagementnetwork.org/pub/cm/management/2003/htc/Knezevic.pdf*

Knezevic, S. Z., Ulloa, S. M., Datta, A., Arkebauer, T., Bruening, C. & Gogos, G. (2011). Weed control and crop tolerance to propane flaming as influenced by time of day, *in* D. C. Cloutier (ed.), *Proceedings 9th EWRS Workshop on Physical and Cultural Weed Control, 28–30 March 2011*, pp. 66–78.

Kurstjens, D. A. G. & Kropff, M. J. (2001). The impact of uprooting and soil-covering on the effectiveness of weed harrowing, *Weed Research* 41(3): 211–228.
URL: *http://dx.doi.org/10.1046/j.1365-3180.2001.00233.x*

Kurstjens, D. A. G. & Perdok, U. D. (2000). The selective soil covering mechanism of weed harrows on sandy soil, *The selective soil covering mechanism of weed harrows on sandy soil* 55(3-4): 193–206.
URL: *http://www.sciencedirect.com/science/article/B6TC6-40NMSHJ-7/2/670aad35b934963ae0cff06c93bde62c*

Llewellyn, R. & Pannell, D. (2009). Managing the herbicide resource: an evaluation on extension for herbicide-resistant weed management, *AgBioForum* 12(3/4): 358–369.
URL: *http://www.agbioforum.org/v12n34/v12n34a11-llewellyn.htm*

Mathiassen, S. K., Bak, T., Christensen, S. & Kudsk, P. (2006). The effect of laser treatment as a weed control method, *Biosystems Engineering* 95: 497–505.
URL: *http://www.sciencedirect.com/science/article/B6WXV-4M57H8K-1/1/d694981c000e8944e8f69701dc0d4de5*

Melander, B. (2006). Current achievements and future directions of physical weed control in europe, *AFPP 3rd International conference on non-chemical crop protection methods*, number 9998, Danish Institute of Agricultural Sciences, Department of Integrated Pest Management, Research Centre Flakkebjerg, Orgprints, Lille, France, pp. 49–58. URL: *http://orgprints.org/9998/*

Melander, B., Rasmussen, I. A. & Bàrberi, P. (2005). Integrating physical and cultural methods of weed control – examples from european research, *Weed Science* 53(3): 369–381. URL: *http://dx.doi.org/10.1614/WS-04-136R*

Midtiby, H. S., Mathiassen, S. K., Andersson, K. J. & Jørgensen, R. N. (2011). Performance evaluation of a crop/weed discriminating microsprayer, *Computers and Electronics in Agriculture* 77: 35–40.
URL: *http://dl.acm.org/citation.cfm?id=1994680*

Mir, S. A. & Quadri, S. M. K. (2009). *Climate change, intercropping, pest control and beneficial microorgansims*, Vol. 2 of *sustainable agriculture reviews*, Springer, Dordrecht, Heidelberg, London, New York, chapter Decision support systems: concepts, progress and issues - a review, pp. 373–399.

Nadimi, E. S., Andersson, K. J., Jørgensen, R. N., Maagaard, J., Mathiassen, S. & Christensen, S. (2009). Designing, modeling and controlling a novel autonomous laser weeding system, *7th World Congress on Computers in Agriculture Conference Proceedings*, number 711P0409e, ASABE - American Society of Agricultural and Biological Engineers, St. Joseph, MI. 22-24 June 2009, Reno, Nevada (electronic only).
URL: *http://asae.frymulti.com/abstract.asp?aid=29077*

Nash, E., Nikkilä, R., Pesonen, L., Oetzel, K., Mayer, W., Seilonen, I., Kaivosoja, J., Bill, R., Fountas, S. & Sørensen, C. (2010). Machine readable encoding for definitions of agricultural crop production and farm management standards, *Deliverable 4.1.1*, FutureFarm. Integration of Farm Management Information Systems to support real-time management decisions and compliance of management standards - Knowledge Management in the FMIS of Tomorrow.
URL: *http://futurefarm.eu/node/183*

Nash, E., Wiebensohn, J., Nikkilä, R., Vatsanidou, A., Fountas, S. & Bill, R. (2011). Towards automated compliance checking based on a formal representation of agricultural production standards, *Computers and Electronics in Agriculture* 78(1): 28–37.
URL: *http://www.sciencedirect.com/science/article/pii/S0168169911001244*

Nikkilä, R., Wiebensohn, J., Nash, E. & Seilonen, I. (2010). Integration of farm management information systems to support real-time management decisions and compliance of management standards, *Deliverable D4.4*, FutureFarm. WP 4 topic: Knowledge Management in the FMIS of Tomorrow.
URL: *http://www.futurefarm.eu/system/files/FFD4.4_Proof_Of_Concept_Final.pdf*

Nikkilä, R., Seilonen, I. & Koskinen, K. (2010). Software architecture for farm management information systems in precision agriculture, *Computers and Electronics in Agriculture* 70(2): 328–336.
URL: *http://www.sciencedirect.com/science/article/B6T5M-4XMC000-1/2/54b46b3c8aa2a9ee60d022eb392b017a*

Oebel, H. & Gerhards, R. (2006). Kameragesteuerte Unkrautbekämpfung – eine Verfahrenstechnik für die Praxis, *Journal of Plant Diseases and Protection* Special Issue XX: 181–187.
URL: *http://www.jpdp-online.com/Artikel.dll/04-Oebel_MTAyNDIy.PDF*

Oerke, E. (2006). Crop losses to pests, *Journal of Agricultural Science* 144(1): 31–43.

Oerke, E.-C., Gerhards, R., Menz, G. & Sikora, R. A. (eds) (2010). *Precision Crop Protection - the Challenge and Use of Heterogeneity*, 1 edn, Springer Verlag, Dordrecht, Heidelberg, London, New York.
URL: *http://springer.com/978-90-481-9276-2*

Owen, M. (2008). Review weed species shifts in glyphosate-resistant crops, *Pest management science* 64: 377–387.
URL: *http://www.ask-force.org/web/HerbizideTol/Owen-Weed-Species-Shifts-2008.pdf*

Parsons, D. J., Benjamin, L. R., Clarke, J., Ginsburg, D., Mayes, A., Milne, A. E. & Wilkinson, D. J. (2009). Weed manager – a model-based decision support system for weed management in arable crops, *Comput. Electron. Agric.* 65(2): 155–167.
URL: *http://portal.acm.org/citation.cfm?id=1501021.1501071*

Reichardt, M., Jürgens, C., Klöble, U., Hüter, J. & Moser, K. (2009). Dissemination of precision farming in Germany: acceptance, adoption, obstacles, knowledge transfer and training activities, *Precision Agriculture* 10(6): 525–545.
URL: *http://www.springerlink.com/content/d64qk22343jl1386/*

Rueda-Ayala, V., Rasmussen, J., Fournaise, N. & Gerhards, R. (2011). Influence of timing and intensity of post-emergence weed harrowing in winter wheat on weed control, crop recovery and crop yield, *Weed Research* . in press.

Rueda, V., Rasmussen, J. & Gerhards, R. (2010). *Precision Crop Protection - the Challenge and Use of Heterogeneity*, in Oerke et al. (2010), 1 edn, chapter Mechanical weed control, pp. 279–294.
URL: *http://springer.com/978-90-481-9276-2*

Rueegg, W. T., Quadranti, M. & Zoschke, A. (2007). Herbicide research and development: challenges and opportunities, *Weed Research* 47: 271–275.

Rydahl, P. (2003). A web-based decision support system for integrated management of weeds in cereals and sugarbeet, *Bulletin OEPP/EPPO Bulletin* 33: 455–460.

Rydahl, P., Berti, A. & Munier-Jolain, N. (2008). O.24 - decision support systems (dss) for weed control in europe - state-of-the art and identification of 'best parts' for unification on an european level, ENDURE. Papers presented at the ENDURE International Conference 2008, diversifying crop protection, La Grande-Motte, France, 12-15 October 2008. URL: *http://www.endure-network.eu/international_conference_2008/proceedings/tuesday_october_14*

Schmitz, M., Martini, D., Kunisch, M. & Mösinger, H.-J. (2009). *agroXML enabling standardized, platform-independent internet data exchange in farm management information systems*, Springer US, Boston, MA, pp. 463–468.
URL: *http://www.springerlink.com/content/kr4755744634g63r/*

Schulze-Lammers, P. & Vondricka, J. (2010). *Precision Crop Protection - the Challenge and Use of Heterogeneity*, in Oerke et al. (2010), 1 edn, chapter Direct Injection Sprayer, pp. 295–310.
URL: *http://springer.com/978-90-481-9276-2*

Singh, K., Agrawal, K. & Bora, G. C. (2011). Advanced techniques for weed and crop identification for site specific weed management, *Biosystems Engineering* in Press.
URL: *http://www.sciencedirect.com/science/article/B6WXV-52DB334-1/2/d827016de58dd4a8c65fb4ef4a9dfe71*

Sökefeld, M. (2010). *Precision Crop Protection - the Challenge and Use of Heterogeneity*, in Oerke et al. (2010), 1 edn, chapter Variable rate technology for herbicide application, pp. 335–347.
URL: *http://springer.com/978-90-481-9276-2*

Srinivasan, A. (2006). *Handbook of precision agriculture: principles and applications*, Food Products Press, an imprint of The Haworth Press, Inc., Binghamton, NY.
URL: *http://www.haworthpress.com/store/product.asp?sku=5627*

Steinberger, G., Rothmund, M. & Auernhammer, H. (2009). Mobile farm equipment as a data source in an agricultural service architecture, *Computers and Electronics in Agriculture* 65(2): 238–246.
URL: *http://portal.acm.org/citation.cfm?id=1501079*

Stigliani, L. & Cosimo, R. (1993). SELOMA: Expert system for weed management in herbicide-intensive crops, *Weed Technology* 7: 550–559.

Storrie, A., Jones, R., Monjardino, M., Preston, C., Stewart, V., Walker, S., Hashem, A., Simpfendorfer, S., Holding, D., Widderick, M., Lemerle, D., Walsh, M., Bowcher, A., Evans, C., Moore, J. & D'Emden, F. (2006). Integrated weed management in australian cropping systems – a training resource for farm advisors, *Technical report*, CRC for Australian Weed Management, University of Adelaide, PMB 1, Glen Osmond, SA 5064 Australia. November.
URL: *http://www.grdc.com.au/director/events/linkpages/weedlinks#Integrated Weed Management manual*

Sui, R., Thomasson, J. A., Hanks, J. & Wooten, J. (2008). Ground-based sensing system for weed mapping in cotton, *Computers and Electronics in Agriculture* 60: 31–38.
URL: *http://www.sciencedirect.com/science/article/B6T5M-4P8R7BS-1/1/ac4921ce458fb93 1f3f9ba2cb9c7cbe6*

Swanton, C. J., Mahoney, K. J., Chandler, K. & Gulden, R. H. (2008). Integrated weed management: Knowledge-based weed management systems, *Weed Science* 56(1): 168–172.

Thorp, K. & Tian, L. (2004). A review on remote sensing of weeds in agriculture, *Precision Agriculture* 5(5): 477–508.
URL: *http://openurl.ingenta.com/content?genre=article&issn=1385-2256&volume=5 &issue=5&spage=477&epage=508*

Tian, L. (2002). Development of a sensor-based precision herbicide application system, *Computers and Electronics in Agriculture* 36: 133–149.

Timmermann, C., Gerhards, R. & Kühbauch, W. (2003). The economic impact of the site-specific weed control, *Precision Agriculture* 4(3): 249–260.
URL: *http://dx.doi.org/10.1023/A:1024988022674*

van der Weide, Y., Bleeker, P. O., Achten, V. T. J. M., Lotz, L. A. P., Folgegerg, F. & Melander, B. (2008). Innovation in mechanical weed control in crop rows, *Weed Research* 48(3): 215–224.
URL: *http://dx.doi.org/10.1111/j.1365-3180.2008.00629.x*

Vondřička, J. (2007). *Study on the process of direct nozzle injection for real-time site-specific pesticide application*, PhD thesis, Rheinische Friedrich-Wilhelms-Universität, Bonn. Forschungsbericht Agrartechnik des Arbeitskreises Forschung und Lehre der Max-Eyth-Gesellschaft Agrartechnik im VDI (VDI-MEG).
URL: *http://www.landtechnik.uni-bonn.de/ifl_research/pp_15/vondricka_dissertation.pdf*

Weis, M. (2010). *An image analysis and classification system for automatic weed species identification in different crops for precision weed management*, PhD thesis, Institute for Phytomedicine, Department of Weed Science, University of Hohenheim, Stuttgart, Germany.
URL: *http://opus.ub.uni-hohenheim.de/volltexte/2010/519/*

Weis, M. & Sökefeld, M. (2010). *Precision Crop Protection - the Challenge and Use of Heterogeneity*, in Oerke et al. (2010), 1 edn, chapter Detection and identification of weeds, pp. 119–134.
URL: *http://springer.com/978-90-481-9276-2*

Wiles, L. (2009). Beyond patch spraying: site-specific weed management with several herbicides, *Precision Agriculture* 10(3): 277–290.
URL: *http://dx.doi.org/10.1007/s11119-008-9097-6*

Wiles, L. & Brodahl, M. (2004). Exploratory data analysis to identify factors influencing spatial distributions of weed seed banks, *Weed Science* 52: 936–947.
URL: *http://dx.doi.org/10.1614%2FWS-03-068R*

Wiles, L. J. (2005). Sampling to make maps for site-specific weed management, *Weed Science* 53: 228–235.
URL: *http://dx.doi.org/10.1614%2FWS-04-057R1*

Wiles, L., King, R., Schweizer, E., Lybecker, D. & Swinton, S. (1996). Gwm: general weed management model, *Agricultural Systems* 50: 355–376.

Wiles, L. & Schweizer, E. (2002). Spatial dependence of weed seed banks and strategies for sampling, *Weed Science* 50(5): 595–606.
URL: *http://dx.doi.org/10.1614%2F0043-17453ASDOWSB%5D2.0.CO%3B2*

Wilkerson, G. G., Modena, S. A. & Coble, H. D. (1991). Herb: Decision model for postemergence weed control in soybean, *Agronomy Journal* 83: 413–417.
URL: *http://agron.scijournals.org/cgi/content/abstract/agrojnl;83/2/413*

Williams, M. & Mortensen, D. (2000). Crop/weed outcomes from site-specific and uniform soil-applied herbicide applications, *Precision Agriculture* 2(4): 377–388.

Zornbach, W. (2011). Leitlinien Integrierter Pflanzenschutz, slides, online. Fachgespräch über die aktuelle Situation bei der Entwicklung von Leitlinien zum integrierten Pflanzenschutz.
URL: *http://nap.jki.bund.de/index.php?menuid=76&downloadid=168&reporeid=0*

7

Managing Weeds with Reduced Herbicide Inputs: Developing a Novel System for Onion

Harlene Hatterman-Valenti
North Dakota State University
USA

1. Introduction

Weeds are a major challenge in crop production. Often weeds cause significant yield losses and even a few weeds producing seeds can cause weed problems in subsequent years. For example, sicklepod (*Senna obtusifolia*) average seed production is 8,000 seeds per plant (English and Oliver, 1981). Chemical weed control methods have been shown to be one of the most cost effective weed control options (Pike et al., 1991). Herbicides dominated the pesticides used in the United States during 2004 and accounted for two-thirds of the approximately $8.5 billion spent on agricultural pesticides (Padgitt et al., 2000). However, with the weed control benefits from herbicide usage also came environmental and health concerns. These concerns have resulted in much research on the safety of each chemical. Most of these environmental and health concerns are dealt with prior to herbicide registration. Manufacturers conduct numerous experiments in order to accurately determine product utility, market value, and regulatory needs. These experiments include toxicity trials to a wide range of organisms to determine the product's safety to plants, animals, and environmental fate. In addition, an enormous amount of testing is done for product quality and efficacy. Considering the vast investment that a manufacturer has incurred prior to product launch and the relatively short period of time to recoup their investment before the product is off patent, it becomes crucial that a product is registered quickly and at the lowest effective use rate. Recommending rates above this rate would potentially lead to widespread rate reductions, while recommending rates below this rate would potentially lead to widespread performance issues. With either scenario, the manufacturer's ability to recoup their investment becomes greatly reduced.

2. Industry perspective

Doyle and Stypa (2004) indicated that herbicide rates are selected on the basis of maximized product value. Therefore, a rate structure is selected which provides an optimum investment return for the conditions of the target market. In other words, rates are selected that will satisfy producer weed control expectations under the environmental conditions where the crop(s) is generally grown. For many of the commodity crops, these growing conditions can vary greatly and are considered when the product rate structure is selected. In addition, manufacturers realize that weed species differ in their susceptibility to a specific herbicide and that the labeled rate for this herbicide may be higher than what is needed for certain weed species, but because the rate range selected needs to be efficacious to as many weeds as possible, rates will be high for some weed species.

3. Weed management decisions

When chemical weed control decisions are made, many questions need to be considered, including the need to spray an herbicide, which product to use, and when, where and how to apply that product. In all of these considerations, there are opportunities to reduce the risks associated with herbicide use. However, a producer will not adopt these practices if there is a resulting crop yield loss, increase in field weed populations, uninsured profitability, or increased environmental risk. Unfortunately, as agricultural profit margins decrease, producers search for ways to control input costs which includes how they manage weeds.

For most field cropping systems, herbicide usage comprises approximately 20 to 30% of the input costs (Derksen et al., 2002). One may wonder if the cost-cutting approach of applying herbicides at reduced rates is worth the risk. However, in Canada, a 10% reduction in herbicide usage, without crop yield reduction or increased field weed populations would save producers $85 million. This 10% herbicide use reduction could occur by either avoiding the need to apply the herbicide because weed densities were kept below economic threshold levels or by reducing herbicide rates. Eliminating herbicide use would alleviate the potential controversy with off-label applications, but would only be successful for the most vigorous and competitive cropping systems (Van Acker et al., 2001).

Deciding when to control weeds requires detailed knowledge of the weed populations in the field, the potential interference from those weeds, and the potential benefit obtained from controlling the weeds. When producers relied on preemergence herbicides for weed control in a specific crop, it was important to scout for weeds prior to harvest so that the weed potential for the following year could be assessed. However, over the past 20 years or so with the introduction of postemergence herbicides, this reliance has changed (Blackshaw et al., 2006).

3.1 Early weed identification

It is critical that weed species be identified early in the season. This can be accomplished by routinely scouting fields, but can also be challenging since many species have similar appearances at the cotyledon stage. Numerous training aids are available to ensure that unfamiliar species are identified correctly and that appropriate management options are employed. Whole fields should be scouted and weed patches, low spots and field margins should be considered separately, since they do not represent the entire field. Scouting these fields later in the season will provide valuable information on the species and numbers of weeds that have escaped control and added to the weed seed reservoir. This information is needed for long-term weed management planning.

4. Yield loss factors

Yield loss from weeds depends on many factors including competitive abilities of the crop and weeds. Adequate weed control with reduced herbicide rates can be successful by increasing the competitiveness of the cropping system and incorporating an integrated weed management system (Mohler, 2001; Mulugeta and Stoltenberg, 1997; Swanton et al., 2008). Fodor et al., (2008) showed that a competitive crop utilizes resources before the weeds. This will only occur if a good crop stand is established for a vigorous growing crop. They concluded that crop rotation, seedbed preparation, crop type and variety selection, seed quality and treatment, seeding rate and stand density, seeding date, fertilizer rate and placement, and pest and disease control influenced crop competitiveness and that the failure to manage all components promoted weed competition with the crop. Similar research has identified cereal traits such as plants taller than their neighbors, with many horizontal leaves and a vigorous root system as traits that would enable these plants to effectively capture light, water and nutrients from neighboring plants and contribute to plant competitiveness (Donald and Hamblin, 1976; Lemerle et al., 2001). The field pea (*Pisum sativum*) 'Jupiter' had the greatest tolerance to competition and the ability to suppress weed growth compared to 10 cultivars ranked low to medium in their tolerance to competition and their ability to suppress weeds (MacDonald, 2002). Unfortunately, cultivar studies have shown to vary considerably between years and locations (Cousens and Mokhtari, 1998).

5. Competitive cropping system components

Components of a competitive cropping system include: diverse crop rotations, competitive crop cultivars, higher seeding rates, reduced row spacing, specific fertilizer placement, and the use of green manures or cover crops (Derksen et al., 2002; Blackshaw et al., 2006). Lemerle et al. (1995) ranked several annual winter crops for their competitiveness against annual ryegrass (*Lolium multiflorum*) in Australia. Oats (*Avena sativa*) was determined to be the most competitive with only 2 to14 % yield reduction from annual ryegrass at a density of 300 plants/m^2. Rye (*Secale cereale*) was the second most competitive crop with a yield reduction of 14 to 20%. Both field pea and narrowleaf lupine (*Lupinus angustifolius*) were the least competitive with 100% yield reduction. In Canada, the competitive ranking of crops from highest to lowest was: barley = rye > oats > canola (*Brassica spp.*) = wheat (*Triticum aestivum*) > peas = flax (*Linum sitatissimum*). Thus the competitiveness of a crop can vary depending upon growing conditions and the weed species.

5.1 Diverse crop rotations

Diverse crop rotations and the use of green manures or cover crops have historically been recognized to be beneficial for crop production. Rotating between distinctly unrelated crops will result in higher grain yields compared to continuous cropping of wheat (Table 1). For example, seeding wheat to an area that was barley (*Hordeum vulgare*) the year before resulted in a 12.5% increase, on average, in wheat yield compared to continuous wheat. However, if wheat was seeded to an area that was soybean (*Glycine max*) the previous year, the average wheat yield increase, compared to continuous wheat, was 42.9%. Some of the benefits from a well-planned, diverse, crop rotation include: reduced insect and disease problems, improved soil fertility, improved soil tilth and aggregate stability, better soil water management, reduced soil erosion, and reduced allelopathic effects. Diverse crop rotations can also discourage weed establishment and reduce weed seed production due to different planting and harvest times that disrupt the weed species lifecycles.

	Wheat yield, t/ha								
Previous crop	1977	1978	1979	1980	1981	1982	1983	1984	8-yr. avg.
Wheat	1.5	1.7	2.4	2.5	2.3	2.6	2.9	1.1	2.1
Barley	1.8	1.7	2.4	2.5	2.8	3.1	3.2	1.2	2.4
Flax	2.1	2.5	2.4	2.4	2.5	3.2	2.9	2.5	2.6
Corn	2.1	2.2	2.9	2.5	3.0	3.6	2.6	2.6	2.6
Soybean	2.8	2.9	2.8	2.8	3.0	3.2	3.6	3.0	3.0
Sunflower	2.0	2.2	3.0	2.8	3.0	2.6	2.9	3.0	2.7
Sugarbeet	2.3	2.3	2.8	2.6	3.0	2.9	3.5	3.2	2.8

Table 1. Wheat yields under conventional tillage when seeded the year following the various previous crops, Fargo, ND. Adapted from Peel, 1998.

5.2 Cover crops and living mulches

Producers have used cover crops to give a crop a competitive edge over weeds. Planting the correct cover crop after the harvest of a crop will help to reduce erosion, reduce nutrient leaching, improve soil structure, and suppress weed emergence. Gallandt (2009) measured common lambsquarters (*Chenopodium album*) weed seed rain for four years in a vegetable rotation of broccoli (*Brassica oleracea*) and winter squash (*Cucurbita moschata*) managed with no cover crop (control), fall cover crop (fall CC), two consecutive years of red clover (2-yr. CC), and alternate years of vegetable and cover crops with a summer fallow (alt.-yr. CC) (Figure 1). It was suggested that the alternate years of vegetable and cover crops with a summer fallow had lower common lambsquarters seed rain because the fallowing periods during the cover crop years depleted the seedbank, thus prevented common lambsquarters from increasing.

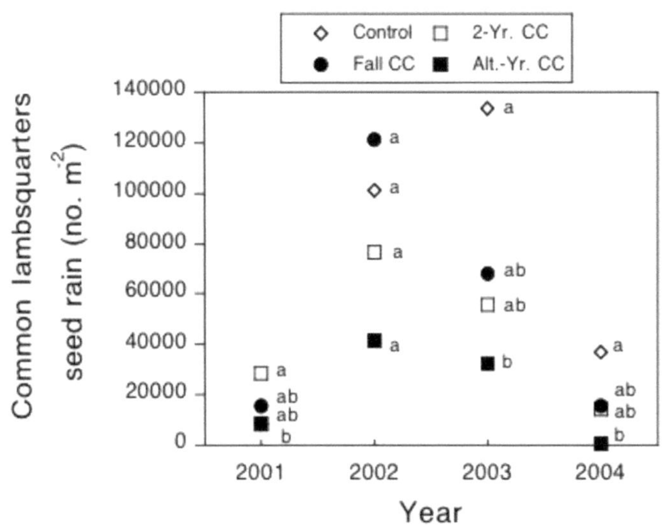

Fig. 1. Effect of cover crop systems on common lambsquarters seed rain in 2001 through 2004. Means within a year with different letters are significantly different from each other at the $P \leq 0.05$ level (Tukey's HSD). Adapted from Gallandt, 2009.

Cover crops have been used as living mulches for weed management. Perennial living mulches such as crownvetch (*Securigera varia*), flatpea (*Lathyrus sylvestris*), birdsfoot trefoil (*Lotus corniculatus*), and white clover (*Trifolium repens*) do not have to be reseeded each year and can be used to conserve nitrogen, reduce soil erosion, and increase soil organic matter, while they reduce weed population and crop yield losses due to weeds (Hartwig and Ammon, 2002).

5.3 Crop density

In general, an increase in crop density will increase the crop's competitiveness against weeds. This increase in crop density can occur by increasing the seeding rate, decreasing the space between rows, or both. Increasing wheat seeding rate from 175 to 280 plants/m^2 increased wheat yield while reducing wild oat biomass and seed production (Stougaard and Xue, 2004). However, Anderson et al. (2004) showed that if higher seeding rates were being used to improve the competitiveness of a wheat crop, it is important to optimize the seeding rate for yield and quality based on pre-seeding rainfall and growing season rainfall (Table 2). There is also an economical seeding rate optimum. Increasing the seeding rate of canola can allow the crop to compete better with weeds, but increasing the seeding rate above 150 seeds/m^2 reduced the profitability of the crop (Upadhyay, 2006).

PSR (mm)	GSR (mm)	Yield expectation (t/ha)	Minimum population needed (plants/m^2)	Approximate sowing rate (kg/ha)
0	150	1.50	60	22
	200	2.25	90	39
	250	3.00	120	56
100	200	2.55	102	47
	250	3.30	132	65
	300	4.05	162	86
200	250	3.60	144	76
	300	4.35	174	92
	350	5.10	204	116

Table 2. Estimates of minimum wheat plant population (plants/m^2) based on pre-seeding rainfall (PSR, mm) and rainfall in the growing period (GSR, mm) in Western Australia. Adapted from Anderson et al., 2004.

Another method to increase the stand density is by reducing the spacing between rows. Reduced row spacing has been shown to increase the crop competitiveness over weeds (Tharp and Kells, 2001; Willingham et al., 2008). Often the narrower-row spacing and reduced herbicide rate had similar weed control as the same crop at the wide-row spacing regime and herbicide applied at the manufacturer's suggested use rate.

The use of the twin-row system is another way to reduce the spacing between rows and has also resulted in increased yields for several row crops (Grichar et al., 2004; Willingham et al., 2008.) The twin-row system resulted in greater ground cover, leaf area indices, light interception at the canopy, and crop growth rate compared to the single wide-row system.

However, Grichar (2007) showed that narrower row spacing or twin-row planting does not always result in higher yields or increased net returns (Table 3). In addition, broadleaf crops seem to be less sensitive to row spacing than cereals. Thus, it is important to match the row spacing and seed rate in order to obtain a plant density that optimizes crop yield and competition against weeds.

Seeding rate (seeds/30.5 cm)	Row spacing	El Campo	Pt. Lavaca	El Campo	Pt. Lavaca
		2003		2004	
6	38-inch	5.4	5.2	5.6	4.8
	twin	11.1	9.8	11.3	8.6
10	38-inch	8.7	9.1	7.7	5.7
	twin	17.1	16.5	16.8	14.1
15	38-inch	7.1	7.4	6.7	5.5
	twin	14.8	14.2	14.1	10.7
$LSD_{0.05}$		1.2	1.8	1.0	2.0

Table 3. Soybean plant populations (plants/30.5 cm) as influenced by row spacing and seeding rates in 2003 and 2004 at El Campo and Pt. Lavaca, TX. Adapted from Grichar, 2007.

5.4 Fertilizer placement

The importance of specific fertilizer placement for a competitive crop was indicated by Fodor et al., (2008) when they concluded that a competitive crop utilizes resources before the weeds. They compared three planting dates for winter wheat and two nitrogen rates as a spring top-dressing application. Results indicated that delayed planting led to reduced wheat growth and greater weed biomass production and that the higher rate of nitrogen resulted in fewer weeds for the early and optimum time seeded plots. In contrast, the higher rate of nitrogen resulted in more weeds for the late seeded treatment.

6. Integrated weed management principles

Integrated weed management systems primarily utilize specific weed assessment; weed population ecology; understanding of economic thresholds; knowing the critical period for control; knowing the competitiveness of the crop; and understanding an herbicide's biologically effective dosage (Knezevic et al., 2002; Liebman and Gallandt, 1997; Swanton et al., 2008). The critical period of weed control is the span of time during the crop growth cycle when weeds must be controlled to prevent yield losses (Mohler, 2001). The best time to control weeds and the length of the critical period depend on a number of variables including weed emergence timing, weed densities, the competitive ability of weeds compared to crops, and environmental factors. Knezevic et al. (2002) suggested a standardized method for data analysis of critical period for weed control trials so that uniform decisions could be made on the weed control need and application timing, and to obtain efficient herbicide use from both biological and economical perspectives.

Unfortunately, most competitive studies have been conducted with agronomic crops. These crops have many weed management options and the ability to utilize several competitive cropping system components. For example, a multiyear study was conducted to compare weed management in wheat, barley, canola, and field pea using full or reduced herbicide rates, crop rotation, seeding date, seeding rate, and fertilizer timing (Blackshaw et al., 2005a, 2005b). They reported that after four continuous years, the weed seed bank did not differ when 50% of the herbicide rate was used as long as the crops were seeded early, at a high crop seeding rate, and with spring-applied banded fertilizer. The most obvious question is what components of a competitive cropping system and integrated weed management methods could be used to reduce herbicide inputs in a noncompetitive crop?

7. Poor crop competitiveness of onion

Onion (*Allium cepa*) is considered a poor competitive crop because the plant generally emerges later than many cool-season weeds and is very susceptible to weed canopy coverage and competition for light (Dunan et al., 1999). Morphological traits of onion include a shallow root system, slow establishment period, and long, narrow, erect leaves. These morphological traits have resulted in blow-out areas or extensive damage to newly emerged onion seedlings when high winds or storms pass through an area (Greenland, 2000). To reduce wind erosion, growers plant barley between the onion rows as a companion crop. The barley emerges quickly in comparison to onion, but also further complicates weed management issues since the grower does not want to reduce barley germination, but will need to kill the barley before it competes with onion. The barley is killed with an application of a postemergence grass herbicide when plants are 4 to 6 in tall. The companion crop has reduced onion establishment issues associated with wind erosion, but also requires additional herbicide input. Additionally, rainfall and wet conditions may delay the grass herbicide application, causing competition between barley and onion, resulting in reduced onion yield (Hatterman-Valenti and Hendrickson, 2006).

Weed competition is a severe problem throughout onion establishment and maturation (Swaider et al., 1992). The inability of onion to morphologically produce a sufficient canopy allows early-season in-row weeds, such as common lambsquarters and redroot pigweed (*Amaranthus retroflexus*), to substantially reduce yield (Boydston and Seymour, 2002).

7.1 Critical period for weed removal

The effect of day length on onion bulb initiation was the most important factor determining the critical period for weed removal (Bond and Burston, 1996). Growth switches from leaf production to bulb development for long-day onion varieties when day length reaches 14 to 16 hours. Weed competition before bulb development slows leaf production, which reduces bulb size at harvest. Weeds uncontrolled in the onion row at emergence and 2 weeks after emergence resulted in complete loss of the onion crop (Wicks et al., 1973). Bond and Burston (1996) concluded that optimum time to control weeds varied from 21 to 56 d after 50% crop emergence, but single and multiple hand-weeding did not consistently prevent yield losses.

Herbicides applied once, either preemergence or postemergence, are not sufficient for season-long broadleaf weed control and adequate onion yields (Ghosheh, 2004). The long season needed to grow large-diameter onion allows for successive flushes of weeds, which makes consecutive weed control activities necessary. Additionally, most herbicides cannot be applied to onion until the two-true-leaf stage due to label restrictions.

8. Herbicide micro-rate introduction

Micro-rate herbicide treatments in onion were developed from the pioneering research of North Dakota State University and University of Minnesota extension specialist Dr. Alan G. Dexter in sugarbeet (*Beta vulgaris*) (Woznica et al., 2004). The micro-rate program uses herbicides applied at reduced rates approximately 50 to 75% compared to recommended rates and reapplied three to five times at 5- to 7-day intervals (Zollinger et al., 2008). Smaller broadleaf weeds were easier to control and required less herbicide to control with the micro-rate program. Also, crop safety increased and less herbicide per area per season was used. Multiple applications also widened the application window allowing the grower to control multiple weed flushes. In addition, the micro-rate program increased the economic return from the purchase of less herbicide (Dale, 2000).

8.1 Micro-rate evaluation on onion

Early season similarities in establishment and herbicide sensitivity for sugarbeet and onion suggested that the micro-rate program may be adapted to onion. Initial testing occurred in the greenhouse to evaluate any postemergence herbicide with activity on annual broadleaf weeds. Reduced rate applications were made to onion, common lambsquarters, and redroot pigweed in the cotyledon to first-true-leaf stage. Any herbicide that caused severe injury to onion was eliminated. Herbicides for field testing were narrowed to four: acifluorfen, bromoxynil, metribuzin, and oxyfluorfen at 0.25X, 0.13X, and 0.06X, where "X" was the lowest labeled herbicide rate, with either two or three sequential applications at a 7-day interval. Initial applications were made when broadleaf weeds reached the first-true-leaf stage. Depending on the year and location, onion were not emerged, in the loop stage, or in the flag-leaf stage when the first micro-rate application was made. A hand-weeded control and grower standard practice were included for comparison.

The grower standard practice consisted of a preemergence application of DCPA immediately following planting and a postemergence application of bromoxynil and oxyfluorfen at the onion two-leaf stage. Dimethamid-P, bromoxynil and oxyfluorfen were applied to the entire study at the onion five-leaf stage. Best management practices were used for planting, fertility, irrigation, and pest control, and were identical for all plots at each location. Weekly weed counts were taken to evaluate weed control compared to the conventional herbicide standard and the hand-weeded check. A visual evaluation was taken approximately 2 weeks after the standard application to evaluate mid-season weed control using a 0 to 100% scale, where 0 is equal to no visible injury or no control and 100 equal to complete kill.

8.2 Weed control evaluations

The high rate of bromoxynil (70 g/ha) applied twice or three times provided the greatest early season control of common lambsquarters (Table 4) (Loken and Hatterman-Valenti, 2010). The similar control between two and three weekly applications suggested that ideally, the producer would have seen that there wasn't an additional weed flush after the second week, thus would not have made the third application.

The high rate of bromoxynil and the high rate of oxyfluorfen provided the greatest early season control of redroot pigweed (Table 4). Three sequential micro-rate applications provided greater control than two sequential micro-rate applications averaged over all herbicides (data not shown). Some redroot pigweed continued to emerge after the last micro-rate application, suggesting that additional micro-rate applications should be considered to control later flushes and may be used to replace a standard bromoxynil plus oxyfluorfen application.

Onion injury was not observed the first year, but in the second year, onion treated with oxyfluorfen displayed approximately 15% injury. These seedling plants (one- to two-leaf) had leaves that were constricted at the soil surface. Constricted leaves occasionally resulted in onion seedling death, but most plants initiated the next true leaf after injury and outgrew the symptoms. Slight injury (approximately 5%) was noticed from bromoxynil at these locations, and all plants outgrew the injury symptoms. Environmental conditions may have contributed to this injury because the average daily temperatures from April to May during the second year were 1.7 °C cooler, with numerous cloudy days that may have enhanced herbicide injury.

		Common lambsquarters		Redroot pigweed	
Herbicide	Rate	Two[a]	Three	Two	Three
	g/ha	%			
Bromoxynil	18	35[c]	49	43	66
Bromoxynil	35	60	82	66	89
Bromoxynil	70	92	99	89	97
Oxyfluorfen	18	31	44	38	49
Oxyfluorfen	35	49	71	55	81
Oxyfluorfen	70	69	78	75	95
Metribuzin	5	38	35	34	38
Metribuzin	10	31	41	34	47
Metribuzin	21	38	54	48	63
Acifluorfen	18	32	32	31	44
Acifluorfen	35	30	38	32	54
Acifluorfen	70	43	56	62	76
DCPA[b]		79		58	
Hand weeded		100		100	
LSD (0.05)		13		12	

[a] Two and three refer to the number of applications in the micro-rate system.
[b] Conventional herbicide management check, DCPA (preemergence) at 11 kg/ha, bromoxynil (postemergence) at 280 g/ha, and oxyfluorfen (postemergence) at 1,120 g/ha.
[c] Visual estimates of weed control using a 0 to 100% scale, where 0 is equal to no visible control and 100 equal to complete kill.

Table 4. Effect of micro-rate herbicide treatments averaged across five locations on common lambsquarters and redroot pigweed percent control 2 weeks after the standard herbicide application to onion at the two-leaf stage. Adapted from Loken and Hatterman-Valenti, 2010.

Onion total yield generally mimicked weed control data, with the greatest total yield from those treatments that provided the greatest early-season broadleaf weed control, namely the three weekly herbicide applications (data not shown). Onion treated with oxyfluorfen (high rate, three applications) had the greatest large-grade and total yield, although total yield was similar to the yield with DCPA (Table 5) (Loken and Hatterman-Valenti, 2010). There was an herbicide by environment interaction for large-grade onion yield, which was attributed to the yield fluctuations in bromoxynil treatments due to common purslane (*Portulaca oleracea*) competition. Common purslane was present at two of the five locations. Bromoxynil does not control common purslane, therefore, at these locations; large-grade onion yield went from comparable or greater yields in comparison with oxyfluorfen treatments to significantly lower yields.

		Cull[a]		Small		Medium		Large		Total	
Herbicide	Rate	Two[b]	Three	Two	Three	Two	Three	Two	Three	Two	Three
	g/ha	t/ha									
Bromoxynil	18	0.1	0.2	3.8	4.5	4.2	3.8	1.9	0.8	9.9	9.3
Bromoxynil	35	0.2	0.1	4.0	4.3	4.3	7.5	1.1	4.7	9.6	17.0
Bromoxynil	70	0.0	0.1	3.9	4.0	10.0	9.9	5.2	6.0	20.0	20.0
Oxyfluorfen	18	0.2	0.3	3.4	3.3	5.6	6.6	1.5	4.8	11.0	15.0
Oxyfluorfen	35	0.2	0.2	3.9	3.8	7.7	9.3	6.2	13.0	18.0	26.0
Oxyfluorfen	70	0.1	0.1	3.7	3.0	8.2	9.1	11.0	18.0	23.0	31.0
Metribuzin	5	0.3	0.1	2.7	2.7	1.4	2.8	0.0	0.3	4.5	5.8
Metribuzin	10	0.2	0.3	2.3	3.8	2.3	3.3	0.9	0.7	5.6	8.1
Metribuzin	21	0.2	0.2	2.5	3.8	3.8	5.1	5.6	4.9	12.0	14.0
Acifluorfen	18	0.3	0.2	2.4	3.2	2.1	3.4	0.3	0.6	5.0	7.4
Acifluorfen	35	0.2	0.2	3.3	4.3	3.1	5.2	0.2	0.6	6.9	10.0
Acifluorfen	70	0.4	0.1	3.9	4.1	7.4	7.6	5.2	6.7	17.0	19.0
DCPA[c]		0.2		3.4		11.0		13.8		28.0	
Handweeded		0.1		4.7		12.0		9.9		26.0	
LSD (0.05)		0.2		NS		2.6		3.3		4.8	

[a] Cull: split or diseased bulbs, small: bulb diameter less than 2.5 cm, medium: bulb diameter 2.5-5.7 cm, large: bulb diameter greater than 7.6 cm.

[b] Two and three refer to the number of applications in the micro-rate system.

[c] Conventional herbicide management check, DCPA (preemergence) at 11 kg/ha, bromoxynil (postemergence) at 280 g/ha, and oxyfluorfen (postemergence) at 1,120 g/ha.

Table 5. Effect of micro-rate herbicide treatments averaged across five locations on cull-, small-, medium-, large-grade, and total onion yield. Adapted from Loken and Hatterman-Valenti, 2010.

9. Strip-tillage in onion

The use of strip-tillage has also been investigated to eliminate the use of a companion crop and the postemergence herbicide application to kill the companion crop. Strip-tillage, leaving wheat stubble between rows, was compared to conventional tillage with two preemergence herbicide treatments and two micro-rate herbicide treatments for two years. The remaining wheat stubble provided the needed structure to reduce wind erosion and the untilled area between onion rows may have reduced hairy nightshade (*Solanum sarachoides*) emergence (Table 6) (Gegner, 2009). Peachey et al. (2006) found at least an 88%

reduction in hairy nightshade populations when spring tillage was eliminated. In addition, the micro-rate treatments provided better hairy nightshade control. Micro-rate treatments also controlled early-season common lambsquarters and redroot pigweed (data not shown). A standard, mid-season herbicide application of bromoxynil and oxyfluorfen at 280.4 g/ha and 1120 g/ha, respectively, when onion reached the three-leaf stage, controlled mid-season, broadleaf weeds as well as many of the broadleaf weed escapes from the preemergence herbicide treatments, and resulted in no yield differences. However, in one of the two years, greater large-diameter onion, marketable onion, and total onion yields occurred with strip-tillage compared to conventional tillage (Table 7) (Gegner, 2009).

Herbicide		2007	2008
1WA2A		------------- plants/m²	----------
	Pendimethalin	100 a[c]	53 b[c]
	DCPA	50 b	132 a
	Oxyfluorfen	0 c	4 c
	Bromoxynil	3 c	11 c
1WA4A			
	Pendimethalin	0 a	33 b
	DCPA	0 a	50 a
	Oxyfluorfen	1 a	0 c
	Bromoxynil	1 a	0 c

[a] 1WA2A, 1 week after 2nd micro-rate herbicide application.
[b] 1WA4A, 1 week after 4th micro-rate herbicide application.
[c] Means for each application timing and year followed by the same letter are not significantly different according to Fisher's Protected LSD (0.05).

Table 6. Effect of herbicide on hairy nightshade density 1WA2A[a] and 1WA4A[b] at Oakes, ND, during 2007 and 2008. Adapted from Gegner, 2009.

Location	Small[a]	Medium	Large	Total	Marketable
Oakes 2007	----------	----------	t/ha	----------	----------
Strip-till	10.8 a[b]	32.8 a	49.3 a	93.0 a	82.1 a
Conventional	14.3 a	33.1 a	31.8 b	79.2 b	65.0 b
Oakes 2008					
Strip-till	7.6 b	25.6 a	25.3 a	58.4 a	50.8 a
Conventional	10.2 a	24.6 a	22.2 a	56.9 a	46.8 a

[a] Small: bulb diameter less than 2.5 cm, medium: bulb diameter 2.5-5.7 cm, large: bulb diameter greater than 7.6 cm.
[b] Means within each column and year followed by the same letter are not significantly different according to Fisher's Protected LSD (0.05).

Table 7. Effect of tillage on onion grade, total yield, and total marketable yield at Oakes during 2007 and 2008. Adapted from Gegner, 2009.

10. Conclusions

These results and other research conducted at North Dakota State University have shown that bromoxynil or oxyfluorfen applied at micro-rates can provide early-season annual broadleaf weed (common lambsquarters and redroot pigweed) control in onion and potentially replace the use of DCPA. The use of micro-rates also reduces the amount of bromoxynil and oxyfluorfen applied to onion. Conservation research results suggest that strip-tillage and bromoxynil or oxyfluorfen applied as micro-rates may be used to eliminate the use of a companion crop and further reduce the amount of herbicides applied to a noncompetitive crop such as onion without sacrificing yield or increasing weed numbers the following year. There is a continuous research effort to investigate ways to further reduce herbicide inputs in a noncompetitive crop such as onion. It is anticipated that adjuvant use and/or tank-mixing herbicide micro-rates would allow even lower herbicide rates and further reduce herbicide inputs when growing onion.

11. References

Blackshaw R.E., H.J. Beckie, L.J. Molnar, T. Entz, and J.R. Moyer. 2005a. Combining agronomic practices and herbicides improves weed management in wheat-canola rotations within zero-tillage production systems. *Weed Science,* 53:528-535.

Blackshaw R.E., J.R. Moyer, K.N. Harker, and G.W. Clayton. 2005b. Integration of agronomic practices and herbicides for sustainable weed management in a zero-till barley field pea rotation. *Weed Technology,* 19:190–196.

Blackshaw, R.T.E., F.T. O'Donovan, K.N. Harker, G.W. Clayton, and R.N. Stougaard. 2006. Reduced herbicide doses in field crops: a review. *Weed Biology and Management,* 6:10-17.

Bond, W. and S. Burston. 1996. Timing of removal of weeds from drilled salad onions to prevent crop losses. *Crop Protection,* 15:205-211.

Boydston, R.A. and M.D. Seymour. 2002. Volunteer potato (*Solanum tuberosum*) control with herbicides and cultivation in onion (*Allium cepa*). *Weed Technology,* 16:620–626.

Cousens, R.D. and S. Mokhtari. 1998. Seasonal and site variability in the tolerance of wheat cultivars to interference from *Lolium rigidum*. *Weed Research,* 38:301-307.

Dale, T.M. 2000. Application method and adjuvant effects on low-dose postemergence herbicide efficacy in sugarbeet (*Beta vulgaris*). MS thesis. Fargo, ND: North Dakota State University. 57 p.

Derksen, D.A., R.L. Anderson, R.E. Blackshaw, and B. Maxwell. 2002. Weed dynamics and management strategies from cropping systems in the northern Great Plains. *Agronomy Journal,* 94:174-185.

Donald, C.M. and J. Hamblin. 1976. The biological yield and harvest index of cereals as agronomic and plant breeding criteria. *Advances in Agronomy,* 28:361-405.

Doyle, P. and M. Stypa. 2004. Reduced herbicide rates - a Canadian perspective. *Weed Technology,* 18:1157-1165.

Dunan, C.M., P. Westra, and F.D. Moore. 1999. A plant process economic model for weed management decisions in irrigated onion. *Journal of the American Society of Horticulture Science,* 124:99-105.

English, L.J. and L.R. Oliver. 1981. Influence of sicklepod (*Cassia obtusifolia*) density on plant growth and seed production. *Proceedings of the Southern Weed Science Society,* 34: 250.

Foder, L., E. Lehoczky, E. Fodorne Feher, P. Nagay, and O. Palmai. 2008. Crop competiveness influenced by seeding dates and top-dress nitrogen rates. *Communications in Agricultural and Applied Biological Sciences*, 73:945-950.

Gegner, S.L. 2009. Effect of reduced tillage systems on sugarbeet (*Beta vulgaris* L.) and onion (*Allium cepa* L.). M.S. thesis. North Dakota State University, Fargo, ND. 93 p.

Greenland, R.G. 2000. Optimum height at which to kill barley used as a living mulch in onions. *HortScience*, 35:853-855.

Ghosheh, H.Z. 2004. Single herbicide treatments for control of broadleaved weeds in onion (*Allium cepa*). *Crop Protection*, 23:539-542.

Grichar, W.J. 2007. Row spacing, plant populations, and cultivar effects on soybean production along the Texas gulf coast. In: *Crop Management*, 10.08.2011, Available from http://www.plantmanagementnetwork.org/pub/cm/research/2007/gulfsoy/default.asp.

Grichar, W.J., F.A. Besler, R.G. Lemon, and D.J. Pigg. 2004. Comparison of conventional and twin-row production systems on cotton growth and development. *Proceedings of the Beltwide Cotton Conference*, p. 2119-2203.

Hartwig, N.L. and H.U. Ammon. 2002. Cover crops and living mulches. *Weed Science*, 50:688-699.

Hatterman-Valenti, H.M. and P.E. Hendrickson. 2006. Companion crop and planting configuration effect on onion. *HortTechnology*, 16:12-15.

Knezevic, S.Z., S.P. Evans, E.E. Blankenship, R.C. VanAcker, and J.L. Linquist. 2002. Critical period of weed control: the concept and data analysis. *Weed Science*, 50:773-786.

Lemerle, D., G.S. Gill, C.E. Murphy, S.R. Walker, R.D. Cousens, and S. Mokhtari. 2001. Genetic improvement and agronomy for enhanced wheat competiveness with weeds. *Australian Journal of Agricultural Research*, 52:527-548.

Liebman, M. and E.R. Gallandt. 1997. Ecological management of crop-weed interactions, p. 291-343. In: L.E. Jackson (ed.). *Ecology in Agriculture*, Academic Press, ISBN 0-12-378260-0, San Diego, CA.

Loken, J.R. and H.M. Hatterman-Valenti. 2010. Multiple applications of reduced-rate herbicides for weed control in onion. *Weed Technology*, 24:153-159.

Mohler, C.L. 2001. Enhancing the competitive ability of crops, p. 269-322. In: M. Liebman, C.L. Mohler, and C.P. Staver (eds.). *Ecological Management of Agricultural Weeds*. Cambridge University Press, ISBN 0521560683, Cambridge.

Mulugeta, D. and D.E. Stoltenberg. 1997. Weed and seedbank management with integrated methods as influenced by tillage. *Weed Science*, 45:706-715.

Padgitt, M., D. Newton, R. Penn, and C. Sandretto. 2000. Production Practices for Major Crops in U. S. Agriculture, 1990-1997. Resource Economics Division, Economic Research Service, *U. S. Department of Agriculture, Statistical Bulletin NO. 969*. 104 p.

Peachey, B.E., R.D. William, and C. Mallory-Smith. 2006. Effect of spring tillage sequence on summer annual weeds in vegetable row crop rotations. *Weed Technology*, 20:204-214.

Peel, M. 1998. Crop rotations for increased productivity. Fargo, ND: North Dakota State University Extension Service EB-48, 10-008-2011, Available from: http://www.ag.ndsu.edu/pubs/plantsci/crops/eb48-1.htm#intro.

Pike D.R., M.D. McGlamery, and E.L. Knake. 1991. A case study of herbicide use. *Weed Technology*, 5:639–646.

Stougaard, R.N. and Q. Xue. 2004. Spring wheat seed size and seeding rate effects on yield loss due to wild oat (*Avena fatua*) interference. *Weed Science*, 52:133-141.

Swaider, J.M., G.W. Ware, and J.P. McCollum, eds. 1992. Producing Vegetable Crops. 4th ed. p. 381-404, Interstate Publishers, Inc., ISBN 9780813429038, Danville, IL.

Swanton, C.J., K.J. Mahoney, K. Chandler, and R.H. Gulden. 2008. Integrated weed management: knowledge-based weed management systems. *Weed Science*, 56:168-172.

Tharpe, B.E. and J.J. Kells. 2001. Effect of glufosinate-resistant corn (*Zea mays*) population and row spacing on light interception, corn yield, and common lambsquarters (*Chenopodium album*) growth. *Weed Technology*, 15.414-418.

Upadhyay, B.M., E.G. Smith, G.W. Clayton, K.N. Harker and R.E. Blackshaw. 2006. Economics of integrated weed management in herbicide-resistant canola. *Weed Science*, 54:138-147.

Van Acker, R.C., D.A. Derksen, M.H. Entz, G. Martens, T. Andrews, and O. Nazarko. 2001. Pesticide - free production (PFP): an idea drawing farmers to implement integrated pest management. *Proceedings of the Brighton Crop Protection Conference – Weeds*. p. 269-276.

Wicks, G.A., D.N. Johnston, D.S. Nuland, and E.J. Kinbacher. 1973. Competition between annual weeds and sweet Spanish onions. *Weed Science*, 21:436-439.

Willingham, S.D., B.J. Brecke, J. Treadaway-Ducar, and G.E. MacDonald. 2008. *Weed Technology*, 22:74-80.

Woznica, Z., K. Adamczewski, and E. Szelezniak. 2004. Application of herbicide micro-rates in sugar beet production. *Progress in Plant Protection*, 44:523–530.

Zollinger, R., ed. 2008. 2008 North Dakota Weed Control Guide. Fargo, ND, p. 135, North Dakota State University Extension Service Rep. W-253.

8

Microbial Weed Control and Microbial Herbicides

Tami L. Stubbs and Ann C. Kennedy
Washington State University and United States Department of Agriculture – Agricultural Research Service
United States of America

1. Introduction

Microbial weed control represents an innovative means to manage troublesome weeds and utilize the naturally occurring biological herbicides produced by soil microorganisms. These compounds kill or hinder the growth of weeds so that beneficial plant species can gain a competitive advantage. The vast diversity of microorganisms in our environment is largely untapped, and the potential discovery and characterization of these microbial compounds represents an opportunity to complement chemical herbicides, or reduce the potential for erosion or soil degradation due to tillage for weed control. Invasive weeds continue to threaten the productivity of agricultural lands and natural areas; however, for many weeds adequate, cost-effective control measures presently are not available (Jones & Sforza, 2007). Discovery of biological controls for invasive plants represents an alternative way to slow the spread of these weeds using natural enemies (Jones & Sforza, 2007). Further advances in microbial genetics will continue to improve our understanding of the wealth of genetic diversity and potential in the soil and to better use plant-microbe interactions. The development of biocontrol agents would lessen the need for chemical herbicides and provide greater options for weed management. Microbes have a place in integrated, ecologically based weed management and their potential is only just being realized.

The concept of utilizing microbial herbicides has been explored for more than a quarter century, but there remain many challenges to overcome before they can be widely used in agricultural, range and forest lands, or waterways. Those challenges include improving the efficacy of the microbial activity, survival of microorganisms, persistence of the suppressive compound, delivery systems, determining host range, and avoiding injury to non-target organisms. Other considerations are interactions with chemical herbicides, regulations, commercialization and mass production, and economic feasibility.

Biological controls for weeds can generally be divided into one of three general types: classical, augmentative or inundative, and cultural. The classical approach involves the introduction of a control agent into an area where it did not previously exist, and where the agent eventually becomes able to sustain itself. An example of this is the release of *Puccinia chondrillina* to control *Chondrilla juncea* (rush skeletonweed; Barton, 2004). Augmentative or inundative biological control refers to repeated application of a foreign agent with the intent to reduce weed densities to a level where beneficial plant species can compete. An example of an augmentative control is *Colletotrichum gloeosporioides* for control of sicklepod (*Senna obtusifolia* L.; Boyette, 2006). Cultural weed control might include crop rotation, fallow periods, sanitation to prevent the introduction and spread of weed seeds, and maintaining soil fertility to produce healthy crop plants. The successful history of insects in weed control and the ever increasing list of successful insect biological controls placed in the hands of land managers offers hope for microbial weed control. Use of biological herbicides requires a shift in thinking from the use of chemical herbicides, as biological controls will most likely not eliminate a weed problem as quickly or as thoroughly as some herbicides on an annual basis; the goal is to inhibit the weed pest below an economically damaging threshold over a long time period in order for beneficial species to gain a competitive advantage (Ghosheh, 2005). Among the criteria that Charudattan (2005) lists for determining which invasive plants make suitable candidates for microbial herbicides are those that have a number of available pathogens that might be suitable for biological control, and where the cost of using that control will be competitive with other control measures. There are a number of reasons for developing microbial herbicides, and those include the potential for herbicide resistant weeds; chemical herbicides may persist in soil for longer than one growing cycle, thus limiting options for crop rotations; there may be limitations on herbicide registrations for certain crops; there is the potential for fewer undesirable effects to the environment than from chemical herbicides; and finally there is the potential for injury to non-target organisms (Guske et al., 2004). Special considerations are needed for management of parasitic weeds (Sauerborn et al., 2007). Herbicide use to control parasitic weeds is difficult, because the greatest damage to the host plant has often occurred before the parasitic weed emerges above the soil surface. Another challenge is that parasitic weeds are closely related to their hosts, and selective control with herbicides is difficult (Sauerborn et al., 2007).

The various regions of the world may be plagued by specific invasive weed problems, while other weeds are problematic worldwide. No matter whether they are worldwide problems or regional challenges, many are elusive to management efforts. These weeds vary by region and ecosystem, and several microorganisms have been studied or are under development in greenhouse or field studies as potential sources for microbial herbicides (Table 1A and B).

Weed Pest	Ecosystem	Region	Biocontrol agent	Reference
Abutilon theophrasti (Velvetleaf)	Croplands	U.S., Canada	*Fusarium oxysporum*; *Colletotrichum coccodes*; *Pseudomonas putida* & *Acidovorax delafieldii*	Kremer & Schulte, 1989; Wymore & Watson, 1989; Owen & Zdor, 2001
Ailanthus altissima (Tree-of-heaven)	Native forests, urban areas	North America, Europe	*Aecidium ailanthi* J.Y. Zhuang sp. nov.; *Coleosporium* sp.; *Fusarium oxysporum* f. sp. *perniciosum*; *Verticillium albo-atrum*	Review by Ding et al., 2006; Schall & Davis, 2009
Alternanthera philoxeroides (Alligatorweed)	Aquatic, upland sites	Worldwide	*Nimbya alternantherae*	Pomella et al., 2007
Amaranthus spp. (Amaranthus)	Croplands	Europe, North America	*Phomopsis amaranthicola*	Rosskopf et al., 2006
Amaranthus retroflexus (Redroot pigweed)	Croplands	Europe, North America	*Alternaria alternata*	Lawrie et al., 2002a

Weed Pest	Ecosystem	Region	Biocontrol agent	Reference
Amaranthus hybridus (Pigweed), *Senna obtusifolia* (Sicklepod), *Crotalaria spectabilis* (Showy crotalaria)	Croplands	North America	*Phomopsis amaranthicola, Alternaria cassia, Colletotrichum dematium, Fusarium udum*	Chandramohan & Charudattan, 2003
Arceuthobium tsugense (Hemlock dwarf mistletoe)	Coniferous forests	Vancouver Island, Canada	*Neonectria neomacrospora*	Rietman et al., 2005
Brunnichia ovata (Redvine), *Campis radicans* (Trumpetcreeper), *Pueraria lobata* (Kudzu)	Croplands, wastelands, natural areas	Southern U.S.	*Myrothecium verrucaria*	Boyette et al., 2006, 2008a
Cannabis sativa (Ditchweed)	Croplands, grasslands	Kazakhstan	*Fusarium oxysporum* f. sp. *cannabis*	Tiourebaev et al., 2001
Carduus pycnocephalus (Italian thistle)	Grasslands, woodlands	Tunisia	*Puccinia carduorum*	Mejri et al., 2010a
Centaurea diffusa (Diffuse knapweed), *Centaurea maculosa* (Spotted knapweed)	Rangelands	Western U.S., Canada	*Fusarium* spp.	Caesar et al., 2002
Chenopodium album (Common lambsquarters)	Croplands	Worldwide	*Ascochyta caulina*	Ghorbani et al., 2002; Vurro et al., 2001
Chrysanthemoides monilifera ssp. *monilifera* (Boneseed)	Natural areas	Southeastern Australia	*Endophyllum osteospermi*	Wood & Crous, 2005
Cirsium arvense (Canada thistle)	Croplands, rangelands, pastures, roadsides	Temperate regions of northern hemisphere	*Phyllosticta cirsii; Stagonospora cirsii; Alternaria cirsinoxia; Pseudomonas syringae* pv. *tagetis;* Mix of *Phoma destructiva, Phoma hedericola, Mycelia sterila, Phoma nebulosa, Phomopsis cirsii*	Evidente et al., 2008; Yuzikhin et al., 2007; Bailey, 2004; Gronwald et al., 2002; Tichich & Doll, 2006; Guske et al., 2004; Leth et al., 2008
Cirsium arvense (Canada thistle), *Ranunculus acris* (Tall buttercup)	Pasture	New Zealand	*Sclerotinia sclerotiorum*	Bourdot et al., 2006a
Damasonium minus (Starfruit)	Croplands (rice), aquatic	Australia	*Plectosporium alismatis*	Jahromi, 2007
Eichhornia crassipes (Water hyacinth)	Aquatic	Tropical, sub-tropical regions	*Alternaria eichhorniae* isolate 5; *Cercospora piaropi*	Shabana & Mohamed, 2005; Shabana, 2005; Tessman et al., 2008
Euphorbia esula/virgata (Leafy spurge)	Rangelands, natural areas	North America	*Uromyces scutellatus*	Caesar, 2006

Weed Pest	Ecosystem	Region	Biocontrol agent	Reference
Euphorbia heterophylla (Wild poinsettia)	Croplands (soybean)	Brazil	*Sphaceloma poinsettiae*	Nechet et al., 2004
Galium spurium (False cleavers)	Croplands	Canada	*Plectosporium tabacinum*	Zhang et al., 2002
Gaultheria shallon (Salal)	Native forests	Canadian and American Pacific Coast	*Phoma exigua; Valdensinia heterodoxa*	Zhao & Shamoun, 2006; Vogelgsang & Shamoun, 2004; Wilkin et al., 2005
Hydrilla verticillata (Hydrilla)	Aquatic	U.S. and worldwide	*Fusarium culmorum; Mycoleptodiscus terrestris*	Shabana et al., 2003; Shearer & Jackson, 2006
Isatis tinctoria (Dyer's woad)	Natural areas	Western North America	*Puccinia thlaspeos*	Kropp et al., 2002; Kropp & Darrow, 2006
Lantana camara (Lantana)	Natural and cultivated lands	South Africa	*Mycovellosiella lantanae* var. *lantanae; Corynespora cassiicola* f. sp. *lantanae*	Den Breeyen & Morris, 2003; Pereira et al., 2003
Matricaria perforata (Scentless chamomile)	Croplands	Canada	*Colletotrichum truncatum*	Graham et al., 2006, 2007; Peng et al., 2005a
Miconia calvescens (Velvet tree; miconia)	Natural areas	Hawaii and French Polynesia	*Ditylenchus drepanocercus*	Seixas et al., 2004
Nasella neesiana (Chilean needle grass)	Pastures, grasslands	Australia, New Zealand	*Uromyces pencanus*	Anderson et al., 2010
Orobanche cumana (Sunflower broomrape)	Croplands	Mediterranean region, southeast Europe	*Fusarium oxysporum* f. sp. *orthoceras*	Müller-Stöver & Sauerborn, 2007
Orobanche ramosa (Broomrape; branched broomrape)	Croplands	Central & western Europe	*Fusarium oxysporum* (FOG); *Fusarium* spp.	Müller-Stöver et al., 2009; Kohlschmid et al., 2009; Boari & Vurro, 2004
Orobanche aegyptiaca (Egyptian broomrape)	Croplands	India, Israel	*Fusarium solani*	Sharma et al., 2011; Dor & Hershenhorn, 2009
Orobanche crenata, Orobanche foetida (Broomrape)	Croplands	Northern Tunisia	*Pseudomonas fluorescens* Bf7-9	Zermane et al., 2007
Papaver somniferum (Opium poppy)	Illicit plants		*Pleospora papaveracea*	Bailey et al., 2004
Portulaca oleracea (Common purslane), *Trianthema portulacastrum* (Horse purslane), *Euphorbia maculata* (Spotted spurge), *Euphorbia supina* (Prostrate spurge)	Tomato fields	Southeastern U.S.	*Myrothecium verrucaria*	Boyette et al., 2007a
Raphanus raphanistrum (Wild radish)	Croplands, vineyards	Australia	*Hyaloperonospora parasitica; Pseudomonas fluorescens*	Maxwell & Scott, 2008; Flores-Vargas & O'Hara, 2006

Weed Pest	Ecosystem	Region	Biocontrol agent	Reference
Salsola kali (Russian thistle)	Croplands, pastures, rangelands	Western U.S., Eurasia	*Uromyces salsolae*	Hasan et al., 2001
Schinus terebinthifolius (Brazilian peppertree)	Forests	Florida, U.S.	*Neofusicoccum batangarum*	Shetty et al., 2011
Senecio vulgaris (Common groundsel)	Croplands (carrots)	Switzerland	*Puccinia lagenophorae*	Frantzen & Müller-Scharer, 2006
Senna obtusifolia (Sicklepod)	Croplands	Southeastern U.S.	*Colletotrichum gloeosporioides*	Boyette et al., 2007b
Sesbania exaltata (Hemp sesbania)	Croplands	Southern U.S.	*Colletotrichum truncatum*	Boyette et al., 2007c, 2008b
Sonchus arvensis (Perennial sowthistle)	Croplands	North America, Europe	*Alternaria sonchi*	Evidente et al., 2009a
Striga hermonthica (Striga, Witchweed)	Croplands	Africa	*Fusarium oxysporum* (Foxy 2; PSM 197); *F. oxysporum* (4-3-B); *F. nygamai; Pseudomonas fluorescens; P. putida*	Venne et al., 2009; Yonli et al., 2004; Ahonsi et al., 2002
Taraxacum officinale (Dandelion)	Turfgrass	Worldwide	*Phoma herbarum; Phoma macrostoma; Sclerotinia minor*	Stewart-Wade & Boland, 2004; Zhou et al., 2004; Abu-Dieyeh & Watson, 2006, 2007
Ulex europaeus (Gorse)	Native areas, pastures, forests	New Zealand	*Chondrostereum purpureum; Fusarium tumidum*	Bourdot et al., 2006b; Yamoah et al., 2008
Xanthium strumarium (Common cocklebur)	Croplands	Southern U.S.	*Alternaria helianthi*	Abbas et al., 2004
Xanthium occidentale (Noogoora burr)	Croplands, rangelands	Australia	*Puccinia xanthii*	VanKlinken & Julien, 2003

Table 1A. List of current biological control research projects on dicotyledonous weed species that have shown promise in greenhouse and/or field trials, including the invasive weed species, ecosystem, region of importance, biological control agent and reference.

Biological herbicides represent a means to reduce dependence on synthetic herbicides; focus on ecologically grounded methods of management; reduce weed seed bank populations through environmentally friendly practices; and potentially reduce costs of weed control in crop production, rangeland restoration, forestry and aquatic systems (Bailey et al., 2010; Kennedy & Stubbs, 2007).

2. Challenges for weed control with microbial herbicides

The list of challenges in developing a successful microbial herbicide is long. Once a potential biological control agent is identified, the first challenge lies in reproducing lab and/or greenhouse results successfully in the field. The potential biological control agent and the toxin responsible for weed inhibition must be able to survive the harsh, and often unpredictable, environmental conditions that exist in the field. Li & Kremer (2006) showed inhibition of several weed species using rhizobacteria isolated from various weed hosts in

Weed Pest	Ecosystem	Region	Biocontrol agent	Reference
Aegilops cylindrica (Jointed goatgrass)	Croplands	Western U.S.	*Pseudomonas* spp.; *Xanthomonas* spp.	Kennedy & Stubbs, 2007
Avena fatua (Wild oat)	Cereal crops, croplands, native areas	North America, Europe, Australia	*Fusarium avenaceum; F. culmorum; Drechslera avenacea* (conidial state of *Pyrenophora chaetomioides); Puccinia coronate* f. sp. *avenae*	deLuna et al., 2011; Ghajar et al., 2006; Carsten et al., 2001
Bromus diandrus (Great brome)	Croplands	Tunisia	*Pseudomonas trivialis strain* X33D	Mejri et al., 2010b
Bromus tectorum (Downy brome; cheatgrass)	Croplands, rangelands, natural areas	Western North America	*Pseudomonas fluorescens strain* D7; *Pyrenophora semeniperda; Ustilago bullata*	Kennedy et al., 1991, 2001; Meyer et al., 2001, 2007
Digitaria sanguinalis (Giant crabgrass; large crabgrass)	Croplands	China	*Curvularia eragrostidis* QZ-2000; *C. intermedia*	Zhu & Qiang, 2004; Jiang et al., 2008; Tilley & Walker, 2002
Elytrigia repens (Quackgrass)	Croplands	Temperate regions of northern & southern hemisphere	*Ascochyta agropyrina* var. *nana*	Evidente et al., 2009b
(Grasses)	Croplands	Australia	*Pyrenophora semeniperda*	Medd & Campbell, 2005
Microstegium vimineum (Japanese stiltgrass)	Forests	Eastern U.S.	*Bipolaris* sp.	Kleczewski & Flory, 2010
Poa annua (Annual bluegrass)	Turfgrass	Japan	*Xanthomonas campestris* pv. *poae* (JT-P482)	Imaizumi et al., 1999
Setaria viridis (Green foxtail)	Croplands	Worldwide	*Drechslera gigantea; Exserohilum rostratum; E. longirostratum; Pyricularia setariae*	Casella et al., 2010; Green et al., 2004

Table 1B. List of current biological control research projects on monocotyledonous weed species that have shown promise in greenhouse and/or field trials, including the invasive weed species, ecosystem, region of importance, biological control agent and reference.

greenhouse studies. They noted the importance of greenhouse studies as a step toward identifying isolates that may be suitable for testing under the more variable conditions of the field, where there is greater competition from indigenous organisms and unpredictable environmental factors. In any case, the conditions under which a microorganism best survives depend on the microorganism itself and it cannot be assumed to be the same for all biocontrol agents. Cool, moist conditions are required for survival of the deleterious rhizobacteria (DRB) used to control jointed goatgrass (*Aegilops cylindrica*) and downy brome (*Bromus tectorum*) in the field (Kennedy & Stubbs, 2007). For greatest biocontrol success, field application of organisms should be timed when rains are expected. On the other hand, Boguena et al. (2007) found that extremely cold temperatures reduced the ability of *Ustilago bullata* to infect downy brome and may limit its use as a biological control. One challenge to the development of fungal bioherbicides is the inability to infect the weed pest without a period of free water retention (Chittick & Auld, 2001). Favorable moisture and temperature conditions are critical to the efficacy of many mycoherbicides. Tichich et al. (2006) found that for optimum population growth of the pathogen *Pseudomonas syringae* pv. *tagetis* targeting Canada thistle (*Cirsium arvense*), periods of wet weather were required. This moisture requirement often limits candidates for biocontrol. After investigating the dew period requirements of three fungal pathogens that infected green foxtail (*Setaria viridis)*, Peng & Boyetchko (2006) found that *Drechslera gigantea* was the best choice for control of green foxtail because this pathogen was the most virulent and it did not require a specific dew temperature for efficacy.

For successful biocontrol, the microorganism's growth stage needs to be matched with the period of greatest vulnerability of the weed. Studies are needed to ensure that the most virulent stage of biocontrol agent growth also coincides with the susceptible host habit and an active growth period of the host. The efficacy of *Plectosporium alismatis* to reduce starfruit (*Damasonium minus* (R. Br.) Buch) in rice was increased by using conidia and chlamydospores (Cliquet & Zeeshan, 2008) and applying the biocontrol agent on juvenile rather than older starfruit (Jahromi, 2007). Qiang et al. (2006) found that the mycelia of *Alternaria alternata* strain 501 were able to infect the host plant, *Eupatorium adenophorum,* in a much shorter time than with conidia. The *Pseudomonas fluorescens* strain D7 has the greatest efficacy on downy brome when the bacterium is applied in the fall. In addition, populations of *P. fluorescens* D7 in soil are greatest in the fall and spring, which coincides with active root growth of downy brome (Kennedy et al., 1991).

Another challenge to successful microbial herbicides is developing strains that are effective against weed populations with high genetic diversity. Diversity within a species can lead to inconsistent results with biocontrol measures (Bailey, 2004; Ward et al., 2008). Biological control isolates may be specific to the region where they were first isolated, for example Ash et al. (2008) studied fungal pathogens from Korea and Australia, and found that Korean isolates showed less pathogenicity on Australian weeds than the Australian isolates. The efficacy of *Sclerotinia minor* on dandelion (*Taraxacum officinale*) was dependent upon weed accession, age and plant competition (Abu-Dieyeh & Watson, 2007). Both physiological and ecological considerations need to be examined for successful biocontrol interactions.

Prior to commercialization of a biological herbicide, extensive host-range studies must be completed to determine effects of the agent on non-target organisms in order to minimize any harmful effects. Potential microbial herbicides must be virulent on the target species, while non-target plants remain disease-free (Bailey, 2004). Wapshere's (1974) concept of concentric spheres of related plant species is a starting point for investigations of non-target plant species. In host-range studies of *Alternaria alternata*, a potential biological control of water hyacinth (*Eichhornia crassipes*), the biocontrol agent also inhibited the weed water lettuce (*Pistia stratiotes* L.; Mohan Babu et al., 2002). They evaluated 29 economically and environmentally significant species that encompassed more than 18 families. After finding that only two plant species were inhibited, they concluded that *A. alternata* would not be harmful to any economically important plants. Kennedy et al. (2001) found that *Pseudomonas fluorescens* strain D7 inhibited the target weed, downy brome, a few other *Bromus* species, jointed goatgrass and medusahead (*Taeniatherum caput-medusae*) in bioassays, greenhouse and field studies. They investigated representatives from a wide array of families that included native, crop and weed species, while concentrating on familes, tribes, subtribes and accessions closely related to *Bromus* species. No dicots and only a few monocots were negatively affected by the bacterium. While a few other grasses were suppressed slightly by *P. fluorescens* D7 in bioassays, they were not suppressed in greenhouse studies using nonsterile soil. These two examples illustrate the importance of extensive testing of other plant species. The lack of host-range testing often leads to the early demise of a potential biocontrol agent.

Other challenges are large-scale production and storage, survival of the organism, requirement of the agent for formulation, shelf-life of the organism, and delivery system of the biocontrol agent to the host plant. The ability to produce large quantities of microbial products, and maintain survival of the organisms, are the major obstacles to field-scale application of biological control agents (Amsellem et al., 1999). Teshler et al. (2007) examined shelf life of the *Sclerotinia minor* bioherbicide and found that storage temperatures less than 11°C increased *S. minor* shelf life, but CO_2 did not affect shelf life. Cooler storage temperatures helped to prolong the shelf life of *Fusarium oxysporum* (Foxy2 and PSM197) for control of *Striga*; however, vacuum packing was not helpful. The fungicide Apron XL enhanced shelf life, but another fungicide, Ridomil Gold (both fungicides Metalaxyl-M, Syngenta, Basle, Switzerland & Germany), did not (Elzein et al., 2009). The authors attributed this difference to the higher recommended application rate for Ridomil Gold, or possibly that Apron XL is ineffective against ascomycetes such as *Fusarium oxysporum*. While most of the microbial herbicides developed to-date have been mycoherbicides, bacterial herbicides are becoming available and have different challenges and advantages. These include simpler fermentation processes, ease of upscaling and mass-production, production of secondary metabolites and lack of spore formation that may require specific growth conditions (Li et al., 2003).

Application technology and cultural practices also affect the efficacy of microbial herbicides applied in field situations. Aerosol applications in the field may need different inoculum levels than what are found to be successful in greenhouse studies. In addition, droplet size, and spray direction may need to be readjusted to reach the weed or soil for greatest weed reduction (Lawrie et al., 2002b). Byer et al. (2006) found that finer droplet size led to greater spray retention of *Colletotrichum truncatum* on scentless chamomile and *Colletotrichum gloeosporioides* f. sp. *malvae* on round-leaved mallow. Peng et al. (2005b) looked at spray retention for efficacy of the biological control agent *Pyricularia setariae* on green foxtail. They looked at sprayer type and nozzle size for varying application rates and droplet sizes. In general, finer droplet size was more advantageous, and the authors suggested further study with other factors such as use of adjuvants and formulations. Boyette et al. (2007b) found that *Colletotrichum gloeosporioides* was an effective biocontrol agent of sicklepod in soybean; however, the wider row spacings required repeat applications of the agent for adequate weed suppression. The challenges to successful biocontrol are many, and continued research and development are needed to offer the alternative of microbial weed management.

3. Successful microbial herbicides

The majority of biological herbicides developed to-date are mycoherbicides; however, several bacterial herbicides are under development as well. A screening procedure for a bacterial biological control agent has been developed by the authors (Figure 1). Our screening procedure takes advantage of the soil microbial community and the diversity within. This screening includes sampling the soil and plant material at the peak of population growth and when the suppression naturally occurs. An initial assay can separate the weed-suppressive microorganisms from the bulk of the native population. Further multiple screenings on various plants are suggested to get rid of those isolates that may also inhibit beneficial plants. While inhibition of the target weed is critical, non-host range must be determined early in the process to ensure the product progresses through the registration process and eventually becomes marketable. Both steps need to be thorough to ensure that no lesser known, but economically important, host is detected later in the process. Souissi and Kremer (1998) utilized a multiple-well plate procedure with leafy spurge (*Euphorbia esula* L.) callus to rapidly determine phytotoxicity of rhizobacterial isolates, and Vidal et al. (2004) have developed a successful method to produce yellow starthistle (*Centaurea solstitialis* L.) calli for bioassay screening of biocontrol pathogens such as *Phoma exigua*. A quick substitute bioassay will reduce the length of time needed for the initial screenings. In many cases, a quick screen is not possible (Kennedy et al., 1991), but automating data recording can often make data collection easier (Doty et al., 1994).

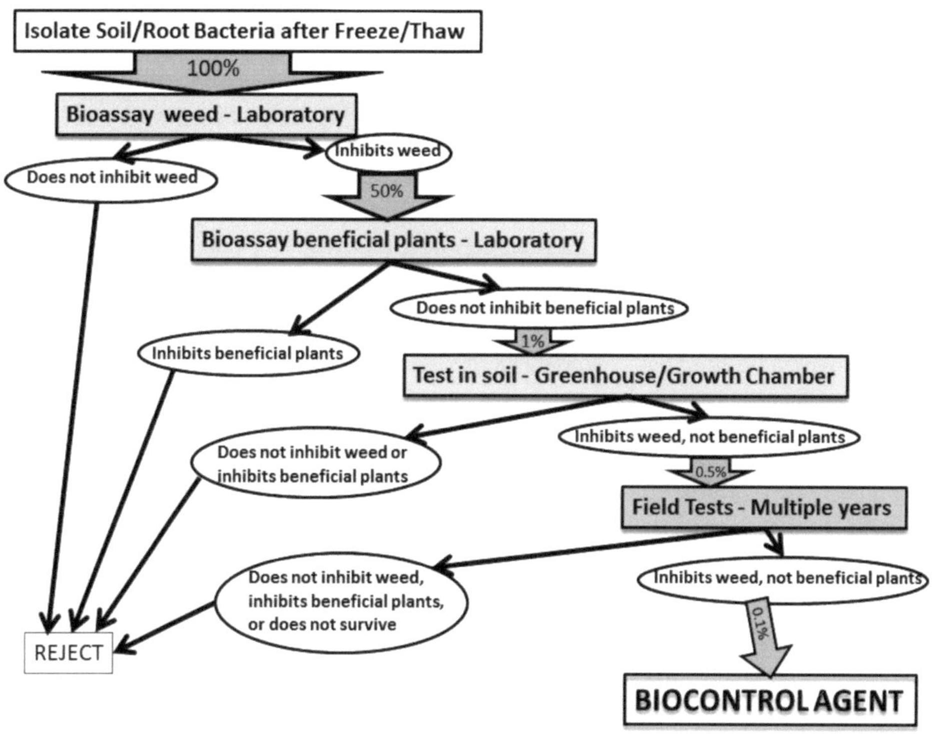

Fig. 1. Flow of assays to obtain a biological control organism from soil that successfully suppresses the growth of target organisms but has a limited host range and does not inhibit beneficial plant species.

Since the early 1980's, there have been several successful biological herbicides released to the market. Biomal® (ATCC 20767) *Colletotrichum gloeosporioides* f. sp. *malvae,* was developed as a weed biocontrol agent against round-leaved mallow (*Malva neglecta*) and registered in 1992. DeVine was marketed as a control for stranglervine (*Morrenia odorata*) in citrus, and Collego was used to control northern jointvetch (*Aeschynomene virginica*) in rice. Velgo is a potential mycoherbicide for control of velvetleaf in the U.S. and Canada (Mortensen, 1998; Owen & Zdor, 2001). DeVine and Collego were registered for use in the United States, and each controls a single weed species. Four bioherbicides released in Canada included two strains of *Chondrostereum perpureum* (HQ1 (Myco-Tech Paste) and PFC 2139 (Chontrol Paste)) for control of trees and shrubs (Becker et al., 2005), *Colletotrichum gloeosporioides* f. sp. *malvae* to control round-leaved mallow (Cross and Polonenko, 1996; no longer available due to small market size), and *Sclerotinia minor* (IMI 3144141) for control of dandelions (Abu-Dieyeh & Watson, 2007; Bailey et al., 2010). CAMPERICO (*Xanthomonas campestris* pv. *poae* isolate JT-P482) was registered in Japan for control of annual bluegrass (*Poa annua* L.) in 1997 (Imaizumi et al., 1999). The fungal pathogen *Alternaria destruens* strain 059 (Smolder G and Smolder WP) was registered in the U.S. in 2005 for control of dodder (*Cuscuta* sp.) in field crops and ornamental plants (USEPA, 2005).

The mode of action of each biocontrol agent is as varied as the microorganisms themselves (de Luna et al., 2011). They range from simple but effective compounds like cyanide (Kremer & Souissi, 2001; Owen & Zdor, 2001) and organic acids to complex molecules with tertiary structure (Bouizgarne et al., 2006; Gurusiddaiah et al., 1994), and from secondary metabolites (Kroschel & Elzein, 2004) to plant growth regulators, such as auxins and ethylene (de Luna et al., 2005). Pedras et al. (2003) isolated toxins from *Pseudomonas fluorescens* strain BRG100. One of the toxins, pseudophomin A, showed greater inhibition of green foxtail than did the other, pseudophomin B, which showed greater activity against several plant pathogens. There are sufficient successful products out in the market or in development that indicate continued efforts will provide more microbial herbicides and better weed management options.

4. Formulations to improve success of microbial herbicides

In order to overcome the obstacles associated with development of a microbial herbicide, and dramatically improve the chances for success of microbial herbicides, numerous researchers have investigated combining the biological control organism with a formulation designed to improve application, survivability, and efficacy. Several research projects have studied the use of a formulation or carrier in combination with a biological control organism (Table 2).

As mentioned earlier, appropriate temperatures and length of dew period are critical to the success of fungal bioherbicides. Formulations can extend the period of time before dew is required, and adjuvants such as unrefined corn oil and Silwet L-77 can improve chances for success of mycoherbicides (Abbas et al., 2004; Boyette, 2006; Boyette et al., 2007a). Elzein et al. (2004) found that *Fusarium oxysporum* 'Foxy 2' could be encapsulated in a pesta formulation to improve shelf life. Amsellem et al. (1999) utilized the 'Stabileze' formulation (containing starch, sucrose, corn oil and silica) to enhance preservation of mycelia from two *Fusarium* spp. and found that they remained viable for over one year. The 'Stabileze' method has also been utilized to enhance survival of bacteria. Zidack & Quimby (2002) used this

Biological Control Agent	Weed Pest	Formulation / Carrier	Reference
Alternaria eichhorniae (Ae5)	Water hyacinth (*Eichhornia crassipes*)	Nine oil emulsions	Shabana, 2005
Alternaria helianthi	Common cocklebur (*Xanthium strumarium*)	Unrefined corn oil; Silwet L-77	Abbas et al., 2004
Bipolaris sp.	Japanese stiltgrass (*Microstegium vimineum*)	Tween 20	Kleczewski & Flory, 2010
Colletotrichum gloeosporioides	Sicklepod (*Senna obtusifolia*)	Adjuvants (unrefined corn oil; invert emulsion-MSG 8.25; Silwet L-77)	Boyette, 2006
Colletotrichum truncatum	Hemp sesbania (*Sesbania exaltata*)	Unrefined corn oil; Silwet L-77	Boyette et al., 2007c
Curvularia eragrostidis (QZ-2000)	Large crabgrass (*Digitaria sanguinalis*)	Tween 80; rapeseed oil	Zhu & Qiang, 2004
Curvularia intermedia	Large crabgrass (*Digitaria sanguinalis*)	Silwet L-77	Tilley & Walker, 2002
Fusarium oxysporum f. sp. *cannabis*	Ditchweed (*Cannibis sativa*)	Birch sawdust; wheat seeds; oat seeds	Tiourebaev et al., 2001
Fusarium oxysporum f. sp. *orthoceras*	Sunflower broomrape (*Orobanche cumana*)	Pesta granules; commercial iron fertilizers	Müller-Stöver & Sauerborn, 2007
Fusarium oxysporum (FOG)	Branched broomrape (*Orobanche ramosa*)	Pesta granules; alginate pellets	Kohlschmid et al., 2009
Fusarium oxysporum; F. arthrosporioides	Broomrapes (*Orobanche* spp.)	Alginate beads; 'Stabileze' (starch, sucrose, corn oil, silica)	Amsellem et al., 1999
Fusarium oxysporum f. sp. *strigae* (Foxy 2)	*Striga (Striga hermonthica)*	Film-coat on sorghum seeds (gum arabic, 40%); Pesta granules; seed treatment (gum arabic, SUET binder)	Elzein et al., 2004, 2006, 2009
Fusarium tumidum	Gorse (*Ulex europaeus*)	Tween 80; 5% Triton X-100	Yamoah et al., 2008
Helminthosporium gramineum subsp. *echinochloae; Curvularia lunata*	Barnyardgrass (*Echinochloa crus-galli*)	Tween 20	Zhang et al., 2007
Mycoleptodiscus terrestris	Hydrilla (*Hydrilla verticillata*)	Diatomaceous earth	Shearer & Jackson, 2006
Myrothecium verrucaria	*Portulaca* spp., *Euphorbia* spp.	Silwet L-77	Boyette et al., 2007a
Myrothecium verrucaria	Kudzu (*Pueraria lobata*)	Silwet L-77; unrefined corn oil	Hoagland et al., 2007
Neonectria neomacrospora	Hemlock dwarf mistletoe (*Arceuthobium tsugense*)	'Stabileze'	Rietman et al., 2005
Phomopsis amaranthicola	*Amaranthus* spp.	16 adjuvants	Wyss et al., 2004
Pseudomonas fluorescens strain G2-11	Green foxtail (*Setaria viridis*); velvetleaf (*Abutilon theophrasti*)	Corn gluten meal; semolina flour	Zdor et al., 2005

Biological Control Agent	Weed Pest	Formulation / Carrier	Reference
Pseudomonas syringae pv. *tabaci*; *P. syringae* pv. *tagetis*	Multiple weeds	'Stabileze' method (bacteria, oil, sucrose; silica)	Zidack & Quimby, 2002
Pseudomonas syringae pv. *tagetis*	Canada thistle (*Cirsium arvense*)	Silwet L-77; Canada thistle sap	Gronwald et al., 2002; Tichich et al., 2006; Tichich & Doll, 2006
Puccinia thlaspeos	Dyer's woad (*Isatis tinctoria*)	Surfactants Sylgard, IFA-S90, Regulaid	Kropp & Darrow, 2006
Sclerotinia minor	Dandelion (*Taraxacum officinale*)	Barley (*Hordeum vulgare*) grits	Teshler et al., 2007

Table 2. Research projects examining formulations or carriers to improve survival and efficacy of microbial herbicides.

method with two *Pseudomonas* spp., and showed that populations were still high after one year, and that components of the formulation could be varied depending on bacterial species response.

Elzein et al. (2006, 2010) studied seed coatings containing *Fusarium oxysporum* isolates to control *Striga*, and found a 40% gum arabic seed coating combined with dried chlamydospores to be the most effective combination of seed coating and inoculum type for causing disease in *Striga* (Elzein et al., 2006). Zhao & Shamoun (2005) tested combinations of gelatin and potato dextrose broth concentrations for optimum efficacy of *Phoma exigua* to control salal *(Gaultheria shallon)*, a perennial evergreen shrub. *Fusarium oxysporum* f. sp. *orthoceras* (FOO) is known to suppress the root parasitic weed broomrape *(Orobanche cumana)* in sunflower. In addition, Benzo(1,2,3)thiadiazole-7-carbothioic acid S-methyl ester (BTH) induces sunflower resistance to *Orobanche cumana.* Using FOO and BTH, Müller-Stöver et al. (2005) improved the efficacy of *Fusarium oxysporum* f. sp. *orthoceras* (FOO) to reduce broomrape infection in greenhouse studies. They also found FOO incorporated into a wheat-kaolin and iron mix further improved efficacy by increasing FOO survival (Müller-Stöver & Sauerborn, 2007).

Chittick & Auld (2001) examined the use of hydrophilic polymers as a formulation for a mycoherbicide to improve efficacy of *Colletotrichum orbiculare* on *Xanthium spinosum* (Bathurst burr) in Australia. Hoagland et al. (2007) studied formulation, application method and growth media for control of kudzu (*Pueraria lobata*) using *Myrothecium verrucaria* fungi. Shabana (2005) found that the efficacy of *Alternaria eichhorniae* isolate #5 could be improved, and the requirement for a dew-period avoided, by applying the fungi using an oil emulsion. With that formulation, complete control of water hyacinth under field conditions in Egypt was achieved. Hurrell et al. (2001) found a granular mycelium-wheat formulation of *Sclerotinia sclerotiorum* controlled *Cirsium arvense* best in pasture lands of New Zealand when the formulation was a water-miscible powder applied as a slurry rather than a dry product. They also found that spring and early summer applications, when some moisture was present, were more effective at reducing the weed than late summer or early autumn. While some moisture was needed for efficacy, too much rain was thought to wash the agent off the leaf. Kohlschmid et al. (2009) found that a combination of alginate pellets and pesta granules formulated with the *Fusarium oxysporum* isolate FOG was more efficacious and reliable for controlling the parasitic weed branched broomrape (*Phelipanche ramosa*) than the untreated control under field conditions. Boyette (2006) examined several adjuvants in combination with the mycoherbicide *Colletotrichum gloeosporioides* for control of sicklepod. These compounds reduced length of the dew period requirement, and improved the performance of this organism in controlling sicklepod. Boyette et al. (2007c) showed that *Colletotrichum truncatum* formulated with unrefined corn oil and the surfactant Silwet L-77 was able to effectively control hemp sesbania in the greenhouse and field by reducing the dew period requirement. Zhang et al. (2007) utilized protoplast fusion as a means to improve the biocontrol efficacy of *Helminthosporium gramineum* subsp. *echinochloae* strain HM1 against barnyardgrass (*Echinochloa crus-galli*) in rice.

Most of the research conducted to develop formulations that improve survival and efficacy of microbial herbicides has been directed at fungal herbicides; however, there have been a few studies aimed at improving delivery of bacterial herbicides. Zdor et al. (2005) studied effects of DRB (*Pseudomonas fluorescens* strain G2-11) in combination with corn gluten meal and semolina flour in soil assays using weed and crop species. Zhang et al. (2010) studied the stability of pyoluteorin, a polyketide metabolite produced by fluorescent pseudomonads that has shown potential to control weeds, among other pests. Tichich and Doll (2006) examined a novel application approach for the pathogen *Pseudomonas syringae* pv. *tagetis* where the sap of infected Canada thistle is extracted, combined with water and Silwet-77, and sprayed. While there was disease expression in the treated plants, it was not enough to control the Canada thistle and further work is needed on this approach.

Various technologies have been used and will continue to be used to enhance biological weed control (Cohen et al., 2002). The protoplast fusion technique was used to create new strains using *Helminthosporium gramineum* subsp. *echinochloae* strain HM1 (high pathogenicity, low spore formation) and *Curvularia lunata* (low pathogenicity, high spore formation) to create strains that effectively control barnyardgrass and other weeds in rice production (Zhang et al., 2007). Hypervirulence selection or manipulation may improve efficacy of biological control agents. Cohen et al. (2002) transformed genes of the indole-3-acetamide (IAM) pathway to cause an auxin imbalance that increased the virulence of *Fusarium oxysporum* and *F. arthrosporioides*, pathogenic on broomrape (*Orobanche aegyptiaca).* Sands and Pilgeram (2009) outline the steps to enhance virulence of the biocontrol agent using amino acid overproduction. They discuss control of the parasitic weeds *Orobanche* and *Striga*, which are especially challenging to control due to the close relationship they develop with their hosts. Economic formulations and genetic manipulations to alter phenotype will assist in the understanding and development of microbial herbicides.

5. Integrating microbial herbicides with other control measures

In many cases, microbial herbicides alone will not be enough to remedy invasive weed problems. Researchers worldwide have shown that an integrated approach utilizing microbial weed management in a synergistic or additive manner with chemical herbicides or in combination with cultural practices or biological controls with insects is more successful than any of the control measures alone. Innovative approaches for microbial management of several different weed species in various regions of the world have been employed to sustainably manage some of the world's worst weed problems in diverse systems.

Denoth et al. (2002) reviewed biological control projects on weeds and insects, and determined that for weeds, those projects where a number of agents were released showed greater success. Wandeler et al. (2008) used weevils as an insect vector to apply *Puccinia punctiformis* to creeping thistles (*Cirsium arvense*) in the field, causing systemic rust infection. Another integrated approach to biological control with microorganisms utilizes mixtures of organisms, rather than a single pathogen to control weed growth. Chandramohan & Charudattan (2003) propose the use of a mixture of four fungal plant pathogens (*Phomopsis amaranthicola, Alternaria cassia, Colletotrichum dematium* f.sp. *crotalariae, Fusarium udum* f.sp. *crotalariae*) to control pigweed (*Amaranthus hybridus* L.), sicklepod and showy crotalaria (*Crotalaria spectabilis* Roth.). In their greenhouse study, they showed that it was possible to control these three weeds together using several fungal strains without losing efficacy or host-specificity. Some potential biological control agents work best as individual applications, and do not exhibit synergy when applied together. Dooley & Beckstead (2010) found no improvement in downy brome inhibition using *Pseudomonas fluorescens* strain D7 in combination with the fungal pathogen *Pyrenophora semeniperda*. These two microorganisms suppress downy brome at different growth stages. Another example of this is *Chondrostereum purpureum*, which is applied in spring and *Fusarium tumidum*, which is applied in early winter to reduce gorse (*Ulex europaeus*) regrowth in New Zealand forests (Bourdot et al., 2006b). These mycoherbicides inhibit gorse at different growth stages, but together they did not further reduce the regrowth of stumps. Further studies are needed to understand the weed reduction and ecological implications of consortia in biocontrol efforts.

In greenhouse studies, Caesar (2003) found that the combined effects of *Fusarium oxysporum* and *Rhizoctonia solani* with flea beetle adults and larvae resulted in greater inhibition of *Euphorbia esula/virgata* than any of the biological control agents alone. In their survey of *Lepidium draba* throughout Europe, Caesar et al. (2010) showed that plants sustaining both insect damage and disease were being colonized by the root pathogen *Rhizoctonia solani*. They concluded that when determining potential biological control agents, the synergistic relationships between plant pathogens and insects should be considered. Likewise, Kremer et al. (2006) found that the most effective biocontrol of leafy spurge occurred with the synergistic effect of "plant-associated microorganisms and root-damaging insects". There was a higher incidence of *Fusarium* and *Rhizoctonia* isolates in *Euphorbia* plants that had injury caused by insect feeding. Rayamajhi et al. (2010) noted the additive effects of using combinations of a weevil, psyllid, and rust fungus (*Puccinia psidii*) to reduce regrowth of tree stumps of *Melaleuca quinquenervia* in southern Florida. The use of multiple enemies against invasive weeds is another combination that could be successful in biocontrol programs.

Babalola et al. (2007) examined the use of trap crops such as cowpea (*Vigna unguiculata*) combined with application of bacteria (*Enterobacter sakazakii* and *Pseudomonas* spp.) to stimulate germination and cause the subsequent death of the parasitic weed *Striga hermonthica*. Similarly, Ahonsi et al. (2003) used ethylene-producing *Pseudomonas syringae* pv. *glycinea* in combination with nitrogen-fixing *Bradyrhizobia japonicum* strains to induce germination and death in *Striga hermonthica* seeds in the presence of cowpea or soybean. No matter how successful a biocontrol agent may be, the control is never considered to be 100% and additional practices and management efforts need to be integrated with these microorganisms or insects to attain weed management or control.

Two methods have been employed to determine whether a synergistic relationship exists between a chemical herbicide and a microbial herbicide (Gressel, 2010): 1) random testing of herbicides, and 2) screening chemical herbicides with the intent to alter the target plant's defenses, leaving it more vulnerable to attack by pathogens. Better control of *Chenopodium album* was achieved by using low rates of herbicides in combination with toxins from *Ascochyta caulina* (Vurro et al., 2001). Jahromi et al. (2006) studied interactions between herbicides and *Plectosporium alismatis* to control starfruit in rice. Kropp & Darrow (2006) found that herbicides and surfactants did not negatively affect teliospore viability of *Puccinia thlaspeos* when it was sprayed on to *Isatis tinctoria* (Dyer's woad) plants in the field. Abu-Dieyeh & Watson (2006) examined the relationship between the fungal pathogen *Sclerotinia minor* Jagger applied with different turfgrass mowing heights compared to herbicide alone to control dandelion. In the greenhouse, *S. minor* caused greater damage to dandelions than the herbicide at all mowing heights; however, under field conditions, close mowing had an unfavorable effect on *Sclerotinia minor* for dandelion control. Magani et al. (2009) demonstrated that the fungal mycoherbicide *Fusarium oxysporum*, when applied in granular form followed by a post-emergence herbicide treatment, was successful in controlling the parasitic plant *Striga* in Nigeria.

Peng and Byer (2005) tested seven herbicides at reduced rates to determine whether there might be synergistic effects with *Pyricularia setariae* for control of green foxtail. Responses were variable and depended on weed growth stage, herbicide and application rates. However, each of the herbicides at a one-quarter rate applied with a sub-lethal dose of the biocontrol organism succeeded in controlling green foxtail in the greenhouse (Peng & Byer, 2005). Jahromi et al. (2006) studied interactions between the fungal pathogen *Plectosporium alismatis* and herbicides for control of starfruit in Australian rice production. In glasshouse experiments, there was no synergistic effect with the pathogen and 2-methyl-4-chlorophenoxyacetic acid (MCPA); however, there was a synergistic effect when the pathogen was applied after the sublethal dose of the herbicide Londax® (bensulfuron methyl, DuPont, Wilmington, DE).

Gressel (2010) notes that glyphosate is most commonly utilized in synergistic relationships with plant pathogens, and hypothesizes that this is due to the capability of glyphosate to affect multiple weed defense mechanisms. Boyette et al. (2008a) in a field study controlled redvine (*Brunnichia ovata*) and trumpetcreeper (*Campis radicans*) using the synergistic relationship between glyphosate and the fungus *Myrothecium verrucaria*. The combination of the chemical herbicide and the microbial herbicide was able to control the weeds better than either treatment alone. Boyette et al. (2008b) concluded that hemp sesbania (*Sesbania exaltata*) control might be augmented by utilizing the biological control agent *Colletotrichum truncatum* in combination with reduced rates of the herbicide glyphosate when the fungus is applied after the herbicide. In a previous experiment under controlled conditions, Boyette et al. (2006) tested *Myrothecium verrucaria* in combination with glyphosate on kudzu (*Pueraria lobata*), redvine and trumpetcreeper at various temperatures. Greatest disease development was achieved at higher temperatures, and weed inhibition was greatest when the fungus was applied after the herbicide rather than prior to or with the glyphosate. Cook et al. (2009) found that dodder (*Cuscuta pentagona*) was more effectively controlled in greenhouse studies using a mixture of the pathogen *Alternaria destruens*, glyphosate, oil and an ammonium sulfate surfactant than when using any of the treatments alone, while not harming the host citrus (*Citrus* spp.) plant.

Care must be taken when integrating biological controls with herbicides so that the herbicide does not affect microbial survival and efficacy. Ray et al. (2008) found that dose and herbicide type may be detrimental to the water hyacinth pathogen *Alternaria alternata*. The effects of four herbicides at recommended and reduced rates were tested on the rust fungus *Puccinia lagenophorae* to be used against *Senecio vulgaris* (common groundsel). Wyss & Müller-Scharer (2001) found that the herbicides they tested were either too toxic to the fungus, or did not increase plant susceptibility to the fungus, and so the combination of the rust fungus and herbicide was not an option for control of *Senecio vulgaris*. Shabana & Mohamed (2005) combined *Alternaria eichhorniae* isolate 5 with 3,6-dichloro-2-pyridinecarboxylic acid (MDCA) to weaken water hyacinth defenses, which increased disease severity and has the potential for greater biocontrol.

The soil possesses a wealth of genetic potential waiting to be discovered and used in weed management systems. Soil quality or the chemical, physical and biological properties of soil can influence plant-microbe interactions and the success of any biocontrol agent (Kennedy & Papendick, 1995). In addition, soil quality investigations often include weed populations as an indicator of soil quality, in part because a healthy soil and healthy desirable plant is the best weed control (Magdoff, 2001; Ryan et al., 2011). Soil investigations are needed in determining the impact of the soil environment on weed populations. The structure and function of soil microbial communities develop depending on the soil, location, climate, slope, aspect and vegetation. Management practices can influence the soil microbial community and plant-microbe interactions (Wander et al., 1995; Lupwayi et al., 1998; Kennedy & Schillinger, 2006). Cultural practices and application of amendments may also play a part in weed-suppressive soils, although the modes of action may be different for each practice. Cropping systems also influence weed-suppressive bacteria (Kremer & Li, 2003; Ryan et al., 2011). Kremer & Li (2003) found that high enzyme activity and greater volume of water stable aggregates correlated with more weed-suppressive bacteria. They found that uncultivated prairie and no-tillage systems contained the highest populations of deleterious rhizobacteria compared to other land use and soil quality indicators, and may be useful in selecting for weed-suppressive practices. In a similar study in Washington state, Kennedy & Stubbs (2007) could not find relationships among management systems and the prevalence of weed-suppressive bacteria.

The concept of weed-suppressive soils can be defined as soils, amendments, or management practices that have the capacity to reduce or limit specific weeds. Management for weed-suppressive soils is the ultimate goal for biological control efforts. The dynamics of weed-suppressive soils may be similar to what is seen with disease-suppressive soils (Mazzola, 2004). The biology, chemistry and physical properties of soil that comprise a weed-suppressive soil need to be characterized in order to manage for suppressive soils. As with all soil quality determinations, no one indicator will explain the complexity of weed suppression.

6. Regulatory issues

Rigorous testing is required prior to the release of a biological herbicide to ensure the safety of humans, animals and the environment. Host-range studies are needed to reduce potential risk and ensure that beneficial, non-target plant species are unaffected by the biocontrol agent. However, the length of time needed to complete assessments of new biological herbicides adds to the costs and the length of time required before an agent can be released (Ghosheh, 2005). Non-host testing is important and the ranges of plant species tested depend on the areas of release, ecosystem variability and potential for dissemination of the biocontrol agent by wind or water. Testing should cover all economically important plant species of the area, and those plants known to be involved in ecosystem maintenance. In agronomic ecosystems, the major crop species are of interest. The U.S. Environmental Protection Agency (EPA) published a list of the top 25 major agricultural crops. Plants were placed on this list because of their economic importance, ecosystem activity or total production values (EPA, 2011). In aquatic systems, several aquatic plants are suggested that include algae, aquatic bacteria, marine and freshwater diatoms. In rangeland ecosystems the non-target species would include native or near native plant species. It is recommended to test six species covering at least four families in the Dicotyledonae, and at least four species of at least two families in the Monocotyledonae. Testing must be performed on all plants of economic importance in agriculture, horticulture or rangeland systems or known to be beneficial to maintenance of the ecosystem that have any reasonable likelihood of serving as hosts. This selection of additional plant species should be based upon a survey of plants closely related (same genus or, if not available, same family) to the target plant and a survey of known hosts of pathogens closely related to the microbial herbicide (EPA, 2011; Wapshere, 1974).

With thorough host-range testing, very few, if any, detrimental effects occur from the release of fungal herbicides to control weeds (Barton, 2004). In a review of pre- and post-release records from 26 projects, Barton (2004) found that there were no reports of a fungal biological control agent striking an unintended plant species. Additional animal, avian, fish and daphnia testing are also required in many countries before bioherbicides can be registered. In addition, as with all research and new products where there are safety concerns, buffer zones are often required to protect animal pastures and other non-target areas (Bourdot et al., 2006a). The risk of applying a microbial herbicide to the environment needs to be considered at the beginning and throughout the development of biocontrol agents.

7. Future prospects for microbial herbicides in sustainable ecosystems

The future of biocontrol is bright and full of possibilities with the many novel, successful biocontrol agents being studied. The advancements in microbial genetics, microbial community analyses and understanding plant-microbe interactions continue to accumulate and will be instrumental in helping microbial biocontrol of weeds move forward.

The area of biological control using soil microorganisms needs further investigations to discover additional isolate-host pairs that are a biocontrol match consisting of a biocontrol agent of highest virulence in contact with the host at its greatest susceptibility. Formulations are needed to increase shelf life of the living organisms to improve survival and efficacy. Research and development of each biocontrol agent are needed so that stakeholders and industry buy in to the marketing, economics and time investments of this approach to weed management. An understanding of microbial community, weed, and soil quality characteristics, and management practices is needed for the development of weed suppressive soils.

Investigations of the structure and function of soil microbial communities are needed to advance the area of biological control. Traditional techniques of microbial analyses to describe the composition and diversity of microbial populations in soils has commonly relied on phenotypic characteristics alone, and molecular investigations add to the information on structure and function of the soils (Mazzola, 2004). Profiling or fingerprinting of soil and soil microbial community structure using substrate utilization and fatty acid methyl ester analyses may be the first step in targeting weed-suppressive potential. There are several nucleic acid-based methods that can be used to probe soil and identify those microbes that produce similar compounds to those already known. Probes will assist field studies of known agents to follow survival in soils and explore soil for additional weed-suppressive factors. Nucleic acid technologies provide greater information on genetics, and possibly function of a given organism. Array, pyrosequencing and metagenomic investigations can provide information on the microbial community and the biological agent within that community. Selection for hypervirulence; construction of molecular probes; understanding the genetic material of the agent, weed-suppressive compounds, and host-microbe interactions can be investigated more thoroughly with these methods. The continual development of novel molecular methods to investigate genetics of a system will provide key information to better understanding of the plant-microbe phenomena. These methods are forever changing and improving to allow us to have increased knowledge of the microbial portion of the ecosystem and the various interactions that can occur.

Soil microbial ecology and the soil microbial community will affect weed ecosystem dynamics, diversity, function, and populations. As with soil quality, the compilation of indicators has been attempted often to hone in on a few indicators of importance. No one approach or method can be used to characterize and follow biocontrol agents, or to isolate and research additional novel plant-microbe interactions. The future is bright for continued development of microbial herbicides to reduce herbicide reliance and provide multiple options in weed management.

8. Conclusions

The wealth of genetic potential of microorganisms on this earth is boundless. There have been many investigations of potential products for weed management. Some have been successful at suppressing weeds in the field and a select few are marketed products that now reduce weed infestations. Further studies are needed to continue to search for additional tools to combat weeds. Increasing our understanding of plant-microbe interactions will assist in this effort. Biocontrol agents need to be specific, competitive and well-matched with the weed of interest. The search for biocontrol agents from the environment entails not only finding microorganisms that inhibit a weed, but that are specific for the weed or related plant species and have an economically viable market. Host-range testing and non-target species testing are needed early in the process. In addition, the development of formulations and delivery systems is necessary to prolong the shelf-life and efficacy of the biocontrol agents in a variety of environments. Biocontrol should not be considered a stand alone option, but may be best if integrated with other methods of control, especially with those that are ecologically sound. Biocontrol agents to reduce or complement chemical herbicides expand options in weed management and tend toward the use of ecologically based systems. They add additional tools in the arsenal of weed management efforts. There is a wealth of genetic potential in the soil and the environment to be explored, screened and tested for weed suppression.

9. References

Abbas, H., Johnson, B., Pantone, D., & Hines, R. (2004). Biological control and use of adjuvants against multiple seeded cocklebur (*Xanthium strumarium*) in comparison with several other cocklebur types. *Biocontrol science and technology,* Vol.14, No.8, (December 2004), pp. 855-860, ISSN 1360-0478

Abu-Dieyeh, M. & Watson, A. (2006). Effect of turfgrass mowing height on biocontrol of dandelion with *Sclerotinia minor. Biocontrol science and technology,* Vol.16, No.5-6, (May 2006), pp. 509-524, ISSN 1360-0478

Abu-Dieyeh, M. & Watson, A. (2007). Efficacy of *Sclerotinia minor* for dandelion control: effect of dandelion accession, age and grass competition. *Weed research,* Vol.47, No.1, (February 2007), pp. 63-72, ISSN 0043-1737

Ahonsi, M., Berner, D., Emechebe, A., Lagoke, S., & Sanginga, N. (2003). Potential of ethylene-producing pseudomonads in combination with effective N_2-fixing bradyrhizobial strains as supplements to legume rotation for *Striga hermonthica* control. *Biological control,* Vol.28, No.1, (September 2003), pp.1–10, ISSN 1049-9644

Ahonsi, M., Berner, D., Emechebe, A., & Lagoke, S. (2002). Selection of rhizobacterial strains for suppression of germination of *Striga hermonthica* (Del.) Benth. seeds. *Biological control,* Vol.24, No.2, (June 2002), pp. 143-152, ISSN 1049-9644

Amsellem, Z., Zidack, N., Quimby, P. Jr., & Gressel, J. (1999). Long-term dry preservation of viable mycelia of two mycoherbicidal organisms. *Crop protection,* Vol.18, No.10, (December 1999), pp. 643-649, ISSN 0261-2194

Anderson, F., McLaren, D., & Barton, J. (2010). Studies to assess the suitability of *Uromyces pencanus* as a biological control agent for *Nassella neesiana* (Poaceae) in Australia and New Zealand. *Australasian plant pathology,* Vol.39, No.1, (January 2010), pp. 69-78, ISSN 1448-6032

Ash, G., Cother, E., McKenzie, C., & Chung, Y. (2008). A phylogenetic and pathogenic comparison of potential biocontrol agents for weeds in the family Alismataceae from Australia and Korea. *Australasian plant pathology,* Vol.37, No.4, (May 2008), pp. 402-405, ISSN 1448-6032

Babalola, O., Sanni A., Odhiambo, G., & Torto, B. (2007). Plant growth-promoting rhizobacteria do not pose any deleterious effect on cowpea and detectable amounts of ethylene are produced. *World journal of microbiology & biotechnology,* Vol.23, No.6, (June 2007), pp. 747–752, ISSN 1573-0972

Bailey, K. (2004). Microbial weed control: an off-beat application of plant pathology. *Canadian journal of plant pathology,* Vol.26, No.3, (July-Sept 2004), pp. 239-244, ISSN 0706-0661

Bailey, B., Hebbar, K., Lumsden, R., O'Neill, N., & Lewis, J. (2004). Production of *Pleospora papaveracea* biomass in liquid culture and its infectivity on opium poppy (*Papaver somniferum*). *Weed science,* Vol.52, No.1, (Jan-Feb 2004), pp. 91-97, ISSN 0043-1745

Bailey, K., Boyetchko, S., & Langle, T. (2010). Social and economic drivers shaping the future of biological control: A Canadian perspective on the factors affecting the development and use of microbial biopesticides. *Biological control: theory and application in pest management*, Vol.52, No.3, (March 2010), pp.221-229, ISSN 1049-9644

Barton, J. (2004). How good are we at predicting the field host-range of fungal pathogens used for classical biological control of weeds? *Biological control: theory and applications in pest management*, Vol.31, No.1, (September 2004), pp. 99-122, ISSN 1049-9644

Becker, E., Shamoun, S., & Hintz, W. (2005). Efficacy and environmental fate of *Chondrostereum purpureum* used as a biological control for red alder (*Alnus rubra*). *Biological control: theory and application in pest management*. Vol.33, No.3, (June 2005) pp. 269-277, ISSN 1049-9644

Boari, A. & Vurro, M. (2004). Evaluation of *Fusarium* spp. and other fungi as biological control agents of broomrape (*Orobanche ramosa*). *Biological control: theory and application in pest management*, Vol.30, No.2, (June 2004), pp. 212-219, ISSN 1049-9644

Boguena, T., Meyer, S., & Nelson, D. (2007). Low temperature during infection limits *Ustilago bullata* (Ustilaginaceae, Ustilaginales) disease incidence on *Bromus tectorum* (Poaceae, Cyperales). *Biocontrol science and technology*, Vol.17, No.1-2, (January 2007), pp. 33-52, ISSN 1360-0478

Bouizgarne, B., El-Maarouf-Bouteau, H., Madiona, K., Biligui, B., Monestiez, M., Pennarun, A., Amiar, Z., Rona, J., Ouhdouch, Y., El Hadrami, I., & Bouteau, F. (2006). A putative role for fusaric acid in biocontrol of the parasitic angiosperm *Orobanche ramosa*. *Molecular plant-microbe interactions*, Vol.19, No.5, (May 2006), pp. 550-556, ISSN 0894-0282

Bourdot, G., Baird, D., Hurrell, G., & de Jong, M. (2006a). Safety zones for a *Sclerotinia sclerotiorum*-based mycoherbicide: Accounting for regional and yearly variation in climate. *Biocontrol science and technology*, Vol.16, No.3-4, (February 2006), pp. 345-358, ISSN 1360-0478

Bourdot, G., Barton, J., Hurrell, G., Gianotti, A., & Saville, D. (2006b). *Chondrostereum purpureum* and *Fusarium tumidum* independently reduce regrowth in gorse (*Ulex europaeus*). *Biocontrol science and technology*, Vol.16, No.3-4, (January 2006), pp. 307-327, ISSN 1360-0478

Boyette, C. (2006). Adjuvants enhance the biological control potential of an isolate of *Colletotrichum gloeosporioides* for biological control of sicklepod (*Senna obtusifolia*). *Biocontrol science and technology*, Vol.16, No.9-10, (October 2006), pp. 1057-1066, ISSN 1360-0478

Boyette, C., Reddy, K., & Hoagland, R. (2006). Glyphosate and bioherbicide interaction for controlling kudzu (*Pueraria lobata*), redvine (*Brunnichia ovata*), and trumpetcreeper (*Campsis radicans*). *Biocontrol science and technology*, Vol.16, No.9-10, (October 2006), pp. 1067-1077, ISSN 1360-0478

Boyette, C., Hoagland, R., & Abbas, H. (2007a). Evaluation of the bioherbicide *Myrothecium verrucaria* for weed control in tomato (*Lycopersicon esculentum*). *Biocontrol science and technology*, Vol.17, No.1-2, (February 2007), pp. 171-178, ISSN 1360-0478

Boyette, C., Hoagland, R., & Weaver, M. (2007b). Effect of row spacing on biological control of sicklepod (*Senna obtusifolia*) with *Colletotrichum gloeosporioides*. *Biocontrol science and technology*, Vol.17, No.9-10, (November 2007), pp. 957-967, ISSN 1360-0478

Boyette, C., Hoagland, R., & Weaver, M. (2007c). Biocontrol efficacy of *Colletotrichum truncatum* for hemp sesbania (*Sesbania exaltata*) is enhanced with unrefined corn oil and surfactant. *Weed biology and management,* Vol.7, No.1, (March 2007), pp. 70-76, ISSN 1445-6664

Boyette, C., Hoagland, R., Weaver, M., & Reddy, K. (2008a). Redvine (*Brunnichia ovata*) and trumpetcreeper (*Campsis radicans*) controlled under field conditions by a synergistic interaction of the bioherbicide, *Myrothecium verrucaria,* with glyphosate. *Weed biology and management,* Vol.8, No.1, (March 2008), pp. 39–45, ISSN 1444-6162

Boyette, C., Hoagland, R., & Weaver, M. (2008b) Interaction of a bioherbicide and glyphosate for controlling hemp sesbania in glyphosate-resistant soybean. *Weed biology and management,* Vol.8, No.1, (2008) pp. 18–24, ISSN 1444-6162

Byer, K., Peng, G., Wolf, T., & Caldwell, B. (2006). Spray retention and its effect on weed control by mycoherbicides. *Biological control: theory and application in pest management,* Vol.37, No.3, (June 2006), pp. 307-313, ISSN 1049-9644

Caesar, A. (2003). Synergistic interaction of soilborne plant pathogens and root-attacking insects in classical biological control of an exotic rangeland weed. *Biological control: theory and application in pest management,* Vol.28, No.1, (September 2003), pp. 144-153, ISSN 1049-9644

Caesar, A. (2006). *Uromyces scutellatus* as a keystone species affecting *Euphorbia* spp. in Europe as shown by effects on density in the field. *Biocontrol science and technology,* Vol.16, No.9-10, (October 2006), pp. 1079-1086, ISSN 1360-0478

Caesar, A., Caesar, T., & Maathuis, M. (2010). Pathogenicity, characterization and comparative virulence of *Rhizoctonia* spp. from insect-galled roots of *Lepidium draba* in Europe. *Biological control: theory and application in pest management,* Vol.52, No.2, (February 2010), pp. 140-144, ISSN 1049-9644

Caesar, A., Campobasso, G., & Terragitti, G. (2002). Identification, pathogenicity and comparative virulence of *Fusarium* spp. associated with insect-damaged, diseased *Centaurea* spp. in Europe. *Biocontrol science and technology,* Vol.47, No.2, (April 2002), pp. 217-229, ISSN 1386-6141

Carsten, L., Maxwell, B., Johnston, M., & Sands, D. (2001). Impact of crown rust (*Puccinia coronata* f. sp. *avenae*) on competitive interactions between wild oats (*Avena fatua*) and stipa (*Nassella pulchra*). *Biological control: theory and applications in pest management,* Vol.22, No.3, (November 2001), pp. 207-218, ISSN 1049-9644

Casella, F., Charudattan, R., & Vurro, M. (2010). Effectiveness and technological feasibility of bioherbicide candidates for biocontrol of green foxtail (*Setaria viridis*). *Biocontrol science and technology,* Vol.20, No.9-10, (October 2010), pp. 1027-1045, ISSN 1360-0478

Chandramohan, S. & Charudattan, R. (2003). A multiple-pathogen system for bioherbicidal control of several weeds. *Biocontrol science and technology,* Vol.13, No.2, (March 2003), pp. 199-205, ISSN 1360-0478

Charudattan, R. (2005). Ecological, practical, and political inputs into selection of weed targets: what makes a good biological control target? *Biological control: theory and application in pest management,* Vol.35, No.3, (December 2005), pp. 183-196, ISSN 1049-9644

Chittick, A. & Auld, B. (2001). Polymers in bioherbicide formulation: *Xanthium spinosum* and *Colletotrichum orbiculare* as a model system. *Biocontrol science and technology,* Vol.11, No.6, (December 2001), pp. 691-702, ISSN 1360-0478

Cliquet, S. & Zeeshan, K. (2008). Impact of nutritional conditions on yields, germination rate and shelf-life of *Plectosporium alismatis* conidia and chlamydospores as potential candidates for the development of a mycoherbicide of weeds in rice crops. *Biocontrol science and technology,* Vol.18, No.7-8, (September 2008), pp. 685-695, ISSN 1360-0478

Cohen, B., Amsellem, Z., Maor, R., Sharon, A., & Gressel, J. (2002). Transgenically enhanced expression of indole-3-acetic acid confers hypervirulence to plant pathogens. *Phytopathology,* Vol.92, No.6, (June 2002), pp.590-596, ISSN 0031-949X

Cook, J., Charudattan, R., Zimmerman, T., Rosskopf, E., Stall, W., & MacDonald, G. (2009). Effects of *Alternaria destruens*, glyphosate, and ammonium sulfate individually and integrated for control of dodder (*Cuscuta pentagona*). *Weed technology,* Vol.23, No.4, (October-December 2009), pp. 550-555, ISSN 1550-2740

Cross, J. & Polonenko, D. (1996). An industry perspective on registration and commercialization of biocontrol agents in Canada. *Canadian journal of plant pathology,* Vol.18, No.4, (December 1996), pp. 446–454, ISSN 0706-0661

de Luna, L., Kennedy, A., Hansen, J., Paulitz, T., Gallagher, R. & Fuerst, E. (2011). Mycobiota on wild oat (*Avena fatua* L.) seed and their caryopsis decay potential. *Plant health progress,* Vol.10, No.1, (February 2011), pp.1-8, ISSN 1535-1025

de Luna, L., Stubbs, T., Kennedy, A., & Kremer, R. (2005). Deleterious bacteria in the rhizosphere. In: *Roots and Soil Management: Interactions between Roots and the Soil,* R. Zobel and S. Wright, (Eds.), 233-261. Monograph no. 48. ISBN 978-0-89118-159-0 ASA. Madison WI.

Denoth, M., Frid, L., & Myers, J. (2002). Multiple agents in biological control: improving the odds? *Biological control: theory and applications in pest management,* Vol.24, No.1, (May 2002), pp. 20-39, ISSN 1049-9644

Den Breeyen, A. & Morris, M. (2003). Pathogenicity and host specificity of *Mycovellosiella lantanae* var. *lantanae,* a potential biocontrol agent for *Lantana camara* in South Africa. *Biocontrol science and technology,* Vol.13, No.3, (May 2003) pp. 313-322, ISSN 0958-3157

Ding, J., Wu, Y., Zheng, H., Fu, W., Reardon, R., & Liu, M. (2006). Assessing potential biological control of the invasive plant, tree-of-heaven, *Ailanthus altissima. Biocontrol science and technology,* Vol.16, No.5-6, (May 2006), pp. 547-566, ISSN 1360-0478

Dooley, S. & Beckstead, J. (2010). Characterizing the interaction between a fungal seed pathogen and a deleterious rhizobacterium for biological control of cheatgrass. *Biological control: theory and application in pest management,* Vol.53, No.1, (January 2010), pp. 197–203, ISSN 1049-9644

Dor, E. & Hershenhorn, J. (2009). Evaluation of the pathogenicity of microorganisms isolated from Egyptian broomrape (*Orobanche aegyptiaca*) in Israel. *Weed biology and management,*Vol.9, No.3, (September 2009), pp. 200-208, ISSN 1445-6664

Doty, J., Kennedy, A., & Pan, W. (1994). A rapid bioassay for inhibitory rhizobacteria using digital image analysis. *Soil science society of America journal,* Vol.58, No.6, (November 1994), pp.1699-1701, ISSN 1435-0661

Elzein, A., Heller, A., Ndambi, B., De Mol, M., Kroschel, J., & Cadisch, G. (2010). Cytological investigations on colonization of sorghum roots by the mycoherbicide *Fusarium oxysporum* f. sp. *strigae* and its implications for Striga control using a seed treatment delivery system. *Biological control: theory and application in pest management,* Vol.53, No.3, (June 2010), pp. 249-257, ISSN 1049-9644

Elzein, A., Kroschel, J., Marley, P., & Cadisch, G. (2009). Does vacuum-packaging or co-delivered amendments enhance shelf-life of Striga-mycoherbicidal products containing *Fusarium oxysporum* f. sp. *strigae* during storage? *Biocontrol science and technology,* Vol.19, No.3-4, (March 2009), pp. 349-367, ISSN 1360-0478

Elzein, A., Kroschel, J., & Leth, V. (2006). Seed treatment technology: An attractive delivery system for controlling root parasitic weed *Striga* with mycoherbicide. *Biocontrol science and technology,* Vol.16, No.1-2, (February 2006), pp. 3-26, ISSN 1360-0478

Elzein, A., Kroschel, J., & Müller-Stöver, D. (2004). Optimization of storage conditions for adequate shelf-life of 'Pesta' formulation of *Fusarium oxysporum* 'Foxy 2', a potential mycoherbicide for Striga: effects of temperature, granule size and water activity. *Biocontrol science and technology,* Vol.14, No.6, (September 2004), pp. 545-559, ISSN 1360-0478

EPA (2011). Microbial pesticide test guidelines, Series 885, Envronmental Protection Agency Chemical Safety and Pollution Prevention, OCSPP, Harmonized test guideline series,
http://www.epa.gov/ocspp/pubs/frs/publications/Test_Guidelines/series885.htm

Evidente, A., Punzo, B., Andolfi, A., Berestetskiy, A., & Motta, A. (2009a). Alternethanoxins A and B, polycyclic ethanones produced by *Alternaria sonchi,* potential mycoherbicides for *Sonchus arvensis* biocontrol. *Journal of agricultural and food chemistry,* Vol.57, No.15, (August 2009), pp. 6656-6660, ISSN 0021-8561

Evidente, A., Berestetskiy, A., Cimmino, A., Tuzi, A., Superchi, S., Melck, D., & Andolfi, A. , (2009b). Papyracillic acid, a phytotoxic 1,6-dioxaspiro(4,4)nonene produced by *Ascochyta agropyrina* var. *nana,* a potential mycoherbicide for *Elytrigia repens. Biocontrol,* Vol.57, No.23, (December 2009), pp. 11168–11173, ISSN 0021-8561

Evidente, A., Cimmino, A., Andolfi, A., Vurro, M., Zonno, M., & Motta, A. (2008). Phyllostoxin and phyllostin, bioactive metabolites produced by *Phyllosticta cirsii,* a potential mycoherbicide for *Cirsium arvense* biocontrol. *Journal of agricultural and food chemistry,* Vol.56, No.3, (February 2008), pp. 884-888, ISSN 0021-8561

Flores-Vargas, R., & O'Hara, G. (2006). Isolation and characterization of rhizosphere bacteria with potential for biological control of weeds in vineyards. *Journal of applied microbiology,* Vol.100, No.5, (May 2006), pp. 946-954, ISSN 0973-7510

Frantzen, J. & Müller-Scharer, H. (2006). Modeling the impact of a biocontrol agent, *Puccinia lagenophorae,* on interactions between a crop, *Daucus carota,* and a weed, *Senecio vulgaris. Biological control: theory and application in pest management,* Vol.37, No.3, (June 2006), pp. 301-306, ISSN 1049-9644

Ghajar, F., Holford, P., Alhussaen, K., Beattie, A., & Cother, E. (2006). Optimising sporulation and virulence in *Drechslera avenacea. Biocontrol science and technology,* Vol.16, No.5-6, (May 2006), pp. 471-484, ISSN 1360-0478Z

Ghorbani, R., Scheepens, P., Zweerde, W., Leifert, C., McDonald, A., & Seel, W. (2002). Effects of nitrogen availability and spore concentration on the biocontrol activity of

Ascochyta caulina in common lambsquarters (*Chenopodium album*). *Weed science*, Vol.50, No.5, (Sept-Oct 2002), pp. 628-633, ISSN 0043-1745

Ghosheh, H. (2005). Constraints in implementing biological weed control: A review. *Weed biology and management*, Vol.5, No.3, (September 2005), pp.83–92, ISSN 1444-6162

Graham, G., Peng, G., Bailey, K., & Holm, F. (2006). Effect of dew temperature, post-inoculation condition, and pathogen dose on suppression of scentless chamomile by *Colletotrichum truncatum*. *Biocontrol science and technology*, Vol.16, No.3-4, (February 2006), pp. 271-280, ISSN 1360-0478

Graham, G., Peng, G., Bailey, K., & Holm, F. (2007). Effect of plant stage, *Colletotrichum truncatum* dose, and use of herbicide on control of *Matricaria perforata*. *BioControl*, Vol.52, No.4, (August 2007), pp. 573-589, ISSN 1386-6141

Green, S., Peng, G., Connolly, T., & Boyetchko, S. (2004). Effect of moisture and temperature on disease of green foxtail caused by *Drechslera gigantea* and *Pyricularia setariae*. *Plant disease*, Vol.88, No.6, (June 2004), pp. 605-612, ISSN 0191-2917

Gressel, J. (2010). Herbicides as synergists for mycoherbicides, and vice versa. *Weed science*, Vol.58, No.3, (July 2010), pp. 324-328, ISSN 0043-1745

Gronwald, J., Plaisance, K., Ide, D., & Wyse, D. (2002). Assessment of *Pseudomonas syringae* pv. *tagetis* as a biocontrol agent for Canada thistle. *Weed science*, Vol.50 No.3, (May-June 2002), pp. 397-404, ISSN 0043-1745

Gurusiddaiah, S., Gealy, D., Kennedy, A., & Ogg, A., Jr. (1994). Isolation and characterization of metabolites from *Pseudomonas fluorescens* strain D7 for control of downy brome (*Bromus tectorum* L.). *Weed science*, Vol.42, No.3, (July - September 1994), pp. 492-501, ISSN 0043-1745

Guske, S., Schulz, B., & Boyle, C. (2004). Biocontrol options for *Cirsium arvense* with indigenous fungal pathogens. *Weed research*, Vol.44, No.2, (April 2004), pp. 107-116, ISSN 0043-1737

Hasan, S., Sobhian, R., & Herard, F. (2001). Biology, impact and preliminary host-specificity testing of the rust fungus, *Uromyces salsolae*, a potential biological control agent for *Salsola kali* in the USA. *Biocontrol science and technology*, Vol.11, No.6, (December 2001), pp. 677-689, ISSN 1360-0478

Hoagland, R., Boyette, C., & Abbas, H. (2007). *Myrothecium verrucaria* isolates and formulations as bioherbicide agents for kudzu. *Biocontrol science and technology*, Vol.17, No.7-8, (September 2007), pp. 721-731, ISSN 1360-0478

Hurrell, G., Bourdot, G., & Saville, D. (2001). Effect of application time on the efficacy of *Sclerotinia sclerotiorum* as a mycoherbicide for *Cirsium arvense* control in pasture. *Biocontrol science and technology*, Vol.11, No.3, (June 2001), pp. 317-330, ISSN 1360-0478

Imaizumi, S., Honda, M., & Fujimori, T. (1999) Effect of temperature on the control of annual bluegrass (*Poa annua* L.) with *Xanthomonas campestris* pv. *poae* (JT-P482). *Biological control*, Vol.16, No.1, (September 1999), pp.13-17, ISSN 1049-9644

Jahromi, F. (2007). Effect of environmental factors on disease development caused by the fungal pathogen *Plectosporium alismatis* on the floating-leaf stage of starfruit (*Damasonium minus*), a weed of rice. *Biocontrol science and technology*, Vol.17, No.7-8, (September 2007), pp. 871-877, ISSN 1360-0478

Jahromi, F., Van De Ven, R., Cother, E., & Ash, G. (2006). The interaction between *Plectosporium alismatis* and sublethal doses of bensulfuron-methyl reduces the growth of starfruit (*Damasonium minus*) in rice. *Biocontrol science and technology,* Vol.16, No.9-10, (October 2006), pp. 929-940, ISSN 1360-0478

Jiang, S.-J., Qiang, S., Zhu, Y.-Z., & Dong, Y.-F. (2008). Isolation and phytotoxicity of a metabolite from *Curvularia eragrostidis* and characterisation of its modes of action. *Annals of applied biology,* Vol.152, No.1, (February 2008), pp. 103-111, ISSN 1744-7348

Jones, W. & Sforza, R. (2007). The European Biological Control Laboratory: an existing infrastructure for biological control of weeds in Europe. *EPPO bulletin,* Vol.37, No.1, (April 2007), pp. 163-165, ISSN 0250-8052

Kennedy, A. & Papendick, R. (1995). Microbial characteristics of soil quality. *Journal of Soil & Water Conservation,* Vol.50, No.3, (May-June 1995), pp. 243-248, ISSN 002-4561

Kennedy, A. & Schillinger, W. (2006). Soil quality and water intake in conventional-till vs. no-till paired farms in Washington's Palouse region. *Soil science society of America journal,* Vol.70, No.3, (May-June 2006), pp.940–949, ISSN 0361-5995

Kennedy, A. & Stubbs, T. (2007). Management effects on the incidence of jointed goatgrass inhibitory rhizobacteria. *Biological control: theory and application in pest management,* Vol.40, No.2, (February 2007) pp.213-221, ISSN 1049-9644

Kennedy, A., Elliott, L., Young, F., & Douglas, C. (1991). Rhizobacteria suppressive to the weed downy brome. *Soil science society of America journal,* Vol.55, No.3, (May-June 1991). pp.722-727, ISSN 1435-0661

Kennedy, A., Johnson, B., & Stubbs, T. (2001). Host range of a deleterious rhizobacterium for biological control of downy brome. *Weed science,* Vol.49, No.6, (Nov-Dec 2001), pp. 792-797, ISSN 0043-1745

Kleczewski, N. & Flory, S. (2010). Leaf blight disease on the invasive grass *Microstegium vimineum* caused by a *Bipolaris* sp. *Plant disease,* Vol.94, No.7, (July 2010), pp. 807-811, ISSN 0191-2917

Kohlschmid, E., Sauerborn, J., & Müller-Stöver, D. (2009). Impact of *Fusarium oxysporum* on the holoparasitic weed *Phelipanche ramosa*: biocontrol efficacy under field-grown conditions. *Weed research,* Vol.49, No.1, (November 2009), pp. 56-65, ISSN 0043-1737

Kremer, R. & Souissi, T. (2001). Cyanide production by rhizobacteria and potential for suppression of weed seedling growth. *Current microbiology,* Vol. 43, No.3, (September 2001), pp. 182–186, ISSN 0343-8651

Kremer, R., & Li, J. (2003). Developing weed-suppressive soils through improved soil quality management. *Soil & tillage research,* Vol.72, No.2, (August 2003), pp. 193-202, ISSN 0167-1987

Kremer, R. & Schulte, L. (1989). Influence of chemical treatment and *Fusarium oxysporum* on velvetleaf (*Abutilon theophrasti*). *Weed technology.* Vol.3, No.2, (April 1989), pp. 369-374, ISSN 0890-037X

Kremer, R., Caesar, A., & Souissi, T. (2006). Soilborne microorganisms of Euphorbia are potential biological control agents of the invasive weed leafy spurge. *Applied soil ecology,* Vol.32, No.1, (May 2006), pp. 27-37, ISSN 1744-7348

Kropp, B. & Darrow, H. (2006). The effect of surfactants and some herbicides on teliospore viability in *Puccinia thlaspeos* (Schub.). *Crop protection,* Vol.25, No.4, (April 2006), pp. 369-374, ISSN 0261-2194

Kropp, B., Hansen, D., & Thomson, S. (2002). Establishment and dispersal of *Puccinia thalaspeos* in field populations of Dyer's woad. *Plant disease,* Vol.86, No.3, (March 2002), pp. 241-246, ISSN 0191-2917

Kroschel, J. & Elzein, A. (2004). Bioherbicidal effect of fumonisin B1, a phytotoxic metabolite naturally produced by *Fusarium nygamai*, on parasitic weeds of the genus Striga. *Biocontrol science and technology*, Vol.14, No.2, (March 2004), pp. 117-128, ISSN 1360-0478

Lawrie, J., Greaves, M., Down, V., Morales-Aza, B., & Lewis, J. (2002a). Outdoor studies of the efficacy of *Alternaria alternata* in controlling *Amaranthus retroflexus*. *Biocontrol science and technology*, Vol.12, No.1, (February 2002), pp. 83-94, ISSN 1360-0478

Lawrie, J., Greaves, M., Down, V., & Western, N. (2002b). Studies of spray application of microbial herbicides in relation to conidial propagule content of spray droplets and retention on target. *Biocontrol science and technology*, Vol.12, No.1, (February 2002), pp. 107-119, ISSN 1360-0478

Leth, V., Netland, J., & Andreasen, C. (2008). *Phomopsis cirsii*: a potential biocontrol agent of *Cirsium arvense*. *Weed research*, Vol.48, No.6, (December 2008), pp. 533-541, ISSN 0043-1737

Li, J. & Kremer, R. (2006). Growth response of weed and crop seedlings to deleterious rhizobacteria. *Biological control: theory and application in pest management*, Vol.39, No.1, (October 2006), pp. 58-65, ISSN 1049-9644

Li, Y., Sun, Z., Zhuang, X., Xu, L., Chen, S., & Li, M. (2003). Research progress on microbial herbicides. *Crop protection*, Vol.22, No.2 (March 2003), pp. 247-252, ISSN 0261-2194

Lupwayi, N., Rice, W., & Clayton, G. (1998). Soil microbial diversity and community structure under wheat as influenced by tillage and crop rotation. *Soil biology & biochemistry*, Vol.30. No.13 (November 1998). pp. 1733-1741, ISSN 0038-0717

Magani, I., Ibrahim, A., & Avav, T. (2009). *Fusarium oxysporum* and post-emergence herbicide for the control of the parasitic plant *Striga hermonthica* in maize. *Biocontrol science and technology*, Vol.19, No.9-10, (October 2009), pp. 1023-1032, ISSN 1360-0478

Magdoff, F. (2001). Concept, components, and strategies of soil health in agroecosystems. *Journal of nematology*. Vol. 33, No.4, (December 2001), pp. 169-172, ISSN 0022-300X

Maxwell, A. & Scott, J. (2008). Pathogens on wild radish, *Raphanus raphanistrum* (Brassicaceae), in south-western Australia - implications for biological control. *Australasian Plant Pathology*, Vol.37, No.5, (2008), pp. 523-533, ISSN 1448-6032

Mazzola, M. (2004). Assessment and management of soil microbial community structure for disease suppression. *Annual Review of Phytopathology*, 2004. Vol.42, No.1 (March 2004), pp. 35–59, ISSN 0066-4286

Medd, R. & Campbell, M. (2005). Grass seed infection following inundation with *Pyrenophora semeniperda*. *Biocontrol science and technology*, Vol.15, No.1, (February 2005), pp. 21-36, ISSN 1360-0478

Mejri, D., Souissi, T., & Berner, D. (2010a). Evaluation of *Puccinia carduorum* for biological control of *Carduus pycnocephalus* in Tunisia. *Biocontrol science and technology*, Vol.20, No.7-8, (August 2010), pp. 787-790, ISSN 1360-0478

Mejri, D., Gamalero, E., Tombolini, R., Musso, C., Massa, N., Berta, G., & Souissi, T. (2010b). Biological control of great brome (*Bromus diandrus*) in durum wheat (*Triticum durum*): specificity, physiological traits and impact on plant growth and root architecture of the fluorescent pseudomonad strain X33d. *Biocontrol*, Vol.55, No.4, (August 2010), pp. 561-572, ISSN 1386-6141

Meyer, S., Nelson, D., & Clement, S. (2001). Evidence for resistance polymorphism in the *Bromus tectorum-Ustilago bullata* pathosystem: implications for biocontrol. *Canadian journal of plant pathology,* Vol.23, No.1, (March 2001), pp. 19-27, ISSN 0706-0661

Meyer, S., Quinney, D., Nelson, D., & Weaver, J. (2007). Impact of the pathogen *Pyrenophora semeniperda* on *Bromus tectorum* seedbank dynamics in North American cold deserts. *Weed research,* Vol.47, No.1, (February 2007), pp. 54-62, ISSN 0043-1737

Mohan Babu, R., Sajeena, A., Seetharaman, K., Vidhyasekaran, P., Rangasamy, P., Som Prakash, M., Senthil Raja, A., & Biji, K. (2002). Host range of *Alternaria alternata*--a potential fungal biocontrol agent for water hyacinth in India. *Crop protection,* Vol.21, No.10, (December 2002), pp. 1083-1085, ISSN 0261-2194

Mortensen, K. (1998). Biological control of weeds using microorganisms. In: *Plant-microbe interactions and biological control,* G. Boland and L. Kuykendall, (Eds.), 223-248, New York, Marcel Dekker, ISBN 0824700430 New York, NY

Müller-Stöver, D., Kohlschmid, E., & Sauerborn, J. (2009). A novel strain of *Fusarium oxysporum* from Germany and its potential for biocontrol of *Orobanche ramosa. Weed research,* Vol.49, No.2, (April 2009), pp. 175-182, ISSN 0043-1737

Müller-Stöver, D. & Sauerborn, J. (2007). A commercial iron fertilizer increases the survival of *Fusarium oxysporum* f. sp. *orthoceras* propagules in a wheat flour-kaolin formulation. *Biocontrol science and technology,* Vol.17, No.5-6, (May 2007), pp. 597-604, ISSN 1360-0478

Müller-Stöver, D., Buschmann, H., & Sauerborn, J. (2005). Increasing control reliability of *Orobanche cumana* through integration of a biocontrol agent with a resistance-inducing chemical. *European journal of plant pathology,* Vol.111, No.3, (March 2005), pp. 193-202, ISSN 0929-1873

Nechet, K. de L., Barreto, R., & Mizubuti, E. (2004). *Sphaceloma poinsettiae* as a potential biological control agent for wild poinsettia (*Euphorbia heterophylla*). *Biological control: theory and applications in pest management,* Vol.30, No.3, (July 2004), pp. 556-565, ISSN 1049-9644

Owen, A. & Zdor, R. (2001). Effect of cyanogenic rhizobacteria on the growth of velvetleaf (*Abutilon theophrasti*) and corn (*Zea mays*) in autoclaved soil and the influence of supplemental glycine. *Soil biology & biochemistry,* Vol.33, No.6, (May 2001), pp. 801-809, ISSN 0038-0717

Pedras, M., Ismail, N., Quail, J., & Boyetchko, S. (2003). Structure, chemistry, and biological activity of pseudophomins A and B, new cyclic lipodepsipeptides isolated from the biocontrol bacterium *Pseudomonas fluorescens. Phytochemistry,* Vol.62, No.7, (April 2003), pp. 1105-1114, ISSN 0031-9422

Peng, G. & Boyetchko, S. (2006). Effect of variable dew temperatures on infection of green foxtail by *Pyricularia setariae, Drechslera gigantea,* and *Exserohilum rostratum. Biological control: theory and application in pest management,* Vol.39, No.3, (December 2006), pp. 539-546, ISSN 1049-9644

Peng, G. & Byer, K. (2005). Interactions of *Pyricularia setariae* with herbicides for control of green foxtail (*Setaria viridis*). *Weed technology,* Vol.19, No.3, (July-September 2005), pp. 589-598, ISSN 550-2740

Peng, G., Bailey, K., Hinz, H. & Byer, K. (2005a). *Colletotrichum* sp: a potential candidate for biocontrol of scentless chamomile (*Matricaria perforata*) in western Canada.

Biocontrol science and technology, Vol.15, No.5, (August 2005), pp. 497-511, ISSN 1360-0478

Peng, G., Wolf, T., Byer, K., & Caldwell, B. (2005b). Spray retention on green foxtail (*Setaria viridis*) and its effect on weed control efficacy by *Pyricularia setariae. Weed technology,* Vol.19, No.1, (January 2005), pp. 86-93, ISSN 550-2740

Pereira, J., Barreto, R., Ellison, C., & Maffia, L. (2003). *Corynespora cassiicola* f. sp. *lantanae*: a potential biocontrol agent from Brazil for *Lantana camara. Biological control: theory and applications in pest management,* Vol.26, No.1, (January 2003), pp. 21-31, ISSN 1049-9644

Pomella, A., Barreto, R., & Charudattan, R. (2007). *Nimbya alternantherae* a potential biocontrol agent for alligatorweed, *Alternanthera philoxeroides. Biocontrol journal of the international organization for biological control,* Vol.52, No.2, (April 2007), pp. 271-288, ISSN 1386-6141

Qiang, S., Zhu, Y., Summerell, B., & Li, Y. (2006). Mycelium of *Alternaria alternata* as a potential biological control agent for *Eupatorium adenophorum. Biocontrol science and technology,* Vol.16, No.7-8, (August 2006), pp. 653-668, ISSN 1360-0478

Ray, P., Kumar, S., & Pandey, A. (2008). Deleterious effect of herbicides on water hyacinth biocontrol agents *Neochetina bruchi* and *Alternaria alternata. Biocontrol science and technology,* Vol.18, No.5-6, (May 2008), pp. 517-526, ISSN 1360-0478

Rayamajhi, M., Pratt, P. , Center T., & Van, T. (2010). Insects and a pathogen suppress *Melaleuca quinquenervia* cut-stump regrowth in Florida. *Biological control: theory and application in pest management,* Vol.53, No.1, (April 2010), pp.1-8, ISSN 1049-9644

Rietman, L., Shamoun, S., & van der Kamp, B. (2005). Assessment of *Neonectria neomacrospora* (anamorph *Cylindrocarpon cylindroides*) as an inundative biocontrol agent against hemlock dwarf mistletoe. *Canadian journal of plant pathology,* Vol.27, No.4, (Oct-Dec 2005), pp. 603-609, ISSN 0706-0661

Rosskopf, E., Yandoc, C., & Charudattan, R. (2006). Genus-specific host range of *Phomopsis amaranthicola* (Sphaeropsidales), a bioherbicide agent for *Amaranthus* spp. *Biocontrol science and technology,* Vol.16, No.1-2, (February 2006), pp. 27-35, ISSN 1360-0478

Ryan, M., Mirsky, S., Mortensen D., Teasdale, J., & Curran, W. (2011). Potential synergistic effects of cereal rye biomass and soybean planting density on weed suppression. *Weed science,* Vol.59, No.2, pp. 238-246, ISSN 0043-1745

Sands, D. & Pilgeram, A. (2009). Methods for selecting hypervirulent biocontrol agents of weeds: why and how. *Pest management science,* Vol.65, No.5, (May 2009), pp. 581-587, ISSN 1049-9644

Sauerborn, J., Müller-Stöver, D., & Hershenhorn, J. (2007). The role of biological control in managing parasitic weeds. *Crop protection,* Vol.26, No.3, (March 2007), pp. 246-254, ISSN 0261-2194

Schall, M. & Davis, D. (2009). *Ailanthus altissima* wilt and mortality: etiology. *Plant disease,* Vol.93, No.7, (July 2009), pp. 747-751, ISSN 0191-2917

Seixas, C., Barreto, R., Freitas, L., Maffia, L., & Monteiro, F. (2004). *Ditylenchus drepanocercus* (Nematoda), a potential biological control agent for *Miconia calvescens* (Melastomataceae): host-specificity and epidemiology. *Biological control: theory and applications in pest management,* Vol.31, No.1, (September 2004), pp. 29-37, ISSN 1049-9644

Shabana, Y. & Mohamed, Z. (2005). Integrated control of water hyacinth with a mycoherbicide and phenylpropanoid pathway inhibitor. *Biocontrol science and technology,* Vol.15, No.7, (November 2005), pp. 659-669, ISSN 1360-0478

Shabana, Y. (2005). The use of oil emulsions for improving the efficacy of *Alternaria eichhorniae* as a mycoherbicide for water hyacinth (*Eichhornia crassipes*). *Biological control: theory and application in pest management,* Vol.32, No.1, (January 2005), pp. 78-89, ISSN 1049-9644

Shabana, Y., Cuda, J., & Charudattan, R. (2003). Evaluation of pathogens as potential biocontrol agents of hydrilla. *Journal of phytopathology,* Vol.151, No.11-12, (Nov.-Dec 2003), pp. 607-613, ISSN 0973-7510

Sharma, P., Rai, P., Siddiqui, S., & Chauhan, J. (2011). First report of *Fusarium* wilt in the broomrape parasite growing on *Brassica* spp. in India. *Plant disease: an international journal of applied plant pathology,* Vol.95, No.1, (January 2011), pp. 75-95, ISSN 0191-2917

Shearer, J. & Jackson, M. (2006). Liquid culturing of microsclerotia of *Mycoleptodiscus terrestris,* a potential biological control agent for the management of hydrilla. *Biological control: theory and application in pest management,* Vol.38, No.3, (September 2006), pp. 298-306, ISSN 1049-9644

Shetty, K., Minnis, A., Rossman, A., & Jayachandran, K. (2011). The Brazilian peppertree seed-borne pathogen, *Neofusicoccum batangarum,* a potential biocontrol agent. *Biological control: theory and application in pest management.* Vol.56, No.1, (January 2011), pp. 91-97, ISSN 1049-9644

Souissi, T., & Kremer, R. (1998). A rapid microplate callus bioassay for assessment of rhizobacteria for biocontrol of leafy spurge (*Euphorbia esula* L.). *Biocontrol science and technology.* Vol.8, No.1, (January 1998), pp. 83-92, ISSN 0958-3157

Stewart-Wade, S. & Boland, G. (2004). Selected cultural and environmental parameters influence disease severity of dandelion caused by the potential bioherbicidal fungi, *Phoma herbarum* and *Phoma exigua. Biocontrol science and technology,* Vol.14, No.6, (September 2004), pp. 561-569, ISSN 1360-0478

Teshler, M., Ash, G., Zolotarov, Y., & Watson, A. (2007). Increased shelf life of a bioherbicide through combining modified atmosphere packaging and low temperatures. *Biocontrol science and technology,* Vol.17, No.3-4, (March 2007), pp. 387-400, ISSN 1360-0478

Tessmann, D., Charudattan, R., & Preston, J. (2008). Variability in aggressiveness, cultural characteristics, cercosporin production and fatty acid profile of *Cercospora piaropi,* a biocontrol agent of water hyacinth. *Plant pathology,* Vol.57, No.5, (October 2008), pp. 957-966, ISSN 0032-08

Tichich, R. & Doll, J. (2006). Field-based evaluation of a novel approach for infecting Canada thistle (*Cirsium arvense*) with *Pseudomonas syringae* pv. *tagetis. Weed science,* Vol.54, No.1, (January 2006), pp. 166-171, ISSN 0043-1745

Tichich, R., Doll, J., & McManus, P. (2006). *Pseudomonas syringae* pv. *tagetis* population dynamics both on and in Canada thistle (*Cirsium arvense*) leaves as affected by rain events. *Weed science,* Vol.54, No.5, (September 2006), pp. 934-940, ISSN 0043-1745

Tilley, A. & Walker, H. (2002). Evaluation of *Curvularia intermedia* (*Cochliobolus intermedius*) as a potential microbial herbicide for large crabgrass (*Digitaria sanguinalis*).

Biological control: theory and applications in pest management, Vol.25, No.1, (September 2002), pp. 12-21, ISSN 1360-0478

Tiourebaev, K., Semenchenko, G., Dolgovskaya, M., McCarthy, M., Anderson, T., Carsten, L., Pilgeram, A., & Sands, D. (2001). Biological control of infestations of ditchweed (*Cannabis sativa*) with *Fusarium oxysporum* f.sp. *cannabis* in Kazakhstan. *Biocontrol science and technology,* Vol.11, No.4, (August 2001), pp. 535-540, ISSN 1360-0478

United States Environmental Protection Agency (2005) http://www.epa.gov/oppbppd1/biopesticides/ingredients/factsheets/factsheet_028301.htm

Van Klinken, R. & Julien, M. (2003). Learning from past attempts: does classical biological control of Noogoora burr (Asteraceae: *Xanthium occidentale*) have a promising future? *Biocontrol science and technology,* Vol.13, No.2, (March, 2003), pp. 139-153, ISSN 1360-0478

Venne, J., Beed, F., Avocanh, A., & Watson, A. (2009). Integrating *Fusarium oxysporum* f. sp. *strigae* into cereal cropping systems in Africa. *Pest management science,* Vol.65, No.5, (May 2009), pp. 572-580, ISSN 1526-498X

Vidal, K., Guermache, F., & Widmer, T. (2004). In vitro culturing of yellow starthistle (*Centaurea solstitialis*) for screening biological control agents. *Biological control: theory and application in pest management,* Vol.30, No.2, (June 2004), pp. 330-335, ISSN 1049-9644

Vogelgsang, S. & Shamoun, S. (2004). Evaluation of an inoculum production and delivery technique for *Valdensinia heterodoxa,* a potential biological control agent for salal. *Biocontrol science and technology,* Vol.14, No.8, (December 2004), pp. 747-756, ISSN 1360-0478

Vurro, M., Zonno, M., Evidente, A., Andolfi, A., & Montemurro, P. (2001). Enhancement of efficacy of *Ascochyta caulina* to control *Chenopodium album* by use of phytotoxins and reduced rates of herbicides. *Biological control: theory and applications in pest management,* Vol.21, No.2, (June 2001), pp. 182-190, ISSN 1049-9644

Wandeler, H., Nentwig, W., & Bacher, S. (2008). Establishing systemic rust infections in *Cirsium arvense* in the field. *Biocontrol science and technology,* Vol.18, No.1-2, (February 2008), pp. 209-214, ISSN 1360-0478

Wander, M., Hedrick, D., Kaufman, D., Traina, S., Stinner, B., Kehrmeyer, S., & White, D. (1995). The functional significance of the microbial biomass in organic and conventionally managed soils. In: The significance and regulation of soil biodiversity. H.P. Collins, G.P. Robertson, M.J. Klug (Eds). pp.87-97, Kluwer Academic Publishers, Netherlands, ISBN 079-233-1389

Wapshere, A. (1974). A strategy for evaluating the safety of organisms for biological weed control. *Annals of applied biology,* Vol.77, No.2, (July 1974), pp. 201–211.ISSN 0003-4746

Ward, S., Reid, S., Harrington, J., Sutton, J., & Beck, K. (2008). Genetic variation in invasive populations of yellow toadflax (*Linaria vulgaris*) in the Western United States. *Weed science,* Vol.56, No.3, (May 2008), pp. 394-399, ISSN 0043-1745

Wilkin, J., Shamoun, S., Ritland, C., Ritland, K., & El-Kassaby, Y. (2005). Genetic diversity and population structure of *Valdensinia heterodoxa*, a potential biocontrol agent for salal in coastal British Columbia. *Canadian journal of plant pathology,* Vol.27, No.4, (Oct-Dec 2005), pp. 559-571, ISSN 0706-0661

Wood, A. & Crous, P. (2005). Epidemic increase of *Endophyllum osteospermi* (Uredinales, Puccinaceae) on *Chrysanthemoides monilifera. Biocontrol science and technology,* Vol.15, No.2, (March 2005), pp. 117-125, ISSN 1360-0478

Wymore, L. & Watson, A. (1989). Interaction between a velvetleaf isolate of *Colletotrichum coccodes* and thidiazuron for velvetleaf (*Abutilon theophrasti*) control in the field. *Weed science,* Vol. 37, No.3, (May 1989), pp.478-483.

Wyss, G., Charudattan, R., Rosskopf, E., & Littell, R. (2004). Effects of selected pesticides and adjuvants on germination and vegetative growth of *Phomopsis amaranthicola,* a biocontrol agent for *Amaranthus* spp. *Weed research,* Vol.44, No.6, (December 2004), pp. 469-482, ISSN 0043-1737

Wyss, G. & Müller-Scharer, H. (2001). Effects of selected herbicides on the germination and infection process of *Puccinia lagenophora,* a biocontrol pathogen of *Senecio vulgaris. Biological control: theory and applications in pest management,* Vol.20, No.2, (February 2001), pp. 160-166, ISSN 1049-9644

Yamoah, E., Jones, E., Bourdot, G., Suckling, D., Weld, R., & Stewart, A. (2008). Factors influencing pathogenicity of *Fusarium tumidum* on gorse (*Ulex europaeus*). *Biocontrol science and technology,* Vol.18, No.7-8, (September 2008), pp. 779-792, ISSN 1360-0478

Yonli, D., Traore, H., Hess, D., Abbasher, A., & Boussim, I. (2004). Effect of growth medium and method of application of *Fusarium oxysporum* on infestation of sorghum by *Striga hermonthica* in Burkina Faso. *Biocontrol science and technology,* Vol.14, No.4, (June 2004), pp. 417-421, ISSN 1360-0478

Yuzikhin, O., Mitina,G., & Berestetskiy, A. (2007). Herbicidal potential of stagonolide, a new phytotoxic nonenolide from *Stagonospora cirsii. Journal of agriculture & food chemistry,* Vol.55, No.19, (September 2007), pp. 7707–7711, ISSN 1520-5118

Zdor, R., Alexander, C. & Kremer, R. (2005). Weed suppression by deleterious rhizobacteria is affected by formulation and soil properties. *Communications in soil science and plant analysis,* Vol.36, No.9-10, (February 2007), pp. 1289-1299, ISSN 1532-2416

Zermane, N., Souissi, T., Kroschel, J., & Sikora, R. (2007). Biocontrol of broomrape (*Orobanche crenata* Forsk. and *Orobanche foetida* Poir.) by *Pseudomonas fluorescens* isolate Bf7-9 from the faba bean rhizosphere. *Biocontrol science and technology,* Vol.17, No.5-6, (June 2007), pp. 483-497, ISSN 1360-0478

Zhang, W., Sulz, M., Bailey, K., & Cole, D. (2002). Effect of epidemiological factors on the impact of the fungus *Plectosporium tabacinum* on false cleavers (*Galium spurium*). *Biocontrol science and technology,* Vol.12, No.2, (April 2002), pp. 183-194, ISSN 1360-0478

Zhang, Z., Burgos, N., Zhang, J., & Yu, L. (2007). Biological control agent for rice weeds from protoplast fusion between *Curvularia lunata* and *Helminthosporium gramineum. Weed science,* Vol.55, No.6, (November 2007), pp. 599-605, ISSN 0043-1745

Zhang, J., Wang, W., Lu, X., Xu, Y., & Zhang, X. (2010). The stability and degradation of a new biological pesticide, pyoluteorin. *Pest management science,* Vol.66, No.3, (March 2010), pp. 248-252, ISSN 1526-4998

Zhao, S. & Shamoun, S. (2005). Effects of potato dextrose broth and gelatin on germination and efficacy of *Phoma exigua,* a potential biocontrol agent for salal (*Gaultheria shallon*). *Canadian journal of plant pathology,* Vol.27, No.2, (Apr-June 2005), pp. 234-244, ISSN 0706-0661

Zhao, S. & Shamoun, S. (2006). Effects of culture media, temperature, pH, and light on growth, sporulation, germination, and bioherbicidal efficacy of *Phoma exigua*, a potential biological control agent for salal (*Gaultheria shallon*). *Biocontrol science and technology*, Vol.16, No.9-10, (October 2006), pp. 1043-1055, ISSN 1360-0478

Zhou, L., Bailey, K., & Derby, J. (2004). Plant colonization and environmental fate of the biocontrol fungus *Phoma macrostoma*. *Biological control: theory and applications in pest management*, Vol.30, No.3, (July 2004), pp. 634-644, ISSN 1049-9644

Zhu, Y. & Qiang, S. (2004). Isolation, pathogenicity and safety of *Curvularia eragrostidis* isolate QZ-2000 as a bioherbicide agent for large crabgrass (*Digitaria sanguinalis*). *Biocontrol science and technology*, Vol.14, No.8, (December 2004), pp. 769-782, ISSN 1360-0478

Zidack, N. & Quimby, P. Jr. (2002). Formulation of bacteria for biological weed control using the Stabileze method. *Biocontrol science and technology*, Vol.12, No.1, (February 2002), pp. 67-74, ISSN 1360-0478

9

Velvet Bean (*Mucuna pruriens* var. *utilis*) a Cover Crop as Bioherbicide to Preserve the Environmental Services of Soil

Angel Isauro Ortiz Ceballos[1], Juan Rogelio Aguirre Rivera[2], Mario Manuel Osorio Arce[3] and Cecilia Peña Valdivia[4]
[1]*Instituto de Biotecnología y Ecología Aplicada (INBIOTECA), Universidad Veracruzana, Xalapa*
[2]*Instituto de Investigaciones de Zonas Desérticas, Universidad Autonoma de San Luis Potosí, San Luis Potosí*
[3]*Colegio de Postgraduados-Campus Tabasco, H. Cárdenas*
[4] *Colegio de Postgraduados-Campus Montecillo, Texcoco*
México

1. Introduction

Cover crops have been used as green manure since the Zhou dynasty (1134-247 BCE) in China. Later on, Greeks and Romans used legumes widely as part of their crops. Pliny, Virgil, Theophrastus and other philosophers (Cato, Varro, Columella and Palladius) wrote about the use of legumes as cover crops (Allison, 1973; Tivy, 1990; Winiwarter, 2006). Also, the Codex Vergara and Florentine Codex,-the most comprehensive textual encyclopedia of Aztec soil knowledge shows the role of application of soil amendments as practice to maintaining or increasing soil fertility (Williams, 2006). The use of legumes as associated or rotation crops or their cultivation as green manure was a strategy to replenish the soil nitrogen that had been used up by crops and to provide organic matter necessary to maintain the soil's physical and chemical conditions favourable for sustained crop production (Mulvaney et al., 2009).

Since the 1940's, economic policies based on the wide availability of cheap fossil fuels and chemical fertilizers encouraged the adoption of maximum-yield production systems, without any consideration of their sustainability or their environmental impact. As a consequence, the use of and research on cover crops, crop rotation and other traditional soil management practices were radically abandoned. However, fertilizers have not replaced the function of organic matter and other management practices; rather, soil erosion and toxic waste have disproportionately increased with the increase of agricultural production, leading to a progressive decline in crop productivity due to soil degradation and contamination of aquifers and surface water bodies. Land productivity has also declined due to increasing problems of weed infestation, pests and diseases (Cox &Atkins, 1979; Mulvaney et al., 2009; Yates et al., 2011).

Farmers in tropical regions use slash-and-burn farming as a common soil-use method. This approach can be sustained under long fallow periods, low population density and strictly for subsistence demands. However, the much higher density of modern populations and their exacting demands for resources make the approach to crumble and decay: Soils start degrading and herbicides, pesticides and chemical fertilizers are needed to sustain crops, with the ensuing and well known economic and environmental problems (Greenland, 1975; Cox & Atkins, 1979; Bandy et al., 1993).

With the widespread use of chemical fertilizers, cover crops and green manures are now disregarded as inefficient, expensive nitrogen supplies, without due consideration of their various biological benefits. Despite the dominant technological trend, some farmers in the tropical regions of Latin America have developed and promoted, on their own initiative, the use of Velvetbean or Picapica mansa (*Mucunu spp.*) as a means to sustainably use clear-cut fields. This way, farmers are able to sustainably and inexpensively grow subsistence and other various crops thanks to the biocontrol of weeds, pests and diseases and the preservation of soil fertility that are provided by using Velvetbean for crop rotation or intercropping (Flores, 1989; Buckles & Perales, 1994; Ortiz-Ceballos & Fragoso, 2004). Clearly, the cover crop concept has to be considered in a broadest sense and not only as a rudimentary way to add nitrogen to the soil.

The Velvetbean technique is becoming better known and has been evaluated by researchers, but no attempts to improve, further develop and disseminate it have been made. Modern criteria and methods to comparatively evaluate the benefits provided by different cover crops are lacking and the properties of *Mucuna* cultivars are still unknown.

2. Green manure and cover crop concepts

Several authors have defined green manure as herbaceous, shrubby or woody plant material that is grown either *in situ* or *ex situ*, and is then incorporated into the soil when still green or before reaching full maturity to maintain and/or improve soil fertility (Allison, 1973; Hauck, 1977; Yost & Evans, 1988; Sarrantonio, 1991). Several different forms to apply green manure to soil do exist:

a. Simultaneous cut and burial
b. Cut and scattering over the soil surface
c. Incorporation of material produced and harvested at some other field
d. Cut to prepare soil improvers and compound manures ("compost")

Cover crops are plant species cultivated as rotation crops or intercropped with annual or perennial crops to control weeds, nutrient loss and weathering, to protect the soil by preventing erosion and to supply biological nitrogen for crop or livestock nutrition (Purseglove, 1987; Yost & Evans, 1988; Sarrantonio, 1991).

However, green manures and cover crops are often considered as expensive, inefficient production elements (as nitrogen sources), disregarding the various environmental services they perform. Farmers, through their long-standing relationship with plants, have incorporated green manures and cover crops into their farming systems as a strategy for sustainable production, as they provide a means for the biocontrol of weeds, pests and diseases and for maintaining soil fertility. For these reasons, the cover crop concept has to be considered in a broadest sense and not only as a rudimentary way to add nitrogen to the soil. This comprehensive concept has been adopted in this work to use velvetbean as bioherbicide.

3. Historical use of cover crops and green manures

Interest to preserve soil fertility for crop production through the use of legumes as cover crops or green manures has been known since the origin of agriculture (Allison, 1973; Tivy, 1990; Sarrantonio, 1991; Winiwarter, 2006; Williams, 2006).

In China and Japan, the use of cover crops and green manures -mainly legumes- that were incorporated directly to the soil as soil improvers or used as rotation, associated or relay crops to support cereal crops has been documented since over 3000 years ago. It was also a common practice in Greece, Rome and Mesoamerica, under conditions of cheap labour, water restrictions and lack of inorganic fertilizers (Allison, 1973; Hauck, 1977; Winiwarter, 2006; Williams, 2006). By 1976, the surface area cultivated with green manures in China was nearly 6.6 million hectares (Hauck, 1977).

Such soil conservation and management practices developed by ancient agricultural civilizations were adopted and remained virtually unchanged in most of the agricultural regions of the world until 1840 when the practice of soil fertilization with guano and superphosphates began. However, widespread use of those materials did not start until 1900. In Germany, the chemical fertilizers industry began in 1850, incentivized by Julius von Liebig's work on fertility and plant nutrition. Overall, it was not until 1945 when the intensive use of synthetic fertilizers in agriculture and livestock raising began in the USA, Europe and Australia (Allison, 1973; Tivy, 1990; Mulvaney et al., 2009).

Traditional agriculture in the USA, Europe, Australia and Mesoamerica consisted of prairie rotation with a wide variety of cereal, root and legume crops, which allowed for a diversity of ploughing methods and nutritional demands over the three to six years rotation cycle. The advantages of this were: a) protection against price volatility and losses due to pests and disease; b) better use of the land, according to climate conditions through the year and, perhaps more importantly, c) maintenance of soil fertility. Thus, by stabilizing the content of nitrogen-rich organic matter, the growth of beneficial microorganisms was promoted and the soil's physical and chemical conditions were improved leading to a sustained crop production (Mckee, 1948; Cobb, 1950; Allison, 1973; Tivy, 1990).

Until then, agricultural research was focused on optimizing and explaining the benefits derived from prairie and crop rotation, as opposed to continuous cultivation, in terms of fertility maintenance, nitrogen supply, improvement of the soil's physical and chemical conditions, and biocontrol of weeds, pests and diseases (Weindling, 1946; Snyder et al.,1959; Mulvaney et al., 2009). Such traditional system integrating agriculture and livestock raising prevailed until the mid twentieth century. However, with the adoption of synthetic nitrogen fertilizers and herbicides, that line of research was abandoned and the use of cover crops, crop rotation and other similar soil management techniques were discontinued, as yields obtained with those were two to four times lower than those obtained with fertilizers (Allison, 1973; Greenland, 1975; Tivy, 1990; Mulvaney et al., 2009; Yamada et al., 2009).

Modern intensive agriculture is characterized by the use of sophisticated technologies, heavy investment of input capital, high crop and animal yields and maximum efficiency.

This demands a large energy input either directly in form of human and animal labour, fossil fuels and electricity, or indirectly in the form of fertilizers, herbicides, pesticides, seed, water and other agrochemicals (Tivy, 1990). The intensification of agriculture and livestock raising relies on larger inputs, the most important of which are those related to plant and animal nutrition. Crop yields depend on fertilizers, pesticides, herbicides and growth promoters; animal yields depend on nutrient-rich fodder (Tivy, 1990; Mulvaney et al., 2009; Yamada et al., 2009).

The intensification of modern agriculture began some 40-60 years ago as a result of scientific and technological developments collectively referred to as the "Green Revolution". The Green Revolution started immediately after the Second World War and was characterized by: a) increased mechanization of soil management, b) increased use of fertilizers, insecticides, pesticides, herbicides, veterinarians and other non-essential additives, and c) a fast and widespread development of genetic improvement programs for plant and animal species (Wade, 1974; Greenland, 1975; Tivy, 1990; Mulvaney et al., 2009; Yamada et al., 2009; Yates et al., 2011). Those scientific and technological developments were rapidly and widely adopted in the USA, Europe and Australia, while socio-economic and cultural factors limited their use in Latin America, Africa and Asia (Greenland, 1975; Pimentel et al., 1980; Yamada et al., 2009). This technological phenomenon was accompanied by a high specialization in production and the separation of agricultural and animal production (Greenland, 1975; Tivy, 1990; Mulvaney et al., 2009).

Such intensification has imposed a considerable cost on the environment, due to: a) increased soil erosion due to the soil exposure to weather effects; the ensuing increased silting has reduced the efficiency of impoundments and irrigation channels and has impaired navigation, with a concomitant loss of nutrients and water quality; b) ecosystem degradation and destruction, as well as drastic changes in the physical environment directly caused by farming activities, followed by the decimation or loss of plant and animal species; c) the use of large amounts of fertilizers has polluted water bodies and aquifers with nitrates and phosphates; d) pesticides have fostered the development of resistance and persistence in pests and pathogens and have affected wild species, particularly birds (Cox & Atkins, 1979; Pimentel et al., 1980; Tivy, 1990; Mulvaney et al., 2009). The ensuing dilemma is that efforts to modify one part of the farming environment have degraded other equally important components: natural ecosystems (Tivy, 1990).

4. Persistence of the slash-and-burn system in maize

The slash-and-burn system has been practiced for millennia in many tropical or warm-humid regions in Latin America, Africa and Asia. Evidence suggests that this system began in New Guinea some 5,000 years BCE. The slash-and-burn system is currently used in approximately 800 to 1,400 million hectares globally, some 11 to 25 million of which are actually cultivated every year, providing basic staples for the subsistence of over 300 million people inhabiting those regions (Mabberley, 1992). A vast literature has been produced on this subject (Greenland, 1975; Cox & Atkins, 1979; Lal, 1987; Tivy, 1990; Mabberley, 1992; Bandy et al., 1993; Willis et al., 2004; Eastmond & Faust, 2006), but a brief synthesis of the main features of this farming system seems appropriate here.

This traditional farming system varies as a function of: a) climate, soil and vegetation; b) farmers' background; c) population density, and d) demand on available land. The system is applied on small clear-cut parcels, sometimes separated one from the other as farmers only have manual tools such as axe, machete and sowing stick available, which restricts the size of the area that can be clear-cut and cultivated every year. Land is clear-cut during the dry season of the year and only partially, as some tree species that facilitate vegetation regeneration during the fallow period are left standing. The dry plant material is burned before plantation starts. Fire is a cheap and powerful tool that humans have been using for over 1.0 - 1.7 million years to quickly and efficiently clear the land, a process which also allows for: a) pest and disease control, b) slowing down the germination of weeds' seeds, c) accelerating the microbial populations growth, d) increasing mineralization of organic nitrogen and other nutrients stored in ashes, and e) lowering the crops' costs.

Nutrients stored in ashes and on the soil surface only last for two to three years of continuous cultivation as they are quickly depleted. The nutrients amount and quality depend on the site potential. Weeds are controlled either manually or using simple tools. A variety of crop species are grown, including maize, rice, manihot, yam, taro, bean, pumpkin, hot pepper, sweet potato and others. Cultivation techniques vary in their spatial distribution, including crop rotation, association or intercropping, which allow for production diversification, an optimal use of land and protecting the soil from the destructive effects of weather. Crop yields rapidly decrease after two to four years of cultivation, due to nutrient depletion, increase in hard-to-control weeds (such as *Imperata cylindrical, Sorghum halepense* and *Cyperus rotundus*) and damages by pests and diseases. As a consequence, a) fertilizers, herbicides and pesticides are required to sustain further cultivation, with well-known economic and environmental disadvantages, or b) the parcel has to be left to fallow for five to twelve years, or c) is abandoned.

When land is left under fallow, new vegetation develops from seeds stored in the soil's seed bank or arriving from the surrounding vegetation and from shoots emerging from previously present species, through a successional process the course of which varies as a function of soil type, climate, cultivation history, soil degradation and presence of still standing species. Second-growth vegetation takes up the few nutrients that still remained in the top soil and that are leached down to lower horizons, to bring them back to the surface in the form of organic matter, thus closing up the nutrient cycle, humus accumulation, weed suppression and the control of pests and diseases. Since ancient times humans have collected and cultivated a number of useful wild species from second-growth vegetation, some of which are commercially valuable nowadays (e.g., mahogany, cedar, rubber, banana, cocoa, palms, etc.). However, as farmer populations in tropical regions become larger and impose more exacting demands on natural resources, the slash-and-burn system tends to break down and become unsustainable, due to the loss of soil fertility and increased erosion resultant from the intensification and protraction of continuous cultivation. In addition, populations immigrating from other environments are often unfamiliar with or do not follow the farming systems traditionally practiced by native populations of tropical regions, which leads to more severe and faster depletion of resources. These factors lead to a shortening of the fallow period which breaks the nutrient cycle down and disrupts the biological control of weeds, pests and diseases.

Nowadays, the slash-and-burn farming system is berated for its impacts on the biosphere as it contributes with over 20-30% of the global CO_2 emissions causing the greenhouse effect

(Kremen et al., 2000). In addition, tropical rain forests harbour a wealth of biological diversity the destruction of which entails a significant loss of plants and animals with actual or potential value for humanity.

5. Velvetbean or "Picapica mansa" (*Mucuna spp.)* cover crop as a means for a sustained use of soil

As discussed in the previous sections, legume cover crops have recently drawn attention in the USA, Europe, Australia, Latin America, Africa and Asia due to a) the unavoidable depletion of fossil fuels and the increase in the cost of nitrogen fertilizers; b) biosphere degradation caused by traditional intensive farming systems; and c) the unsustainability of the slash-and-burn system due to population growth. In addition, cover crops are also attractive for biocontrol of weeds, pests and diseases and for protecting the soil from erosion and weathering (Allison, 1973; Greenland, 1975; Hauck, 1977; Reddy et al., 1986; Lal, 1987; Yost & Evans, 1988; Smyth et al., 1991; Carsky et al., 2001; Elittä et al., 2003; Kaizzi et al., 2004; Baijukya et al., 2005; Yamada et al., 2009; Mulvaney et al., 2009; Olorunmaiye, 2010; Odhiambo et al., 2010; Odhiambo, 2011).

Several authors have described various farming systems including legume cover crops in various tropical regions of the world. Those crops are used for rotation, intercropping or associated with rice, maize, plantations, root, tuber crops, tomato, sorghum, cassava, etc. (Whyte et al., 1955; Gray, 1969; Warriar, 1969; Kay, 1985; Yost & Evans, 1988; López, 1993; Flores, 1989; Kaizzi et al., 2004; Wang et al., 2009; Olorunmaiye, 2010).

5.1 Characteristics of *Mucuna* spp.

The genus *Mucuna*, belonging to the Fabaceae family, covers perhaps 100-150 species of annual and perennial legumes. The study taxon (*Mucuna* spp.) is known by various names in different tropical regions of the world: "Picapica mansa" (Veracruz, México) and "Nescafé" (Tabasco, México), as it is occasionally used as coffee substitute; "frijol de abono" (manure bean in Honduras); "frijol de mula" (mule's bean in Guatemala); "haba de terciopelo" (velvet broadbean in Puerto Rico); "poroto aterciopelado" (velvetbean in Argentina); "ojo de venado" (deer's eye in Spain); "haricot velouté" (Francia); "makhmali sem" (India); "Stizolobia" (Italia); velvetbean" (USA); and banana stock pea (Australia).

This legume is native to Malaysia, South China, China and India but nowadays is widely distributed in many tropical regions. Cultivated and wild varieties from America and Africa were originally introduced and propagated by humans along various commercial routes. The main differences among cultivated species are in the character of the pubescence on the pod, the seed colour, and number of days to harvest of the pod. So far, improved cultivars have only been produced in: a) Australia (White, Mauritius, Black Mauritius, Somerset, Marbilee, Smith and Jubilack), b) USA (Georgia, Alabama, Osceola, Yokohama and Florida, and c) Zimbabwe (Bengal, White Stigless, SES 30, SES 45, SES 68 SES 74 and SES 108). Throughout its distribution range, the species has also been given various scientific names, among them: *M, deeringianum* Bort., *M. utilis* (Wall) Baker ex Burck, *M. pruriens* (L.) DC, *M. cochichinensis* (Lour.) Burk, *M. nivea* (Roxb.) Kuntze, *M. capitata, M. aterrima* Piper & Tracy, *M. Hassjo, M. diabolica, M. cinerum, M. haltonii,*and *M. sloanei* Fawcett & Rendle, some of which are just synonyms but others may represent valid names referring to different taxa

(Whyte et al., 1955; Duke, 1981; Göhl, 1982; Kay, 1985). In addition, an argument among taxonomists on whether the taxon is *Mucuna* or *Stizolobium* still exists; some recognize morphological features that set them apart while others maintain that they are synonyms (Whyte et al., 1955; Duke, 1981; Göhl, 1982; Kay, 1985; Pugalenthi et al., 2005; Zaim et al., 2011).

The genus *Mucuna* includes species of annual and perennial plants with vigorous indeterminate growth that, under favourable conditions, can produce vines 3 to 18 m long (occasionally longer). These plants might grow under short days, with a life-cycle length ranging from 120 to 330 days. Leaves are trifoliate, with oblique lateral folioles 5-20 cm long, 3-15 cm wide. Flowers white to purple, four to six flowers arranged in hanging racemes 2-3 cm long, flowers with wing and keel 3-4 cm long, much longer than the 2 cm long banner. Pods are 5 to 15 cm long, 1-2 cm wide, with three to six seeds, covered by a velvety pubescence, black to white or absent. Seeds are 1-2 cm long, 5-6 cm thick, colour cream, bright black or mottled brown, hilum 3-5 mm long and long aril. Numerous roots 7-10 m long, with abundant nodules near the soil surface. The plants accumulate between 2.2 and 10.9 t/ha of dry biomass and produce between 0.24 and 6.12 t/ha of seed (Duggar, 1899; Tracy & Coe, 1918; Scott 1919; Watson, 1922; Whyte et al., 1955; Duke, 1981; Göhl, 1982; Kay, 1985; Purseglove, 1987; Pugalenthi et al., 2005).

Mucuna grows better in warm humid climates, with annual precipitation from 3.8 to 31.5 dm and temperatures between 18.7 and 30 °C; night temperatures of 21 °C promote flowering. The plants are sensitive to frost during the growth season, drought tolerant once established but do not tolerate excess moisture. They grow on various soil types, with pH between 4.5 and 7.7 (Whyte et al., 1955; Duke, 1981; Kay, 1985; Pugalenthi et al., 2005).

The dry weight composition of *Macuna* green forage, pods and seeds is as follows: a) Foliage: 10.8 to 23.5% protein; 2.1% fat; 48.6% nitrogen-free extract; 19.3% fibre; 14.9% ash; 10.7% digestible protein; 49.6% digestible carbohydrates; 63.4% total digestible nutrients; b) dry pods: 10.0% moisture; 13.4 to 18.1% protein; 13.4% digestible protein; 73.8% total digestible nutrients; 13.0% raw fibre; 4.4% fat and 4.2% ash; c) Seeds: 10.0 % moisture; 19.0 to 37.5% protein; 4.7-9.0% fat; 51.5% nitrogen-free extract; 81.7% total digestible nutrients; 5.3-11.5% raw fibre and 2.9-5.7% ash (Duke, 1981; Göhl, 1982; Kay, 1985; Pugalenthi et al., 2005). The studies on mineral composition of *Mucuna* seeds reveal that they contain potassium, calcium, iron, manganese, zinc, copper, magnesium, phosphorous and sodium with 778-1846, 104-900, 1.3-15, 0.6-9.3, 1.0-15.0, 0.3-4.3, 85-477, 98-498 and 12.7-150.0 mg/100 g, respectively. Among the amino acids found in seeds, the aspartic and glutamic acids are found to be predominant (8.9-19 and 8.6-14.4%, respectively), whereas the levels of other amino acids are found to be low (Pugalenthi et al., 2005).

The seeds of *Macuna* also contain many antinutritional factors such as total free phenolics, tannins (3.1-4.9%), L-Dopa (4.2-6.8%), lectins (0.31-0.71%), protease inhibitors (trypsin and chymotrypsin), phytic acid, flatulence factors (Oligosaccharides), saponins (1.15-1-31%), and hydrogen cyanide (58 mg/kg), alkaloids (Pugalenthi et al., 2005).

Seeds are used for human consumption in some parts of Africa and India, which is just reasonable given their high contents of protein and minerals and low fibre content. For this same reason, their use as part of the diet of monogastric animals is also attractive, although in limited amounts due to their alkaloid contents. Experiments showed that rats fed with "picapica mansa" seeds died after 72 hr, while pigs and milking cows fed with seed-rich diets showed an impoverished quality of fat and milk, respectively (Duggar, 1899; Duke, 1981; Göhl, 1982; Kay, 1985; Pugalenthi et al., 2005). Thus, *Mucuna* seeds seem to constitute a valuable but still underutilized resource in tropical regions as the seed's toxic compounds have first to be eliminated or reduced either through conventional methods or by breeding *ad hoc* cultivars (Pugalenthi et al., 2005).

Mucuna is resistant and tolerant to many pests and diseases, probably due to its content of 3-4-dihydroxyphenylalanine (L-Dopa) and N,N-Dimethyltryptamine (DMT) in leaves and seeds, which provide a chemical barrier to the attack of insects and small mammals. However, it is severely attacked by a) fungi such as *Cercospora stizolobii, Mycospharella cruenta, Phyllostica mucunae, Phymatotrchum omnivorium, Phytophthora dreschsleri, Rhizoctonia solani, Sclerotium rolfsii* and *Pestalotiopsis versicolor*; b) bacteria such as *Xanthomonas stizolobiicola, Pseudomonas stizolobii, P. syringae* and *Striga gesnerioides*; c) virus: Bean common mosaic virus (BCMV), Bean pod mottle virus (BPMV), Bean yellow mosaic virus (BYMV), Cowpea mosaic virus (CoMV), Soybean mild mosaic virus (SMMV), Soybean mosaic virus (SMV), Soybean stunt virus (SSV), True broad bean mosaic virus (TBBMV), Tobacco ringspot virus (TRV), Tobacco streak virus (TSV), Watermelon mosaic virus-II (WMV-II) and Velvetbean severe mosaic virus (VbSMV) as new species of genus *Begomovirus*; and d) nematodes: *Meloidogyne thamesi, M. hapla, M. incognita* and *M. javanica* (Duke, 1981; Kay, 1985; Zaim et al., 2011).

5.2 History of the use of *Mucuna*

Since some 50 yr ago, farmers in some regions of Latin America have developed and disseminated the use of "Velvetbean" as rotation or associated crop to make a continuous use of clear-cut lands for production of subsistence crops (Duke, 1981; Kay, 1985; Triomphe 1996; Ortiz-Ceballos & Fragoso, 2004). Little is known about the introduction of "Velvetbean" to Latin America. Bort (1909) and Duke (1981) claim that this species was initially introduced to the USA in 1876, where it was cultivated and improved and, later on (in the 1920's), was introduced to Mexico and Central America by banana companies. However, the question remains whether it may have been introduced earlier than that through some other commercial routes between Asia and other Latin American countries, as it has happened with other species.

"Velvetbean" was grown in Southern USA because it constituted an important resource as a protein-rich seed forage for feeding cows (for meat and milk production) and pigs, as well as for the nitrogen supply it provided when used as a cover or rotation crop, to maintain the fertility of land cultivated with cereals, cotton and citric crops. For these reasons, the development of short-cycle cultivars through artificial selection and genetic improvement of naturalized and newly introduced varieties was promoted (Whyte et al., 1955; Duke, 1981; Bort, 1909). By 1918, over three million hectares were cultivated with "Velvetbean" in Florida, Mississippi, Alabama and Georgia, using technologies developed in research centres and based on its use as green manure (Tracy & Coe, 1918; Scott, 1919; Watson, 1922; Bort, 1909). The use of "Velvetbean" decreased in the early 1920's but remained important until the mid 1940's. Afterwards, the crop just disappeared from agricultural production statistics as a consequence of the increase in the use of nitrogen fertilizers and the cultivation of soybean as a commercial crop.

This species has also been grown in the tropical regions of over 20 countries as cover crop or green manure, either as a rotation or associated crop, to maintain soil fertility, as forage and for biological control of weeds, pests and diseases, in association with crops such as maize, rice, sorghum, sugar cane, manihot, banana, coconut, citric crops, coffee, rubber and prairies (Whyte et al., 1955; Pugalenthi et al., 2005).

6. Advantages and disadvantages of the use of *Mucuna*

At present, the use of *Mucuna* is a traditional technology that is becoming better known, described and evaluated by researchers both, in experimental fields and in farmers' parcels in various tropical regions of Latin America and Africa. In the following paragraphs, we present a summary review of the advantages and benefits that can be obtained in the biological control of weeds, pests and diseases and the maintenance of soil fertility by using *Mucuna* as a rotation or associated crop.

With regard to dry biomass accumulation, average yields of 7.9 ± 4.7 t/ha are obtained, with 19.0, maximum and 1.5 t/ha, minimum (Table 1).

Biomass (dry matter)	Location	Reference
2.9 - 7.3	Veracruz, Mexico	Eilitta et al. (2003)
18.0 - 19.0	Veracruz, Mexico	Buckles & Perales (1994)
3.4 - 5.3	Tabasco, Mexico	Ortiz-Ceballos et al. (2004)
2.0 - 3.9	Veracruz, Mexico	Ruiz & Laird (1964)
7.0 - 10.7	Manaus, Brasil	Smyth et al. (1991)
4.4 - 7.1	Florida, USA	Reddy et al. (1986)
3.1 - 8.3	Kaduna, Nigeria	Carsky et al. (2001)
7.0 - 16.3	Tela and Jutiapa, Honduras	Triomphe (1996)
10.6 - 12.4	Bulegeni and Kibale, Uganda	Kaizzi et al. (2004)
2.0 - 13.8	Kaduna, Nigeria	Franke et al. (2004)
3.3 - 11.2	Hwedza, Zimbadwe	Whitbread et al. (2004)
1.6 - 9.8	Limpopo, South Africa	Odhiambo (2011)
3.7 - 9.8	Limpopo, South Africa	Odhiambo et al. (2010)
1.5 - 8.7	Southern, Benin Republic	Vanlauwe et al. (2001)
7.4 - 10.2	Florida, USA	Wang et al. (2009)
9.9 - 10.6	Vihiga, Kenya	Kiwia et al. (2009)

Table 1. Biomass yield (t/ha) of different cultivars of Velvetbean (*Mucuna pruriens* var. *utilis*) in the system rotation with maize.

Seed yields average 2.6 t/ha (maximum 6.12 t/ha and minimum 0.24 t/ha). Accumulated dry biomass contributes an average of 189.3 ± 112.2 kg/ha (range 3.0 - 430 kg/ha) of inorganic nitrogen when is incorporated into the soil (Table 2). This wide variation can be explained by the influence of the sowing season and density, as well as the phenological stage at the time of incorporation which influences the biomass content of organic carbon, nitrogen and structural carbohydrates (Odhiambo, 2011). Temperature, humidity and actual evapotranspiration in the habitat are also important factors influencing the rate of biomass accumulation and its decomposition by soil microorganisms. The best time to incorporate the cover crop into the soil is at the onset of pod filling, when some 50 to 75% of the growth cycle has been completed, as maximum yields of dry biomass and accumulated nitrogen can be obtained this way (Allison, 1973; Gerónimo et al., 2002).

Nitrogen	Location	Reference
272 - 316	Tela and Jutiapa, Honduras	Triomphe (1996)
68 - 111	Veracruz, Mexico	Ruiz & Laird (1964)
130 - 330	Tabasco, Mexico	Ortiz-Ceballos et al. (2004)
100 - 190	Florida, USA	Reddy et al. (1986)
168 - 254	Manaus, Brasil	Smyth et al. (1991)
50 - 147	Veracruz, Mexico	Eilitta et al. (2003)
13 - 279	Limpopo, South Africa	Odhiambo (2010)
3 - 279	Limpopo, South Africa	Odhiambo et al. (2011)
150 - 430	Bulegeni and Kibale, Uganda	Kaizzi et al. (2004)
50 - 150	Bukoba District, Tanzania	Baijukya et al. (2005)
127 - 281	Kaduna, Nigeria	Carsky et al. (2001)
101 - 348	Hwedza, Zimbadwe	Whitbread et al. (2004)
22 - 193	Southern, Benin Republic	Vanlauwe et al. (2001)
190 - 262	Florida, USA	Wang et al. (2009)
305 - 329	Vihiga, Kenya	Kiwia et al. (2009)

Table 2. Nitrogen supply (kg/ha) of Velvetbean (*Mucuna pruriens* var. *utilis*) evaluated through the rotation with maize.

In some cases, increases (from 2.2 to 3.8%, for example) in the soil organic matter content have been observed as the age of the *Mucuna*-maize rotation system increases and the soil's chemical conditions are also improved. However, in some cases, no or little significant increases in organic matter content have been found, probably because the time when the rates of organic matter accumulation and decomposition reach an equilibrium is not known and also perhaps because the seasonal effects of weather and soil conditions on organic matter decomposition are not taken into account when the sampling scheme is designed (Triomphe, 1996; Gerónimo et al., 2002; Odhiambo, 2011). Barthès et al. (2004) studied changes in soil carbon (0-40 cm) in a soil sandy loam Ultisol in Benin (Africa), which involved a 12-experimentation on three maize cropping systems under manual tillage. In traditional no-input cultivation, mineral fertilized and association with Velvetbean changes in soil carbon were -0.2, +0.2 and +1.3 t C/ha/yr, with residues carbon to 3.5, 6.4 and 10.0 t/ha/yr, respectively. The carbon originating from maize and Velvetbean in litter-plus-soil represented less than 4% and more 50% of both total and overall residue carbon, respectively.

In those experiments where a significant increase in organic matter was observed, the soils also increased 20 to 30% in moisture content, and showed a higher cation exchange capacity, lower pH, lower apparent density and a reduction in micronutrient recycling. At the same time, with the use of *Mucuna*, reductions in the damage and mortality caused by *Pythium*, *Rhizoctonia* and *Fusarium* on maize seedlings have been documented, probably due to the type of organic matter that is incorporated into the soil, the effects of this on the soil's microclimate and/or its allelopathic effects (Versteeg & Koudokpon, 1990). Rotation of *Mucuna* with maize or banana crops reduces *Radopholus similis*, *Criconemella*, *Scutellonema*,

Melodogyne and root-knot populations, but increases those of *Helicotylenchulus, Rotylenchulus, Rhabditidae, Cephalobidae* and *Pratylenchus* (Watson, 1922; Reddy et al., 1986; Figueroa et al., 1990; Blanchart et al., 2006).

Also, *Mucuna* in a cropping system modified the structure, composition, diversity and interactions of soil biota (earthworms, millipedes, centipedes, Coleoptera adults, Diptera larvae and Isopoda) that can promote soil structure and nutrient availability (Ortiz-Ceballos & Fragoso 2004; Blanchart et al., 2006; Ortiz Ceballos et al., 2007a; Ortiz Ceballos et al., 2007b).

Continuous cultivation of maize on slope-side parcels increases splash and sheet erosion, with soil losses of 127 t/ha/yr, or even as high as 200-3600 t/ha/yr. A *Mucuna*-maize rotation system on 30 to 65% slopes had soil losses of 52.3 t/ha/yr and in a no-burn grazing system with *Mucuna* coverage, soil loss was only 3.9 t/ha/yr (López, 1993). Some evidence indicates that this crop grows well on acidic soils, that the development of its root system is influenced by the availability of phosphorus and magnesium and that its yield increases with soil pH (Halriah et al., 1991). Rotation or intercropping with *Mucuna* has promoted fertility restoration and improvement of the physical conditions of soils that had been compacted by heavy machinery or degraded after intensive slash-and-burn cultivation, thus allowing their reincorporation to food production (Hulugalle et al., 1986; Lal, 1987).

For its vigorous, explosive growth and its allelopathic effect, "picapica mansa" has shown to be effective in weed suppression, particularly gramineous weeds which compete for light, water and space with annual and perennial crops. This has been shown in parcels infested by *Imperata cylindrical, Paspalum fasciculatum, Striga hermonthica* and *S. Asiatica* and *Cyperus rotundus* in Africa (Whyte et al., 1955; Buckles & Perales, 1996; Tarawali et al., 1999; Akobundu et al., 2000; Udensi et al., 1999; Carsky et al., 2001; Chikoye & Ekeleme 2001; Whitbread et al., 2004; Kiwia et al., 2009; Odhiambo et al., 2010; Olorunmaiye, 2011). Thus, this practice reduces the costs of weed control and releases hand labour that can then be devoted to other productive activities (Versteeg & Koudokpon, 1990; Odhiambo et al., 2010; Olorunmaiye, 2011). Planting distances that have been effective for weed control are 1.0 and 1.5m equidistant, equivalent to sowing some 15 kg of "picapica mansa" seed per hectare or some 15,000 - 16,000 plants/ha (Versteeg & Koudokpon, 1990). This provides an ample potential for the restoration of the 11 to 22 million hectares that have been infested by *Imperata cylindrica* in Indonesia (Tempany, 1951; Coomans, 1976). For example, Tarawali et al. (1999) document that, in the southern Benin Republic, *Mucuna* raised the interest of 3000 farmers (1988-1993) mainly for controlling *Imperata* and the numbers testers of the innovation rose up to 10000 farmers by 1996.

Finally, "picapica mansa" seems to be tolerant to the attack of pests and diseases due to its content of toxic secondary metabolites, and is able to outcompete weeds partly due to the production of allelopathic compounds (Duke, 1981; Kay, 1985). However, a survey conducted in tropical regions of developing countries showed only a limited acceptance of using cover crops as green manure; most of the advantages recognized through the survey were of agronomic character, while the disadvantages were mostly economic (Yost & Evans, 1988). In fact, economic hardships lead farmers to be far more concerned about having more land available to produce more subsistence crops to feed a larger population than about preserving soil fertility (Yost & Evans, 1988; Flores, 1989; Versteeg & Koudokpon, 1990). Water competition with the main crop, an increase in hand-labour and costs and the fact that, sometimes, its use is not profitable due to the low price of chemical fertilizers, have been identified as additional disadvantages (Warriar, 1969; Gray, 1969; Yost & Evans, 1988).

Some agronomic disadvantages of using "picapica mansa" are that it: a) is susceptible to burn during the dry season of the year, when grown in the vicinity of parcels managed by slash-and-burn; b) provides shelter for poisonous snakes and rats; c) is defoliated by rabbits, and d) attracts bean slugs *Sarasinula plebeia* Fischer (Versteeg & Koudokpon, 1990; Buckles & Perales, 1996). When legume plants are grown in association or intercropped with maize, a reduction in maize yield often occurs during the first cycle, the severity of such reduction depends on the legume species but also on its density and management. Finally, soil nitrogen losses through leaching and volatilization have also been recorded due to the absence of a crop able to absorb the nitrogen being released through decomposition, the immobilization of nitrogen coincident with the time when the crop makes the highest demands of this nutrient or to the increase in soil acidity (Triomphe, 1996). However, Jensen et al. (2011) indicates that the ability of the legumes to fix N_2 reduces emissions of fossil energy-derived CO_2 and results in lower N_2O fluxes compared to agroecosystems that are fertilizer with mineral N.

7. Effects of *Mucuna* spp. on maize yield

Farmers and researchers alike have found that the use of "picapica mansa", either as rotation crop or intercropped with maize, has beneficial effects on maize yields as a result of the several advantages and benefits described above. Thus, average yields of 3.2 ± 2.39 t/ha (range: 0.3 - 8.3 t/ha) have been reported, which compare favourably with the average yield of 2.2 ± 1.95 t/ha (range: 0.4 - 7.5 t/ha) that is obtained in monoculture (Table 3). The response of maize to improved fallows of mucuna was linearly related to the amount of biomass produced from the mucuna returned to the system. With an agronomic use efficiency of 11.3 kg grain/kg applied N and apparent N recoveries in the range of 25-53%, there were large quantities of N no utilised by the subsequent maize phase.

With *Mucuna*	Without *Mucuna*	Location	Reference
1.0 - 1.5	1.2 - 1.3	Veracruz, Mexico	Eilittä et al. (2003)
1.9 - 4.5	1.4 - 2.5	Tela and Jutiapa, Honduras	Triomphe (1996)
2.6 - 3.2	0.8 - 1.8	Tabasco, Mexico	Ortiz-Ceballos et al. (2004)
4.5 - 5.1	3.8	Veracruz, Mexico	Ruiz J. & Laird (1964)
2.4	1.3	Manaus, Brasil	Smyth et al. (1991)
0.3 - 1.6	0.7 - 2.2	Veracruz, Mexico	Buckles & Perales (1994)
4.8 - 8.3	4.0 - 7.4	Limpopo, South Africa	Odhiambo et al. (2010)
6.3 - 8.2	2.3 - 4.2	Limpopo, South Africa	Odhiambo (2011)
0.7 - 0.9	0.7 - 0.8	Kwara, Nigeria	Olorunmaiye (2010)
3.8 - 7.0	2.9 - 6.3	Lomé, Togo	Sogbedji et al. (2006)
1.4 - 2.6	0.6 - 0.9	Yucatán, Mexico	Eastmond & Faust (2006)
1.4 - 4.2	1.4 - 4.3	Bukoba District, Tanzania	Baijukya et al. (2005)
0.7 - 2.5	0.6 - 1.0	Kaduna, Nigeria	Carsky et al. (2001)
0.8 - 1.0	0.4 - 0.5	Kaduna, Nigeria	Franke et al. (2004)
1.3 - 8.2	3.0 - 7.5	Bulegeni and Kibale, Uganda	Kaizzi et al. (2004)
2.2 - 5.8	1.2 - 2.6	Hwedza, Zimbadwe	Whitbread et al. (2004)
1.1 - 2.8	0.6 - 1.7	Southern, Benin Republic	Vanlauwe et al. (2001)

Table 3. Effect of the presence and absence of Velvetbean (*Mucuna pruriens* var. *utilis*) in maize grain yield (t/ha).

In hot, humid regions, the intensive cultivation of maize monocultures tends to break down and become unsustainable as the soil is degraded and/or herbicides, pesticides and chemical fertilizers are required to sustain the crop, with well-known economic and environmental problems (Cox & Atkins, 1979; Buckles & Perales, 1996; Bandy et al., 1993). By contrast, the *Mucuna* - maize intercropping or rotation system allows the affordable production of subsistence food staples as the system helps preserve soil fertility and biologically control weeds, pests and diseases.

8. Criteria for comparing and evaluating cover crops as bioherbicide

The general features that a cover crop should possess and the advantages of using this soil management practice are often listed. However, suitable methods to evaluate and compare cover crops are still lacking. Therefore, in this section we attempt to provide a summary view of the attributes that should be considered to evaluate and compare cover crops (Whyte et al., 1955; Allison, 1973; Sarrantonio, 1991; Versteeg & Koudokpon, 1990; Triomphe, 1996; Yost & Evans, 1988):

a. Growth rate. Species with vigorous, fast, indeterminate growth and high capability for interspecific competition can quickly protect the soil against weathering and erosion and possess good capacity to suppress weeds.
b. Biomass accumulation. The potential productivity of many species can be evaluated in terms of dry biomass production and nitrogen accumulation, both expressed in terms of kg/ha, which can then be related to the *in situ* evaluation of the Rhizobia capacity and effectiveness.
c. Carbon/nitrogen ratio (C/N). In general, if the C/N ratio of organic materials incorporated into the soil is higher than 30, nitrogen becomes immobilized. When the C/N ratio is lower than 20, nitrogen is released. The C/N ratio affects the action, rate and type of microorganisms involved in organic matter decomposition. However, temperature, humidity and actual evapotranspiration are also determining factors of decomposition rate in natural conditions.
d. Structural carbohydrates. The decomposition rate of organic materials and the action of soil microorganisms on them depend on the chemical nature of the plant tissues contained in the green manure that is incorporated into the soil and on their quality as fodder. Organic compounds such as lignin, hemicelluloses, cellulose and pectic substances are resistant to decomposition and/or digestion. The abundance of these materials influences nitrogen mineralization and organic matter digestibility. This is why models have been developed that predict decomposition rates or digestibility percentages based simply on the percentage content of lignin in foliage.
e. Seed quality and quantity. Plant species with high seed yields are always preferable as this facilitates their establishment and repopulation of new areas. Also important is the time, vigour and synchrony of germination to achieve a rapid establishment, features which are intrinsic to the seed and might be related to the presence of dormancy.
f. Resistance to the attack of pests and diseases. It is well known that wild, cultivated and some domesticated species release organic compounds that have allelopathic or toxic effects on herbivores and pathogens.

g. Other features. Finally, to choose a particular species as cover crop, attention should also be given to characteristics such as its adaptation to climate and soil conditions, low water and nutrient requirements, competition with the main crop, ease of management and low cost of incorporating it into the land's management plan.

9. Conclusions

Based on the above, we can conclude that *Mucuna* as bioherbicide may increase the functional properties of agroecosystem and allow a better agricultural ecosystem productivity: a) biocontrol of weeds and diseases, b) reduce the fossil energy used in the production of food, c) incorporation of OM and N into the soil (sequestration of carbon and lower emission of nitrous oxide), d) preservation of the soil biota, e) regulation of soil moisture and temperature, f) protection from soil erosion, g) *in situ* conservation and improvement of local maize cultivars, and h) sustained harvests.

10. References

Akobundu, I.O.; Udensi, E.U. & Chikoye, D. (2000). Velvetbean (*Mucuna* spp.) suppresses speargrass (*Imperata cylindrica* (L) Raeuschel) and increases maize yield. *International Journal Pest Management* 46, 103-108

Allison, F.E. (1973). *Soil organic matter and its role in crop production*, Elsiever, Amsterdam, The Netherlands, p. 637

Baijukya, F.P.; der Ridder, N. & Giller, K.E. (2005). Managing legume cover crops and residues to enhance productivity of degraded soils in the humid tropics: a case study in Bukoba District, Tanzania. *Nutrient Cycling in Agroecosystems* 73, 75-87

Bandy, D.E.; Garrity, D.P. & Sanchez, P.A. (1993). The worldwide problem of slash-and-burn agriculture. *Agroforestry Today*, 5, 2-6

Barthès, B.; Azontonde, A.; Blanchart, E.; Girardin, C.; Villanave, C.; Lesaint, S.; Oliver, R. & Feller, C. (2004). Effect of a legume cover crop (*Mucuna prurines* var. *utilis*) on soil carbon in an Ultisol under maize cultivation in southern Benin. *Soil Use Management*, 20, 231-239

Blanchart, E.; Villenave, C.; Viallatoux, A. & Barthès, B. (2006). Long-term effect of a legume cover crop (*Mucuna pruriens* var. *utilis*) on the communities of soil macrofauna and nematofauna, under maize cultivation, in southern Benin. *European Journal of Soil Biology*, 42 (Supplement 1), S136-S144

Bort, K.S. (1909). *The Florida velvet beans and its history*. Bureau of Plant Industry, pp. 25-32, Bulletin 141, USDA, Washington, D.C.

Buckles, D. & Perales, H.R. (1994). *Experimento con el frijol terciopelo basados en los agricultores: la innovación dentro de la tradición*, CIMMYT, México, p. 25

Carsky, R.J.; Oyewole, O. & Tian, G. (2001). Effect of phosphorus application in legume cover crop rotation on subsequent maize in the savanna zone of West Africa. *Nutrient Cycling in Agroecosystems* 59, 151-159

Chikoye, D. & Ekeleme, F. (2001). Growth characteristics of ten *Mucuna* accessions and their effect on dry matter of speargrass. *Biological Agriculture and Horticulture* 18, 191-201

Cobb, G.R. (1950). Farming with green manures. *Better Crop With Plant Food*, 34, 24

Coomans, P. (1976). Controle chemique de I´Imperata. *Oleagineux*, 31,109-110

Cox, G.W. & Atkins, M.D. (1979). *Agricultural ecology*, Freeman, San Francisco, California, USA, p.721

Duggar, J.T. (1899). *Velvet bean*. Alabama Agric. Expt. Stat. pp. 113-125, Bulletin 104, Auburn, Alabama. USA

Duke, J.A. (1981). *Handbook of legumes of world economic importance*, Plenum, New York, USA, p. 345

Eastmond, A. & Faust, B. (2006). Farmers, fires, and forests: a green alternative to shifting cultivation for conservation of the Maya forest? *Lanscape and Urban Planning* 74, 267-284

Eilittä A.M.; Sollenberger L.E.; Littell R.C. & Harrington L.W. (2003). On-farm experiments with Maize-Mucuna systems in the Los Tuxtlas region of Veracruz, Mexico. I. Mucuna biomass and maize grain yield. *Experimental Agriculture*, 39, 5-17

Figueroa, A.; Molina, M.E. & Pérez, L. (1990). Cultivos alternos para controlar nematodos en renovación de plantaciones bananeras. *ASBANA-CORBANA*, 14, 19-26

Flores, M. (1989). Velvetbean: An alternative to improve small farmer´s agriculture. *ILEIA*, 5, 8-9

Franke, A.C.; Schulz, S.; Oyewole, B.D. & Bako, S. (2004). Incorporating short-season legumes and green manure crops into maize-based systems in the moist Guinea savanna of West Africa. *Experimental Agriculture* 40, 463-479

Gerónimo, C.A.; Salgado, S.G.; Catzin, R.F.J. & Ortiz-Caballos, A.I. (2002). Descomposición del follaje de nescafé (*Mucuna* spp.) en la época seca. *Interciencia* 27, 625-630

Göhl, B. (1982). *Piensos tropicales: Resúmenes informativos sobre piensos y valores nutritivos*, Producción y Sanidad Animal 12, FAO, Roma, Italia, p. 550

Gray, B.S. (1969). Ground covers and performance. *Journal of the Rubber Research Institute of Malaya*, 21,107-112

Greenland, D.J. (1975). Bringing the green revolution to the shifting cultivator. Better seeds, fertilizers, zero or minimum tillage, and mixed cropping are necessary. *Science*, 190, 841-844

Halriah, K.; Noordwijk, M.V. & Setijono, S. (1991). Tolerance to acid soil conditions of the velvet beans *Mucuna pruriens* var. *utilis* and *M. deeringiana*. *Plant and Soil*, 34, 195-105

Hauck, F.W. (1977). *China: Reciclaje de desechos orgánicos en la agricultura*, Boletín de Suelos Núm. 40, FAO, Roma, Italia, p. 105

Hulugalle, N.R.; Lal, R. & ter Kuile, C.H.H. (1986). Amelioration of soil physical properties by *Mucuna* after mechanized clearing of tropical rain forest. *Soil Science*, 141, 219-224

Jensen, E.S.; Peoples, M.B.; Boddey, R.M.; Gresshoff, P.M.; Hauggaard-Nielsen, H.; Alves, B.J.R. & Morrison, M.J. (2011). Legumes form mitigation of climate change and the provision of feedstock for biofuels and biorefineries: A review. *Agronomy for Sustainble Develoment* DOI10.1007/S13593-011-0056-7

Kaizzi, C.K.; Ssali, H. & Vlek, P.L.G. (2004). The potential of velvet bean (*Mucuna pruriens*) and N fertilizers in maize production on contrasting soils and agro-ecological zones of East Uganda. *Nutrient Cycling in Agroecosystems* 68, 59-73

Kay, D.E. (1985). *Legumbres alimenticias*, Acribia, Zaragoza, España, p. 437

Kiwia, A., Imo, M.; Jama, B. & Okalebo, J.R. (2009). Coppicing improved fallow are profitable for maize production in striga infested soils of western Kenya. *Agroforestry Systems* 76, 455-465

Kremen, C.; Niles, J.O.; Dalton, M.G.; Daily, G.C.; Ehrlich, P.R.; Fay, J.P.; Grewal, D. & Guillery, R.P.(2000). Economic incentives for rain forest conservation across scales. *Science*, 288, 1828-1832

Lal, R. (1987). Management of soil compaction and soil-water after forest clearing in upland soils of humic tropical Asia. In: *Soil management under humid conditions in Asia and Pacific*, Latham M. (Ed.), pp. 273-296, Proceeding 5, IBSRAM, Ibadan, Nigeria

López M., J. (1993). Conservación y productividad de suelos en ladera de la Fraylesca, Chiapas. Tesis de maestría en ciencias. Colegio de Postgraduados. Montecillo, México. 177 p.

Mabberley, D.J. (1992). *Tropical rain forest ecology*, 2nd. ed., Chapman, New York, USA, p. 300

McKee, R. (1948). The other pasture legumes. In: *Grass: Yearbook of agriculture*, USDA, pp. 363-366, Washington, D.C., USA

Mulvaney, R.L.; Khan, S.A. & Ellsworth T.R. (2009). Synthetic nitrogen fertilizers deplete soil nitrogen: a global dilemma for sustainable cereal production. *Journal of Environmental Quality* 38, 2295-314

Odhiambo, J.J.O. (2011). Potential use of green manure legume cover crops in smallholder maize production systems in Limpopo province, South Africa. *African Journal of Agricultural Research* 6, 107-112

Odhiambo, J.J.O.; Ogola, J.B.O. & Madzivhandila, T. (2010). Effect of green manure legume - maize rotation on maize grain yield and weed infestation levels. *African Journal of Agricultural Research* 5, 618-625

Olorunmaiye, P.M. (2010). Weed control potential of five legume cover crops in maize/cassava intercrop in a Southern Guinea savanna ecosystem of Nigeria. *Australian Journal of Crop Science* 4, 324-329

Ortiz-Ceballos, A.I. & Fragoso, C. (2004). Earthworm populations under tropical maize cultivation: the effect of mulching with Velvetbean. *Biology and Fertility of Soils* 39, 438-445

Ortiz-Ceballos, A.I.; Fragoso, C. & Brown, G.G. (2007a). Synergistic effects of the tropical earthworm *Balanteodrilus pearsei* and *Mucuna pruriens* as green manure in maize growth and crop production. *Applied Soil Ecology* 35, 356-362

Ortiz-Ceballos, A.I.; Fragoso, C.; Brown, G.G. & Peña-Cabriales, J.J. (2007b) Mycorrhizal colonization and uptake of nitrogen in maize crop: Combined effect the tropical earthworm and velvetbean mulch. *Biology and Fertility of Soils* 44, 181-186

Pimentel, D.; Andow, D.; Dyson-Hudson, R.; Gallahan, D.; Jacobson, S.; Irish, M.; Kroop, S.; Moss, A.; Schreiner, I.; Shepard, M.; Thompson, T.; Vinzant, B. (1980). Environmental and social costs of pesticides: a preliminary assessment. *OIKOS*, 34, 126-140

Pugalenthi, M.; Vadivel, V. & Siddhuraju, P. (2005). Alternative food/feed perspectives of an underutilized legume *Mucuna pruriens* var. *utilis* - A review. *Plant Foods for Human Nutrition 60*, 201-218

Purseglove, J.W. (1987). *Tropical crop: dicotyledons*, Logman, New York, USA, p. 719

Reddy, K.C.; Soffes, A.R. & Pine, G.M. (1986). Tropical legumes for green manure. I. Nitrogen production and the effects on succeeding crop yields. *Agronomy Journal*, 78, 1-4

Ruiz, J.M. & Laird, R.J. (1964). Tres leguminosas tropicales para abono verde. *El Campo* 29, 45-47

Sarrantonio, M. (1991). *Methodologies for screening: soil-improving legumes*, Rodale Institute, Kutztown, P.A., USA, p. 312

Scott, J.M. (1919). *The velvet bean*, Florida Agric. Expt. Stat., p. 43-58, Bulletin 102, Gainsmeville, Florida, USA

Smyth, T.J.; Cravo, M.S. & Melgar, R.J. (1991). Nitrogen supplied to corn by legumes in a central Amazon oxisol. *Tropical Agriculture (Trinidad)*, 68, 366-372

Snyder, W.C.; Schroth, M.N. & Christon, T. (1959). Effect of plant residues on root rot of bean. *Phytopathology*, 49, 755-756

Sogbedji, J.M.; van Es, H.M. & Agbeko K.L. (2006). Cover cropping and nutrient management strategies for maize production in Western Africa. *Agronomy Journal*, 98, 883-889

Tarawali, S.A. (1994). Evaluating selected forage legumes for livestock and crop production in the subhumic of Nigeria. *Journal of Africultural Science* 123, 5-50

Tempany, H.A. (1951). *Imperata* grass, a major menace on the wet tropics. *World Crops*, 3, 143-146

Tivy, J. (1990). *Agricultural Ecology*, Wiley, New York, USA, p. 287

Tracy, M. & Coe, H.S. (1918). *Velvet beans*, Farming Bulletin 962, United States Department of Agriculture, Washington, D.C., USA

Triomphe, B. (1996). *Evaluation of the long-term effects of green manure use in the maize-based cropping, systems of northern Honduras*, SCAS, University Cornell. Ithaca, New York. USA. 7 p.

Udensi, E.U.; Akobundu, I.O.; Ayeni, A.O. & Chikoye, D. (1999). Management of congograss (*Imperata cylindrica*) using velvetbean (*Mucuna pruriens* var. *utilis*) and herbicides. *Weed Techonology* 13, 201-208

Vanlauwe, B.; Aihou, K.; Houngnandan, P.; Diels, J.; Sanginga, N. & Merckx, R. (2001). Nitrogen management in 'adequate' input maize-based agriculture in the derived savanna benchmark zone of Benin Republic. *Plant and Soil* 228, 61-71

Versteeg, M.N. & Koudokpon, V. (1990). *Mucuna helps control imperata in shuthern Benin*. pp. 7-10, Bulletin 7, West African Farming Systems Research Network, Benin, Africa

Wade, N. (1974). Green revolution (I): A just technology, often unjust in use. *Science*, 186, 1093-1096

Wang, Q.; Klassen, W.; Li, Y. & Codallo, M. (2009). Cover crops and organic mulch to improve tomato yields and soil fertility. *Agronomy Journal*, 101, 345-351

Warriar, S.M. (1969). Cover plant trials. *Journal of the Rubber Institute of Malaya*, 21, 158-164

Watson, J.R. (1922). *Bunch velvet beans to control root-knot*, Florida Agric. Expt. Stat., pp. 55-59, Bulletin 163, Gainsmeville, Florida, USA

Weindling, R. (1946). Microbial antagonism and disease control. *Soil Science*, 61, 23-30

Whitbread, A.M.; Jiri, O. & Maasdorp, B. (2004). The effect of managing improved fallows of *Mucuna pruriens* on maize production and soil carbon and nitrogen dynamics in sub-humid Zimbabwe. *Nutrient Cycling in Agroecosystems* 69, 59-71

Whyte, R.O.; Nilsson-Leissner, G. & Trumble, H.C. (1955). *Leguminosas en la agricultura*, FAO, Estudios Agropecuarios Núm. 21, Roma, Italia, p. 344

Williams, B.J. (2006) Aztec soil knowledge: Classes, management, and ecology. In: *Footprints in the soil: People and ideas in soil history*, Warkentin, B.P. (Ed.), Elsevier, pp. 16-42, The Netherlands

Willis, K.J.; Gillson, L. & Brncic, T.M. (2004). How "Virgin" is virgin rainforest? *Science*, 304, 402-403

Winiwarter, V. (2006). Soil scientists in ancient Rome. In: *Footprints in the soil: People and ideas in soil history*, Warkentin, B.P. (Ed.), Elsevier, pp. 3-16, The Netherlands

Yamada, T., Kremer, R.J., Camargo, P.R. & Wood, B.W. (2009). Glyphosate interactions with physiology, nutrition, and diseases of plants: threat to agricultural sustainability. *European Journal of Agronomy* 31, 111-113

Yates, S.R.; McConnell, L.L.; Hapeman, C.J.; Papiernik, S.K.; Gao, S & Trabue, S.L. (2011). Managing agricultural emissions to the atmosphere: state of the science, fate and mitigation, and identifying research gaps. *Journal of Environmental Quality* 40, 1347-1358

Yost, R. & Evans, D. (1988). *Green manures and legume covers in the tropics*, University of Hawaii, HI, USA, Hawaii, Institute of Tropical Agriculture and Human Resources, Research Series No. 055, p. 37

Zaim, M.; Kumar, Y.; Hallam, V. & Zaidi, A.A. (2011). Velvet bean severe mosaic virus: a distinct begomovirus species causing severe mosaic in *Mucuna pruriens* (L.) DC. *Virus Genes* 43, 138-146

10

Organic Weed Control

Charles L. Webber III[1], James W. Shrefler[2] and Lynn P. Brandenberger[3]
[1]*USDA, ARS, WWARL, Lane, Oklahoma,*
[2]*Oklahoma State University, Lane, Oklahoma*
[3]*Oklahoma State University, Stillwater, Oklahoma*
USA

1. Introduction

Organic vegetable producers rank weeds as one of their most troublesome, time consuming, and costly production problems (OFRF, 1999). Because there are only a few organically approved herbicides, optimizing their application may increase their potential usefulness for organic production systems. Inter-row cultivation for the purpose of weed control is not always the ideal choice for organic vegetable production or due to soil and weather conditions may not always be an available option. Additional cultivations can decrease soil organic matter (Dick, 1983; Gallaher & Ferrer, 1987; Johnson, 1986) and soil water holding capacity (Johnson, 1986), increase soil erosion (Logan et al., 1991; McDowell & McGregor, 1984) and nutrient loss (McDowell and McGregor, 1984), and stimulate new weed growth (Pekrun et al., 2003). In conventional, non-organic production systems, herbicides are increasingly used to avoid the detrimental impact of soil erosion from weed control from cultivation. Preventing soil, nutrient, and organic matter losses due to tillage are a fundamental tenant of certified organic production (USDA National Organic Standards Board, 2010).

The primary source for the majority of organic herbicides are natural, produced by or from plants. Allelopathy is the biochemical interference, inhibitory or stimulatory, between one plant species and another (Rice, 1984). Certain weeds and crops can release chemicals by exudation, leaching, volatilization, and from plant tissue (leaves, stems, roots, flowers) decomposition into the environment. When these biochemcals (allelochemicals) come contact with other plants they can influence (inhibit or simulate) another species growth. Vegetables are not immune to allelopathic effects of other crops. Early development of vegetables may be the most vulnerable part of the life cycle to be exposed to allelopathic chemicals (Russo et al., 1997). The allelopathic impact of one species upon another is more obvious when the allelochemcials prevent germination and establishment of an affected species, but the impact exerted can affect established neighboring plants by limiting their optimum growth or causing death. Allelochemicals can persist in the soil, and therefore, impact the subsequent crop growth. Allelochemicals can also work in concert with competition for light, water, and nutrients between weeds and crop plants. Microbial weed control is another method for managing weeds (see chapter "Microbial weed control and microbial herbicides" by Tami L. Stubbs and Ann C. Kennedy) with potential application within an organic cropping system.

Organic certification in the United States was developed in recognition of the necessity for the use of consistent standards across the U.S. to benefit producers, processors, wholesalers, retailers, and consumers. Prior to establishment of USA's, National Organic Program federal guidelines for organic certification in 2002, a multitude of agencies and associations throughout the U.S. maintained a diverse list of acceptable inputs, production methods, and policies to determine organic certification, which were consistently reviewed with materials being added or removed from use. Differences in certification standards invited marketing inconsistencies, misunderstandings, and misrepresentations concerning organic products. Certified organic crop production is more than a list of acceptable and prohibited inputs or practices that can and can not be used; rather, it is a holistic approach to sustainable and healthy food production that enhances the well-being of the consumer and protects natural resources. One important aspect of certified organic crop production is the prevention of non-organic substances from intentionally, or inadvertently, being brought into the production area. This would include the intentional application of irrigation water, or natural water flow containing disallowed materials, including herbicides, from a pond or lake onto a certified organic production area.

The approval process for herbicides for certified organic production initially follows a similar registration process as conventional herbicides, through US government (United States Environmental Protection Agency) and state approval. A potential organic herbicide must also be cleared by the National Organic Program (NOP) and the individual farmer's certifying agency. The Organic Materials Review Institute (OMRI) is a nonprofit organization founded in 1997 to provide a independent review of potential organic products. OMRI is not a regulating agency, but the agencies or companies which certify organic producers tend to place a great importance on whether a material has been placed on the OMRI approval list. A particular product or material may be appropriate and safe for use on certified organic land even if the product is not on the OMRI list, but the individual farmer's certifying agency must agree that the product can be used in connection with certified organic land and produce. The producer should always check with their certifying agency prior to using any product or material. As with conventional herbicide labels, regulations, approvals, and labels for organic herbicides can and often do change often. The term "organic" has its detractors. The term "natural" is preferred by some. As the legislation stands only the term "organic" is recognized.

Human safety should always remain a primary concern when using any herbicide, even organically approved herbicides. For example, eye exposure to vinegar with acetic acid concentrations greater than 10% can cause blindness. In this chapter the most common organic herbicides and their methods of application are described. The reader should be aware that herbicide labels and regulations can and do change. The herbicides described are not meant to be all inclusive, but to provide information on the more common organic herbicides. Certain herbicides or formulations may have been dropped or added, so all herbicide labels should be read and followed closely. Sometimes organic herbicides fail, the same can be true for synthetic herbicides, in which case it is necessary to use hand tools or cultivation to control weeds. Although hoeing is labor intensive and tedious early preemptive use of tools for weed control can provide reduced time for weeding later in the growing season.

2. Holistic weed control

Organic producers typically understand better than others the importance of a holistic approach to crop production and weed control. The initial site selection for a certified organic production area can have tremendous long term benefits, or adverse consequences, for crop yields and weed control. Although any site that is selected may have some indigenous weed species, either present or stored in the soil profile, it is critical to select land that minimize these weed sources, and then pursue a diligent program to control existing weeds and prevent the introduction of new weeds. Many of the certification requirements favor site selections and processes that help reduce weed growth and the introduction of new weeds (e.g., avoiding drainage into the certified area, good soil health and conservation, cover crops and mulches, and cleaning equipment entering from not certified areas).

2.1 Weed monitoring

Weed monitoring is an essential aspect for the successful and economic control of weeds. It is important to determine the weed species and their locations in the production area. Weed monitoring should not be limited to the current growing season, but throughout the year, and from year to year, so that trends in weed populations can become known. Ideally, weed surveys should be conducted twice during each growing season, once following planting but prior to the first weeding/post-emergence application and then just prior to harvest. The survey results will provide valuable information concerning the weeds missed from previous crops and current preemergence and post-emergence weed control efforts and potential weeds from crop rotations. Weed monitoring throughout the year will help the producer to realize the importance of controlling weeds year long. A single uncontrolled weed can produce 10s of thousands of weed seed. Take for example one of many weed species that can inflect total yield loses, if left uncontrolled: redroot pigweed (*Amaranthus retroflexus* L.). Researchers have reported that a typical redroot pigweed can produce from 9,254 (Pawlowski *et al.*, 1970) to 117,400 (Stevens, 1932) seeds per plant. A vigorous redroot pigweed plant may produce 100,000 seeds (Mitich, 1997), a large plant can produce as many as 230,000 seeds (Stevens, 1957), and closely spaced plants 34,600 seeds/plant (Hauptli & Jain, 1978; Weaver & McWilliams, 1980). In addition to the large number of seeds produced from a single plant, the seeds can overwinter in the flowers (Mohler & Callaway, 1995), and on or below the soil surface (Georgia, 1942), and are known to survive in the soil for 30 (Mitich, 1997) to 40 years (Darlington & Steinbauer, 1961). Therefore, a few escape plants even at the end of growing season can have long-term detrimental impact on future weed infestations, weed control costs, and yield reductions. Redroot pigweed is only one of many weed species that have these tremendous competitive tendencies. Producer should be especially concerned with aggressive and hard to control perennial weeds such as nutsedges (*Convolvulus arvensis* L. and bindweeds [*Calystegia sepium* (L.) R. Br.], for once they are established they are very difficult to eliminate. Weed monitoring will also assist the producers in making informed decisions concerning crop rotations, rotating herbicides, and cultural practices.

2.2 Weeds: opportunistic plant species

Research and grower experience have shown the importance of crop rotations, cover crops, and mulches for crop and soil health, and reduced weed competition. Weeds are opportunistic plant species that will occupy voids between crop sequences, crop rows, and crop plants within rows. The judicious selection of beneficial plant species that will fill these voids between growing seasons, and cover the soil surface as mulches, can promote crop productivity by increasing nutrients, organic matter, preserving soil moisture, and reducing weed competition. Great care must be taken when using plant mulches from locations other than your own certified land to prevent the introduction of additional weeds to your production area. In the same manner, the use of animal manures from outside sources may introduce new weeds. If a new weed species does appear, take immediate action to eliminate the plants, minimize seed dispersal, and determine the source of infestation.

2.3 Cultural practices

2.3.1 Crop rotations, and stale seedbeds

Crop rotations, including winter cover crops, can influence the quantity and species of weed present in spring-planted crops. The previous cropping systems may influence weed pressure during the current growing season depending on the type of crop produced, the herbicides used, and the weeds controlled or not controlled. The presence of a winter cover crop prior to establishment of a spring crop can reduce the weed pressure in the following crop by not allowing weeds to become established during non-production times. Clark and Panciera (2002) reported that rolling down a winter rye crop suppressed weeds in the following spring no-till planted corn crop, eliminating the need for herbicides. Rolling winter rye or a winter rye/hairy vetch mixture suppressed weeds in spring no-till planted bell peppers by 96% for 8 to 10 weeks, while rolling hairy vetch alone reduced weeds by 80% for 2 to 8 weeks (Leavitt et al., 2011). When rolling is used a machine comprised of a drum roller, to which blunt metal strips were welded either horizontally or in a chevron pattern, is drawn over the standing cover crop. The metal strips crush and crimp the stems without chopping. The cover is killed and the residue left after use of a roller-crimper is laid down flat in a uniform direction and layered so that the space through which weeds can emerge is reduced (Teasdale & Mohler, 2000).

Crop rotation can also help prevent domination of any one weed species. The herbicides used in a previous crop may eliminate potentially hard to control weeds or inadvertently select for more troublesome weeds in future crops. Weed management, monitoring and control, should focus not only on the current crop, but also on the weed management systems prior to the current crop. The previous herbicides may not only impact the weeds in the current crop, but may also have detrimental herbicide carry over.

Stale seedbeds and reduced tillage practices can be used in conjunction with many of the other organic weed control practices. In a stale seedbed, the planting area is prepared earlier than normal to allow for the germination, growth and control of weeds, or the killing of an established cover crop to serve as a mulch. The weeds or the cover crop might be killed with a mower, a chopper, organic herbicides, or by some other means. The crop is then direct-seeded, or transplanted into the seedbed using a minimal amount of soil disturbance in order to not promote weed seed germination and growth.

2.3.2 Solarization and tillage

Solarization and fallow tillage are two other approaches that use the weeds' aggressive growth tendencies as a method to help control future weed competition. Solarization is the use of solar radiation to kill weeds, normally using clear polyethylene mulch on moist soil surfaces. Solar radiation passes through the clear plastic, heating the soil. The moist hot environment initiates weed seed germination and stimulates weed growth, but then the hot humid confined environment becomes detrimental to continued weed growth and survival (Johnson et al., 2007). Solarization can also benefit the future crop by adversely impacting other plant pests such as nematodes, fungi, and insects (Johnson et al., 2007). Fallow tillage is the repeated use of soil tillage on fallow land to reduce future weed populations (Johnson et al., 2007). The weeds are repeatedly allowed to germinate and grow, but tilled prior to seed production. Fallow tillage can be used independently or in conjunction with solarization or other organic weed control methods (Johnson et al., 2007). During the cropping season, paper, fiber, and colored plastic mulches serve as weed barriers while promoting crop growth by warming the soil and conserving soil moisture. Soil cultivation during the growing season is also an effective method for controlling weeds between crop rows until canopy closure between rows. Hoeing and hand-weeding is also an option, depending on the production area, labor impacts on return on investment, and the removal of newly induced weed species.

2.3.3 Mulches

The term mulch has an expansive degree of understanding depending on by whom, and how, it is being used. Mulch can be defined as use of a material that covers soil for a variety of uses. The term was at one time used for the application of organically based plant residue. The term mulch has grown to incorporate the use of paper and plastics applied from a roll or polymers applied to the soil (Russo, 1995). Plastic mulch can be applied from a tractor mounted implement that lays the material on to the soil and covers the edge so that the plastic is secured to a bed. A drip irrigation system can be applied at the same time as the plastic mulch with the same implement. Plastic of several colors have been tested with crops and it was determined that they affected plant development and yield (Decoteau et al., 1990; Kaul & Kasperbauer, 1992).

Plastic and natural (dead or living) mulches are used to control weeds (Law et al., 2006). Mulches are typically applied to a prepared bed prior to transplanting. However, if stale seed beds are used the mulch can be applied in the autumn before a next spring's planting. Plastic mulches used in conjunction with an irrigation system under the plastic have the advantage that the soil temperature is generally increased and the soil moisture conserved which benefits crop growth while serving as a barrier to weed establishment (Law et al., 2006). Most weeds will germinate and emerge under the plastic mulches but will die due to the lack of sunlight and/or excessive temperatures. However, nutsedges (e.g., *Cyperus esculentus* L. and *Cyperus rotundus* L.) may actually pierce the thinner plastic mulches and reduce pepper yields through competition. Certain plastic mulches can be used in both conventional and organic production systems. Dead natural mulches can reduce weed populations and increase crop yields. In a 2-yr study, a cowpea mulch reduced weeds at harvest by 80 and 90% and increased bell pepper yields by 202 and 165% compared to plots with the plastic mulch (McGiffen & Hutchinson, 2000). Living mulches can also provide weed suppression for spring-transplanted crops (Paine & Harrison, 1993), but can also reduce pepper yields (Biazzo & Masiunas, 1998).

2.3.4 Flaming

Flaming uses propane gas (LP, LPG, LP-gas) to control weeds with a directed flame (Johnson & Mullinix, 2008). Flaming equipment can either be LP hand-held devices or full size field flamers with multiple flamers across the width of the boom flame (Johnson & Mullinix, 2008; OFRF, 2006). Flaming research has produced mixed results depending on the equipment used, the weed species and size, and the exposure time to the flaming treatment. Research investigating organic weed control methods on stale seedbeds determined that a hand-held LP flamer produced better and more convenient weed control than a number of organic herbicides containing either clove oil and/or acetic acid (OFRF, 2006). The researchers did state that flaming was their least favorite method "due to it being a non-renewable resource." Other researchers have reported unsatisfactory results with full size flaming equipment due to safety and operational issues, and ineffective weed control, especially with consistent long term control of grasses.

2.4 Organic herbicides

2.4.1 Preemergence organic herbicides

2.4.1.1 Corn gluten meal

Corn gluten meal (CGM) is an organic herbicide (Bingaman and Christians, 1995; Christians, 1991). CGM is the by-product of the wet-milling process of corn (Bingaman and Christians, 1995; Quarles, 1999). The protein fraction of CGM is approximately 60% protein and 10% nitrogen (Quarles, 1999). CGM (Alliance Milling Company, Denton TX), normally a yellow powder (McDade, 1999), has been used as a component in dog, fish, and livestock feed (Christians, 1991, 1995; Quarles, 1999). CGM can be purchased in the form of a powder, a pellet, and a granulated material (McDade, 1999; Webber and Shrefler, 2007a).

Christians (1993) investigated the weed control efficacy of broadcast soil applied, non-incorporated, applications of corn starch, corn germ, corn seed fiber, corn meal, and CGM. CGM produced the greatest inhibitory effect and reduced root formation in several weed species, including creeping bentgrass (*Agrostis palustris*) and crabgrass (*Digitaria* ssp.). Bingaman and Christians (1995) in greenhouse research determined that CGM applied at 324 $g{\cdot}m^{-2}$ reduced plant survival, shoot length, and root development for the twenty-two weed species tested, whether the CGM was applied to the soil surface as a preemergence herbicide or mixed into the top 2.54 cm as a preplant-incorporated herbicide. Although plant development was reduced for all weeds tested, the extent of susceptibility differed across species. Plant survival and root development were reduced by at least 70% and shoot length by at least 50% for the weeds: black nightshade (*Solanum nigrum*), common lambsquarters (*Chenopendium album*), creeping bentgrass, curly dock (*Rumex crispus*), purslane (*Portulaca oleracea*), and redroot pigweed (*Amaranthus retroflexus*). When CGM was applied preplant-incorporated, survival and shoot length of the following weeds were reduced at least 50% and root development reduced by at least 80%: catchweed bedstraw (*Galium aparine*), dandelion (*Taraxacum officinale*), giant foxtail (*Setaria faberi*), and smooth crabgrass (*Digitaria ischaemum*). Barnyardgrass (*Echinochloa crus-galli*) and velvetleaf (*Abutilon theophrasti*) were more tolerant to CGM and plant survival reductions were less than 31%. Field studies with three planting dates (3 July 1998, 20 Aug. 1998, and 8 June 1999) demonstrated that CGM incorporated into the top 5-8 cm of soil at 100, 200, 300, and 400

$g \cdot m^{-2}$ reduced weed cover by 50%, 74%, 84%, and 82%, respectively, compared to an untreated check at 3 weeks after treatment (McDade & Christians, 2000).

Crop safety with CGM is a major concern because it is a non-selective organic herbicide. CGM applications for organic weed control did not adversely affect established turf (Christians, 1993). Nonnecke and Christians (1993) did report a decrease in strawberry (*Fragaria xananassa*) fruit number and weight from four applications of CGM, but it was unclear whether the yield reductions were due to the CGM phytotoxicity or excess nitrogen applications associated with CGM (10% nitrogen). Strawberry leaf area was not reduced as a result of CGM applications (Nonnecke & Christians, 1993). In onions, CGM applications of 400 $g \cdot m^{-2}$ to spring-transplanted onions produced fair (72.1%) overall weed control and good (82.7%) broadleaf weed control through the first 46 d after planting (DAP) (Webber et al., 2007a), without reductions in yields from crop injury (Webber et al., 2007b).

The impact of CGM applications on the plant safety of direct-seeded crops has been investigated by McDade and Christians (2000) and Webber and Shrefler (2007b). McDade and Christians (2000) determined that CGM rates of 100, 200, 300, and 400-$g \cdot m^{-2}$ CGM rates reduced average seedling survival for eight vegetables by 48%, 65%, 73%, and 83%, respectively. 'Daybreak' sweet corn (*Zea mays*) was the most tolerant to CGM, requiring at least 300 $g \cdot m^{-2}$ of CGM to produce a 26% reduction in stand. CGM applications of 100 $g \cdot m^{-2}$ reduced seedling survival by 35% for 'Ruby Queen' beet (*Beta vulgaris*), 41% for 'Red Baron' radish (*Raphanus sativus*), 59% for 'Maestro' pea (*Pisum sativum*), 67% for 'Comanche' onion (*Allium cepa*), 68% for 'Black Seeded Simpson' lettuce (*Lactuca sativa*), 71% for 'Provider' bean (*Phaseolus vulgaris*), and 73% for 'Scarlet Nantes' carrot (*Daucus carota*) compared to the control. These findings resulted in a recommendation not to apply CGM even at the lowest application rate (100 $g \cdot m^{-2}$) to direct-seeded vegetables (McDade & Christians, 2000). Webber and Shrefler (2007b) determined that broadcast applications of CGM as low as 100 $g \cdot m^{-2}$ significantly decreased the establishment of direct-seeded black bean 'Black Knight', pinto bean 'Apache', cantaloupe (*Cucumis melo*) 'Magnum 45' and watermelon (*Citrullus lanatus*) 'Allsweet' by 66%, 58%, 50%, and 58%, respectively. Webber and Shrefler (2007b) suggested the potential usefulness of CGM application for direct-seeded vegetables by restricting CGM to the interrow area while leaving a CGM-free application area for the direct-seeding of vegetable crops.

A mechanized applicator was developed and evaluated to apply CGM in a banded configuration (Webber & Shrefler, 2007a). The applicator was constructed using various machinery components (fertilizer box, rotating agitator blades, 12-V motor, and fan shaped, gravity-fed, row banding applicators). The equipment was evaluated for the application of two CGM formulations (powdered and granulated), three application rates (250, 500, and 750-$g \cdot m^{-2}$), and two application configurations (solid and banded). Differences between CGM formulations affected flow rate within, and between, application configurations. The granulated formulation flowed at a faster rate, without clumping, compared to the powdered formulation, while the CGM in the banded configuration flowed faster than the solid application. Webber and Shrefler (2007a) demonstrated the feasibility of using equipment, rather than manual applications, to apply CGM to raised beds for organic weed control. The development of equipment to apply CGM in banded configurations created an opportunity to investigate whether banded CGM applications would provide significant crop safety for direct-seeded vegetables.

As a result of the development of a mechanized application system for the banded placement of CGM between crop rows (seed row not treated) Webber et al. (2010)

investigated the impact of CGM applications (formulations, rates, incorporation, and banded applications) on direct-seeded squash (*Cucurbita pepo*) plant survival and yields. It was determined that neither CGM formulation (powdered or granulated), nor incorporation method (incorporated or non-incorporated), resulted in significant differences in plant survival or squash yields. When average across all other factors (formulations, incorporation method, and banding), CGM rates of 250 to 750-g·m^{-2} reduced squash survival from 70% to 44%, and squash yields from 6402 to 4472-kg·ha^{-1}. However, the banded application (CGM placed between rows) resulted in significantly greater crop safety, 75% survival, and yields, 6402 kg·ha^{-1}, than the broadcast (non-banded) applications, 35% survival and 4119 kg·ha^{-1} yields. It was demonstrated that banded applications of CGM can be useful in direct-seeded squash production and other organic direct-seeded vegetables.

2.4.1.2 Mustard seed meal

Mustard seed meal (MSM) is the by-product of the seed oil pressing process. Research has shown that MSM added to the soil inhibited weed emergence and growth (Ascard & Johansson, 1991; Boydston, 2008; Boydston et. al., 2008, 2011; Miller, 2006, Webber et al., 2009a). As with CGM, MSM is a non-selective natural herbicide that will not discriminate between weeds and crop plants, therefore, care must be taken to control the target species (weeds) and provide sufficient crop safety. Among other crops, MSM has been used to control weeds in turf (Earlywine et al., 2010), onion (Boydston et al., 2011), ornamentals (Boydston et al., 2008), potato, Solanum tuberosum L., (Boydston, 2008), and peppermint, Mentha × piperita L., (Boydston, 2008). Research has shown that the range of weeds controlled or suppressed by MSM is very extensive (Ascard and Johansson, 1991; Boydston, 2008; Boydston et al., 2011; Earlywine, 2010; Handiseni et al., 2011; Vaughn et al., 2006; Webber 2009a; Yu & Morishita, 2011). A partial list of weeds inhibited by MSM would include redroot pigweed (Amaranthus retroflexus L.), green foxtail (Setaria viridis (L.) Beauv.), kochia (Kochia scoparia (L.) Schrad.), Russian-thistle (Salsola tragus L.), common lambsquarters (Chenopodium album L.), barnyardgrass (Echinochloa crus-galli L. Beauv.), annual sowthistle (Sonchus oleraceus L.), buckhorn plantain (*Plantago lanceolata* L.), common chickweed (*Stellaria media* (L.) Vill.), large crabgrass (*Digitaria sanguinalis* L. Scop.), Italian ryegrass (*Lolium perenne* L. spp. multiflorum Lam. Husnot), prickly lettuce (*Lactuca serriola* L.), and wild oat (*Avena fatua* L.) (Boydston et al. 2011; Earlywine et al., 2010; Handiseni et al., 2011; Yu & Morishita, 2011).

Yu and Morishita (2011) compared weed control efficacy of MSM and CGM at 3 rates (2240, 4480, and 6720 kg ha^{-1}) at two locations for five broadleaf and two grass weed species. Yu and Morishita concluded that MSM provided, in general, had a greater weed control efficacy than CGM. The species of mustard can also influence the weed control efficacy. Hoagland et al. (2008) determined that MSM from *Sinapis alba* L. produced as good or greater phytotoxicy and was more consistent than MSM produced from *Brassica napus* L. and *B. Juncea* L.

It is essential to understand that as non-selective herbicides, CGM and MSM can injure or kill germinating and emerging crop seedlings. Crop safely is greater when these substances are applied to established annual or perennial plants. Although CGM and MSM can provide effective early preemergence weed control of germinating weed seeds, supplemental weed control measures will be required to control escaped weeds, established perennial weeds, or weeds emerging in the mid- to late-growing season. Most, if not all, organic certificating entities prohibit the use CGM and MSM derived from genetically modified organisms

(GMO). Although organic materials are naturally derived, care should always be taken to safely handle and apply the materials. For example, MSM can cause extreme dermal reaction in humans and should be used with suitable protective equipment.

Researchers have identified other seed and plant components that natural inhibit plant growth (allelopathy) (Chung et al., 1997; Kuk et al., 2001; Rice, 1984; Tamak et al., 1994a, 1994b). Hopefully, research will continue to develop suitable organic herbicides from not only seed meals, but from other plant components.

2.4.2 Post-emergence, post-directed, and burndown organic herbicides

The organic post-emergence, post-directed, and burndown herbicides are all non-selective, non-translocated, contact herbicides which need to be applied, either prior to crop emergence or transplanting, or post-directed to established crops to assure the herbicides do not injure the crop plant. In general these contact herbicides control broadleaf weeds better than grasses, smaller weeds better than larger weeds, and annual weeds better than perennial weeds. These herbicides destroy the plant's waxy cuticle and cell walls causing desiccation and rapid wilting, which is further improved with uniform application of the material. Depending on the herbicide, an adjuvant may increase the herbicidal activity by increasing the destruction of the cuticle and cell wall, or by providing a more uniform application. Read and follow the label to determine whether an adjuvant is required and, if so, what type of adjuvant and mixture rate.

2.4.2.1 Ammonium nonanoate

Racer® (40% ammonium Nonanoate) is a soap formulation of pelargonic acid with a changing registration history. It is a non-selective contact herbicide for controlling small (2.5 to 5 cm tall) annual broadleaf and grass weeds (Webber et al., 2011a). Repeated applications may be needed to control most grasses or larger (5 cm) broadleaf weeds. It has been cleared for non-crop use in organic crop production and with addition of new formulations may be cleared for use in organic crop production. Organic producers should receive clearance from their certifying agency prior to using Racer.

2.4.2.2 Fatty acids

A recent National Organic Program (NOP) ruling decided that pelargonic acid is a prohibited substance for organic crop production. Until the recent ruling, pelargonic acid, a fatty acid, had tremendous potential as an organic herbicide. It had proven effective as a non-selective post-emergent contact herbicide (Webber et al., 2011b). It provided excellent weed control at low application rates and volumes, but has not been cleared due to its manufacture by synthetic methods. In addition to pelargonic acid, other fatty acids are under consideration and development as potential organic herbicides.

2.4.2.3 Vinegar (5, 10, 15, and 20% acetic acid)

There are a number of organically approved products that contain vinegar (e.g., Weed Pharm®, 20% Acetic Acid) that contain vinegar (e.g., 5%, 10%, and 20% acetic acid). Vinegar (acetic acid) is a non-selective contact herbicide. In general, weed control increases as acetic acid content and application volume increase (e.g., 20, 40, 80, and 100 gpa). Typically, vinegar is less effective in controlling grasses than broadleaf weeds and more effective on annual species than perennials (Webber and Shrefler, 2007c, 2008a, 2008b, 2009b; Webber et al., 2009b). In addition to application volumes and concentration, weed control is also dependent on the weed size and the species. Carpetweed (*Mollugo verticillata* L.) is very sensitive to acetic acid at very low concentrations and application volumes, while yellow nutsedge (*Cyperus esculentus* L.) is able to tolerate high acetic acid concentrations and application volumes. Repeated applications of acetic acid may be necessary for satisfactory weed control depending on weed size, weed species, and whether it is an annual or perennial plant. There is also a difference between non-synthetic and synthetic acetic acid and approval for use in organic production. If the material is intended for use on certified organic land, check for approval of your specific product with your organic certifying agency. Also keep in mind that clearance for organic use does not mean a product can not cause personal injury, if handled in an unsafe manner. Vinegar with greater than 10% acetic acid can cause severe eye damage or even blindness.

2.4.2.4 Clove oil

Clove oil is the active ingredient in a number of organically approved post-emergent non-selective herbicides (e.g. Matratec®, Matran® EC and Matran®, 50% Clove Oil). Clove oil is a post-emergence, non-selective, contact herbicide for the control of actively growing emerged annual and perennial grass and broadleaf weeds. As a contact, non-translocated herbicide its effectiveness increases with application rate and decreasing weed size. As with the other contact herbicides, when weeds are of similar size, the broadleaf weeds are easier to control than the grasses (Webber and Shrefler, 2009a; clove oil weed control efficacy can be as good, or better than acetic acid herbicides, and can be applied at lower application volumes and remain effective. There is evidence that adding certain organically approved adjuvants (e.g., garlic and yucca extracts) will increase weed control with clove oil. Repeated applications may be necessary because larger annual grass weeds may grow back.

2.4.2.5 D-limonene

GreenMatch® (55% d-limonene) is a post-emergence, non-selective, contact herbicide for control on actively growing emerged annual and perennial grass and broadleaf weeds. Shrefler et al. (2011) used d-limonene as a post-directed control of weeds in organic cantaloupe and Lanini et al (2010) conducted greenhouse and field studies using d-limonene. As with the other organic contact herbicides, d-limonene, in general, has greater efficacy on younger smaller weeds than larger older weeds, and greater control of broadleaf weeds than grass weeds (Lanini et al., 2010).

2.4.3 Organic herbicide conclusions

Additional active ingredients and formulations are also being developed. Approval of these involves conducting greenhouse screenings, and progressing to extensive field evaluations. Even if all these active ingredients and their commercial formulations are registered by EPA and approved for organic use, the application technology and timing will play an essential element in their successful integration into existing certified organic systems. Research with post-directed applications of non-selective contact herbicides is showing promise. The height and plant maturity differences between the crop and target weeds are important factors in controlling weeds and protecting the crop from herbicide damage. The post-directed technique is especially effective when used in combination with either preemergence corn gluten meal applications or transplanted crops.

2.4.4 Herbicide precautions

Always read and follow the herbicide labels, take appropriate safety precautions, and don't hesitate to contact your certifying agency prior to applying any substance.

3. Summary

Controlling weeds is an essential aspect of successful crop production. The lack of weed control can result in the total yield loss due to weed competition and with weeds acting as a reservoir for pathogens through disease and insect damage. Weed control should be considered a continuous endeavor not just a seasonal effort. It is more cost effective to prevent an infestation than eliminating a weed species once the production area is infested. Weed control should start in the previous crop, by monitoring, controlling, and managing the weeds. Successful weed management uses a multifaceted approach (rotating crops and herbicides, cover crops, mulches, cultivation) rather than relying solely on herbicides to control the weeds. Knowing which weeds will be present and understanding their growth habits will enable the producer to achieve greater weed control by the wise application of the many weed control methods available.

4. References

Ascard, J. & Johansson, T. (1991). White mustard meal interesting for weed control, In: Weeds and Weed Control Reports. *Proceedings of 32nd Swedish Crop Protection Conference.* Swedish University of Agricultural Sciences, Stockholm, Sweden. p. 139-155.

Biazzo, J. & Masiunas, J.B. (1998) Using living mulches for weed control in Hungarian wax pepper (*Capsicum annuum*) and okra (*Abelmoschus esculentus*). *Illinois Fruit and Vegetable Crop Research Report.* University of Illinois - Champaign.

Bingaman, B.R. & Christians, N.E. (1995) Greenhouse screening of corn gluten meal as a natural control for broadleaf and grass weeds. *HortScience* 30:1256-1259.

Boydston, R.A. (2008) Mustard meal suppresses weeds in potato and peppermint. *Proceedings Western Society of Weed Science,* Anaheim, CA, USA. March 2008.61:37.

Boydston, R.A., Anderson, T. & Vaughn, S.F. (2008) Mustard (*Sinapis alba*) seed meal suppresses weeds in container-grown ornamentals. *HortScience* 43:800-803.

Boydston, R.A., Morra, M.J., Borek, V., Clayton, L. & Vaughn, S.F.(2011) Onion and Weed Response to Mustard (Sinapis alba) Seed Meal. Weed Science In-Press. Online. doi: 10.1614/WS-D-10-00185.1

Christians, N.E. (1991) Preemergence weed control using corn gluten meal. U.S. Patent No. 5,030,268 p.63-65. *United States Patent and Trademark Office.* Washington D.C.

Christians, N.E. (1993) The use of corn gluten meal as a natural preemergence weed control in turf. *Intl. Turfgrass Soc. Res. J.* 7:284-290.

Christians, N.E. (1995) A natural herbicide from corn meal for weed-free lawns. *IPM Practitioner* 17(10):5-8.

Chung, I.M., K.H. Kim, K.H., Ahn, J.K. & Ju, H.J. (1997) Allelopathic potential evaluation of rice varieties on *Echinochloa crus-galli. Kor. J. Weed Sci.* 17:52–58.

Clark, S. and Panciera, M. (2002) Cover crop roll-down for weed suppression in no-till crop production. *Fruit and Vegetable Crops Research Report.* Univ. Kentucky Agr. Expt. Sta. p. 56-57.

Darlington, H.T. & Steinbauer, G.P. (1961) The eighty-year period for Dr. Beal's seed viability experiment. *American J. of Botany* 48: 321-325.

Dick, W.A. (1983). Organic carbon, nitrogen and phosphorus concentrations and pH in soil profiles as affected by tillage intensity. *Soil Sci. Soc. Am. J.* 47:102–107.

Decoteau, D.R., Kasperbauer, M.J. & Hunt, P.G. (1990) Bell pepper plant development over mulches of diverse color. *HortScience* 25:460-462.

Earlywine, D.T., Smeda, R.J., Teuton, T.C., Smas, C.E., & Xiong, X. (2010) Evaluation of oriental mustard (*Brassica juncea*). *Weed Tchnology* 24:440-445.

Gallaher, R.N. & Ferrer, M.B. (1987) Effect of no-tillage vs. conventional tillage on soil organic matter and nitrogen contents. *Commun. Soil Sci. Plant Anal.* 18:1061-1076.

Georgia, A. (1942) *Manual of weeds*. Macmillan, New York. 593 p.

Handiseni, M., Brown, J., Zemetra, R. & Mazzola, M. (2011) Herbicidal activity of brassicaceae seed meal on wild oat (*Avena fatua*), Italian ryegrass (*Lolium multiflorum*), redroot pigweed (*Amaranthus retroflexus*), and prickly lettuce (*Lactuca serriola*) *Weed Technology* 25:127-134 doi: 10.1614/WT-D-10-00068.1

Hauptli, H. & Jain, S.K. (1978) Biosystematics and agronomic potential of some weedy and cultivated amaranthus. *Theoretical and Applied Genetics* 52, 177-185.

Hoagland, L., Carpenter-Boggs, L., Reganold, J.P. & Mazzola, M. (2008) Role of native soil biology in Brassicaceous seed meal-induced weed suppression. *Soil Biology and Biochemistry* 40:1689-1697.

Johnson, A.E. (1986) Soil organic matter, effects on soil and crops. *Soil Use Management*. 2:97-105.

Johnson, W.C., III, Davis, R.F. & Mullinix, B.G., Jr. (2007). An integrated system of summer solarization and fallow tillage for *Cyperus esculentus* and nematode management in the southeastern coastal plain. *Crop Protection* 26:1660-1666.

Johnson, W.C., III, & Mullinix, B.G., Jr. (2008) Potential weed management systems for organic peanut production. *Peanut Sci.* 35:67-72.

Kaul, K. & Kasperbauer, M.J. (1992) Mulch color effects on reflected light, rhizosphere temperature, and pepper yield. *Transactions Kentucky Academy of Science* 53, 109-112.

Kuk, Y., Burgos, N.R., & Talbert, R.E. (2001) Evaluation of rice by-products for weed control. *Weed Science* 49:141-147.

Lanini, W.T., Capps, L., & Roncoroni, J.A. (2010) Field testing of organic herbicides. *Proceedings of the Western Society of Weed Science*. Waikoloa, Hawaii, USA. March 2010. 63:13.

Law, D.M., Rowell, A.B., Snyder, J.C. & Williams, M.A. (2006) Weed control efficacy of organic mulches in two organically managed Bell Pepper production systems. *HortTechnology* 16:225-231.

Leavitt, M.J., Sheaffer, C.C., Wyse, D.L. & Allen, D.L. (2011) Rolled winter rye and hairy vetch cover crops lower weed density but reduced vegetable yields in no-tillage organic production. *HortScience* 46(3), 387-395.

Logan, T.J., Lal, R. & Dick, W.A. (1991). Tillage systems and soil properties in North America. *Soil Tillage Res.* 20:241-270.

McDade, M.C. (1999) Corn gluten meal and corn gluten hydrolysate for weed control. *M.S. Thesis*, Iowa State Univ. Dept. of Hort., Ames, IA.

McDade, M.C. & Christians, N.E. (2000) Corn gluten meal - a natural preemergence herbicide: effect on vegetable seedling survival and weed cover. *Amer. J. Alternative Agr.* 15(4):189-191.

McDowell, L.L. & McGregor, K.C. (1984) Plant nutrient losses in runoff from conservation tillage corn. *Soil and Tillage Research* 4:79-91.

McGiffen, Jr., M.E. & Hutchinson, C. (2000) Cowpea cover crop mulch controls weeds in transplanted Peppers. *HortScience* 35:462.

Miller, T.W. (2006) Natural herbicides and amendments for organic weed control In: *Crop Protection Products for Organic Agriculture. ACS Symposium Series 947,* A.S. Felsot and K.D. Racke (eds.). p.174-175. American Chemical Society, Washington, D.C.

Mitich, L.W. (1997) Intriguing world of weeds - Redroot pigweed (*Amaranthus retroflexus*). *Weed Technology* 11, 199-202.

Mohler, C.L. and Callaway, M.B. (1995) Effects of tillage and mulch on weed seed production and seed banks in sweet corn. *J. of Applied Ecology* 32, 627-639.

Nonnecke, G.R. & Christians, N.E. (1993) Evaluation of corn gluten meal as a natural weed control product in strawberry. *Acta Hort.* 348:315-320.

Organic Farming Research Foundation (OFRF) (1999). *Final results of the third biennial national organic farmers' survey*. Santa Cruz, CA, USA. 161 pages.

Organic Farming Research Foundation (OFRF). (2006) On-farm testing of organic weed control strategies in Indiana. *Project report summary*. Santa Cruz, CA, USA.

Paine, L.K. & Harrison, H. (1993) The historical roots of living mulch and related practices. *HortTechnology* 3(2), 137-143.

Pawlowski, F., Kapeluszny, J., Kolasa, A. & Lecyk, Z. (1970) The prolificacy of weeds in various habitats. *Annales Universitatis Mariae Curie, Sklodowska Lublin, Polonia,* 25:61-75.

Pekrun, C., El Titi, A. & Claupein, W. (2003). Implications of soil tillage for crop and weed seeds. In: *Soil Tillage in Agroecosystems,* A. El Titi, Editor, pp. 115–146. CRC Press, Boca Raton, FL.

Quarles, W. (1999). Corn gluten meal: a least-toxic herbicide. *The IPM Practitioner* 21:1-7.

Rice, E.L. (1984) *Allelopathy*. Orlando, Florida: Academic Press (Second Edition).

Russo, V.M. (1995) Bedding, plant population, and spray-on mulch tested to increase dry bean yield. *HortScience* 30:53-54.

Russo, V.M., Webber, C.L., III, & Myers, D.L. (1997) Kenaf extract affects germination and post-germination development of weed, grass and vegetable seeds. *Industrial Crops and Products* 6:59-69.

Shrefler, J., Webber III, C.L., Taylor, M.J., & Roberts, B.W. (2011). Weed management options for organic cantaloupe production. *Proceeding of the Weed Science Society of America.* Portland, OR, USA. Feb. 2011. Poster # 51. URL http://wssaabstracts.com/public/4/proceedings.html

Stevens, O.A. (1932) The number and weight of seeds produced by weeds. *American J. of Botany* 19, 784-794.

Stevens, O.A. (1957) Weights of seeds and numbers per plant. *Weeds* 5:46-55.

Tamak, J.C., Narwal, S.S., Singh, L. & Ram, M. (1994a) Effect of aqueous extracts of rice stubbles and straw 1 stubbles on the germination and seedling growth of *Convolvulus arvensis, Avena ludoviciana,* and *Phalaris minor. Crop Res.* 8:186-189.

Tamak, J.C., Narwal, S.S., Singh, L. & Singh, I. (1994b) Effect of aqueous extracts of rice stubbles and straw 1 stubbles on the germination and seedling growth of wheat, oat, berseen, and lentil. *Crop Res.* 8:180-186.

Teasdale, J. & Mohler, C. (2000) The quantitative relationship between weed emergence and the physical properties of mulches. *Weed Science* 48:385-392.

United States Department of Agriculture (USDA) National Organic Standards Board (2010) *Policy and Procedures Manual.* United States Department of Agriculture, Washington, D.C.

Vaughn, S.F., Palmquist, D.E., & Duval, S.M. (2006) Herbicidal activity of glucosinolate-containing seedmeals. *Weed Science* 54:743-748.

Weaver S.E. & McWilliams, E.L. (1980) The biology of Canadian weeds: *Amaranthus retroflexus* L., *A. powellii* S. Wats. and *A. hybridus* L. *Canadian J. of Plant Science* 60: 1215-1234.
Webber, C.L. III & J.W. Shrefler. (2007a) Corn gluten meal applicator for weed control in organic vegetable production. *J. Veg. Crop Production* 12:19-26. 2007.
Webber, C.L. III & Shrefler, J.W. (2007b) Impact of corn gluten meal applications on direct-seeded vegetable seedling establishment. *Intl. J. Veg. Sci.* 13:5-15.
Webber, C.L. III & Shrefler, J.W. (2007c) Organic weed control with vinegar: Application volumes and adjuvants. *Proceedings of Horticultural Industries Show*. Ft. Smith, AR, USA. Jan. 2007. 26:149-151.
Webber, C.L. III, Shrefler,J.W. & Taylor, M.J. (2007a) Corn gluten meal as an alternative weed control option for spring-transplanted onions. *Intl. J. Veg. Sci.* 13:17-33.
Webber, C.L. III, Shrefler, J.W. & Taylor, M.J. (2007b) Impact of corn gluten meal applications on spring-transplanted onion (*Allium cepa*) injury and yields. *Intl. J. Veg. Sci.* 13:5-20.
Webber, C.L. III & Shrefler, J.W. (2008a) Acetic acid and weed control in onions (*Allium cepa* L.). *Proceedings of National Allium Research Conference*. Dec. 2008. Savannah, GA, USA. p. 49-54.
Webber, C.L. III & Shrefler, J.W. (2008b) Acetic acid: Crop injury and onion (*Allium cepa* L.) yields. *Proceedings National Allium Research Conference*. Dec. 2008. Savannah, GA, USA. p. 55-59.
Webber, C.L. III & Shrefler, J.W. (2009a) Broadcast application of Matran for broadleaf weed control in spring-transplanted onions. In: *2008 Weed Control Report*. Brandenberger, L. and Wells, L. (eds.). MP-162:17-19. Oklahoma State University, Division of Agricultural Sciences and Natural Resources, Department of Horticulture & Landscape Architecture. Stillwater, OK, USA.
Webber, C.L. III & Shrefler, J.W. (2009b) Broadcast application of vinegar for broadleaf weed control in spring-transplanted onions. In: *2008 Weed Control Report*. Brandenberger, L. and Wells, L. (eds.). MP-162:26-28. Oklahoma State University, Division of Agricultural Sciences and Natural Resources, Department of Horticulture & Landscape Architecture. Stillwater, OK, USA.
Webber, C.L. III, Shrefler, J.W., & Boydston, R.A. (2009a) Mustard Meal as an Organic Herbicide. *Proceedings of the Joint 49th Weed Science Society of America and 62nd Southern Weed Science Society*. Orlando, FL, USA. Feb. 2009. p. 80.
Webber, C.L. III, Shrefler, J.W., Brandenberger, L.P., Taylor, M.J. & Boydston, R.A. (2009b) 2008 Organic herbicide update. *Proceedings of the Horticultural Industries Show*. Ft. Smith, AR, USA. Jan. 2009. p. 237-239.
Webber, C.L. III, Shrefler, J.W., & Taylor, M.J. (2010) Influence of Corn Gluten Meal on Squash Plant Survival and Yields. *HortTechnology* 20:696-699.
Webber, C.L. III, Brandenberger, L.P., Shrefler, J.W., Taylor, M.J., Carrier, L.K. & Shannon, D.K. (2011a) Weed control efficacy with ammonium nonanoate for organic vegetable production. *Int. J. Veg. Sci.* 17:1-8.
Webber, C.L. III, Shrefler, J.W. & Brandenberger, L.P. (2011b) Scythe® (pelargonic acid) weed control in squash. In: *2010 Weed Control Report*. Brandenberger, L. and Wells, L. (eds.). MP-162:20-23. Oklahoma State University, Division of Agricultural Sciences and Natural Resources, Department of Horticulture & Landscape Architecture. Stillwater, OK, USA.
Yu, J. & Morishita, D.W. (2011) Greenhouse screening of corn gluten meal and mustard seed meal as natural weed control products. *Proceedings of the Western Society of Weed Science*. Spokane, WA, USA. March 2011, Abstract #60. (http://wssaabstracts.com/public/6/abstract-60.html)

11

Opportunities for Weed Control in Dry Seeded Rice in North-Western Indo-Gangetic Plains

Gulshan Mahajan[1], Bhagirath Singh Chauhan[2] and Jagadish Timsina[3]
[1]*Punjab Agricultural University, Ludhiana*
[2]*International Rice Research Institute, Los Baños*
[3]*International Rice Research Institute , IRRI-Bangladesh Office, Dhaka*
[1]*India*
[2]*Philippines*
[3]*Bangladesh*

1. Introduction

North-western Indo-Gangetic Plain (NW-IGP) of India has played a leading role in the agricultural transformation of India and is considered as the most fertile plain for the livelihood of millions of peoples of India (Dhillon et al., 2010). Its contribution to the national food basket over years has also registered a remarkable increase compared to other regions. Food security of the nation is highly dependent on the NW-IGP as evident from the contribution of this region to the national buffer stock of food grains, which has generally been 50–75% in wheat and 30–48% in rice (Timsina and Connor, 2001). Sustainable production of rice and wheat in the region is very important for the food security of India. Cultivation of rice requires huge labour and large amount of water. Water and labour, however, are becoming increasingly scarce in the region raising the questions of environmental sustainability and sustainability of rice production systems. In NW-IGP, increasing use of groundwater for rice cultivation has led to declines in water table by 0.1 to 1.0 m yr^{-1}, resulting in water scarcity and increased cost for pumping water (Hira, 2009; Rodell et al., 2009; Humpherys et al., 2010). Implementation of the Mahatma Gandhi National Rural Employment Guarantee Act, introduced by the Indian government in 2005 (GOI, 2011), promising 100 days of paid work in people's home village, has been creating a labour scarcity in Punjab and Haryana which are the cereal bowl of northwest India. Rice transplanting in northwest India, particularly in the Punjab and Haryana, is dependent on millions of migrant labourers from eastern Uttar Pradesh and Bihar (Kumar & Ladha, 2011).

Since rice is primarily grown by transplanting seedlings in puddled fields it requires a large amount of water (~150 cm), of which 20-25 cm is used for puddling (intensive cultivation in wet conditions) only. This suggests that alternatives to puddled transplanted rice are required to save water and increase crop and labour productivity. One way to reduce water and labor demand is to grow dry seeded rice (DSR) instead of the puddled transplanted rice (Mahajan et al., 2009; Sudhir-Yadav et al., 2010). In DSR crop, dry seeds are sown into a prepared seedbed after tillage or under zero-till conditions and depending on water availability soils are kept aerobic, continuously saturated, or flooded. A DSR crop grown without standing water, intended to use less irrigation water than conventional flooded rice, is referred as aerobic rice. Dry seeding of rice with subsequent aerobic soil conditions eliminates the need of puddling and maintains submerged soil conditions, thus reducing the overall water demand and providing opportunities for water and labour savings (Bouman, 2001; Sharma et al., 2002). However, weeds are a serious problem because dry tillage practices and aerobic soil conditions are conducive for germination and growth of weeds, which can cause grain yield losses from 50 to 90% (Rao et al., 2007; Chauhan & Johnson, 2011; Chauhan et al., 2011; Prasad, 2011). The development and adoption of DSR may enable good crop growth but the lack of sustained flooding will greatly increase potential losses from weeds. These systems may integrate direct seeding and herbicide use, yet, to be sustainable, effective weed management strategies are required. A multitude of prerequisites, including level land, effective weed control, efficient water management, and timely water supply in relation to crop water demand, need to be met to ensure a successful DSR crop. In most places, insufficient attention is given to the importance of weeding. In DSR fields, it is not uncommon to see fields full of weeds, mainly grasses. When weed control in rice is neglected, there is a decrease in yield because of weeds, even if other means of increasing production, including application of fertilizers, are practiced. In the NW-IGP, DSR is an emerging production system. The transition from the puddle transplanted rice to DSR can therefore only be successful if accompanied by effective integrated weed management practices.

1.1 Weed species association

Weed flora in DSR consists of various kinds of grasses, broad leaf weeds and sedges (Mahajan et al., 2009) . Community composition of these weeds varies according to crop establishment methods, cultural methods, crop rotation, water and soil management, location, weed control measures, climatic conditions, and inherent weed flora in the area. *Echinochloa colona* and *E. crus-galli* are the most serious weeds affecting DSR. The density of these weeds in DSR depends upon moisture condition in the field. *E. colona* requires less water, so it is more abundant in DSR. *Cyperus rotundus* and *Cynodon dactylon* may be major problems in poorly managed fields or where un-decomposed farm yard manure has been applied. The other weeds of major concern in DSR are *Leptochloa chinensis, Digitaria sanguinalis, Dactyloctenium aegyptium, Caesulia axillaris, Paspalum spp., Alternanthera spp., Trianthema monogyna, Cyperus iria, Cyperus difformis*, and *Fimbristylis miliacea.*

1.2 Losses and critical duration of weed crop competition

Weeds in DSR adversely affect yield, quality and cost of production as a result of competition for various growth factors. Extent of loss may vary depending upon cultural methods, rice cultivars, weed species association, and their density and duration of competition. *Trianthema monogyna* was found to grow faster than other weeds during early stage due to shorter life cycle and contributed much more to the competition as compared to other weeds (Singh, 2008). Relativity yield losses are higher for DSR than for transplanted rice as the rice and weed seedlings emerge simultaneously in DSR. The competitive advantage of transplanted rice over DSR is due to the use of 4-5 weeks old seedlings (20-30 cm tall) in transplanted rice and also that the weeds emerging after rice transplanting are controlled by flooding after transplanting in transplanted rice compared to DSR. In Asia, yield losses due to uncontrolled weeds in DSR are between 45 and 75% and in transplanted rice of approximately 50%. In a recent study, yield losses due to weeds (with one weeding at 28 days after sowing) in aerobic rice were about 50% relative to weed-free rice (Chauhan et al., 2011). In order to formulate an effective and economical weed management system for DSR, it is essential to establish critical duration of weed-crop competition and a limit for an acceptable presence of weeds. In DSR, the longer the presence of competition during the initial period the lower the yield, while at later stages yield might not change since the maximum damage has already accurred. The effective period of weed-crop competition in DSR occurs in two phases; i.e. between 15-30 days, and 45-60 days after seeding. The competition beyond 15 days after seeding may cause significant reduction in the grain yield. However, competition for the first 15 days only may not have much adverse effect on crop (Singh, 2008). In a recent study, critical periods for weed control, to obtain 95% of a weed-free yield, were estimated as between 17 to 56 days after sowing of the DSR crops at 15-cm row spacing (Chauhan & Johnson, 2011).

2. Opportunity

2.1 Prevention

Sowing of clean seed is perhaps the most important weed management technique in any crops. Rice seed contaminated with weed seeds may introduce a new species to a given field or add to an existing weed population. Preventing weeds from entering an area may be easier than trying to control them once they have established. Weeds which mature at the same time as rice are harvested and threshed with the crop resulting in the contamination of rice seed with weed seeds. Therefore, by using clean seed, infestation of weeds can be reduced to a considerable extent. Irrigation water is one of the major means of spread of weed seeds and vegetative propagules. Flowing water moves millions of weed seeds from one place to another. The amount and the type of weed seeds moved depend on the volume and the velocity of the water and the size and weight of the seed or the vegetative propagules. In case of heavy weed infestation, it is advisable to rotate the DSR crop with transplanted rice or other methods of rice planting. A continuous DSR crop may result in a shift to emergence of new weeds like *D. sanguinalis* and *L. chinensis* (Mahajan & Chauhan, 2011a).

2.2 Stale seed bed technique

In this technique, after pre-sowing irrigation, fields are left as such and weeds are allowed to germinate and thereafter are killed through cultivation or with the use of non-selective herbicide (e.g., glyphosate) application. This technique is quite effective in DSR, especially for controlling weeds such as *C. rotundus*. Herbicides may destroy weeds without disturbing the soil, which would be advantageous and hence reducing the possibilities of bringing new seeds to the upper soil surface. The rice seeds should be sown with minimum soil disturbance after destroying the emerged weeds. The use of zero-till-ferti-seed drills may be useful to serve this purpose. A reduction of 59% in the density and by 78% in the fresh weight of *E. colona* was recorded after using stale bed technique in the Philippines (Moody, 1982). Information on the effectiveness of stale seedbed technique is, however, very limited in the literature on DSR.

2.3 Land preparation

In DSR, the fields should be well levelled, if possible by using a laser leveller. Uneven land levelling may result in poor crop establishment (Jat et al., 2009). Developing rice seedlings can be killed or greatly retarded in their growth if ponding occurs due to unevenness of field or due to accumulation of toxic concentrations of applied herbicides. Also, weeds invade the vacant spaces where there are no rice plants, and hence there could be yield losses due to weed competition. Therefore, if the field is not levelled, the rice seedlings cannot establish quickly in the low spots and weeds will grow abundantly in the high spots (Singh et al., 2009). These conditions may result in stunted rice plants with low tiller production. Good field drainage and good water control are essential for the successful crop establishment and reduction of pre-emergence herbicide phytotoxicity. Tillage, however, serves only as a temporary means of weed control because the soil contains many ungerminated weed seeds. Ploughing buries weed seeds at a depth from where they cannot emerge but also brings deeply buried seeds to the surface where the conditions are conducive for germination (Chauhan & Johnson, 2010).

2.4 Choice of cultivar

Cultivars play an important role in crop-weed competition because of their diverse morphological traits, canopy structure and relative growth rate. A quick growing and early canopy cover enables a cultivar to compete better against weeds. Research evidences have shown that traditional tall cultivars like Nerica rice exert effective smothering effect on weeds (Prasad, 2011). Further, it has been observed that early maturing rice cultivars and rice hybrids also have a smothering effect on weeds due to improved vigour and having the tendency of early canopy cover (Mahajan et al., 2011). Little is known about the relative importance of shoot and root competition of hybrids and inbreds in field conditions in DSR and this may be the subject of future research. Most importantly, there is a need to develop rice varieties suitable for DSR as currently, varieties bred for transplanted rice are being used in DSR.

2.5 Seeding rate

For direct drilling of rice, 30-60 kg ha^{-1} seed rates are enough for a healthy crop. Only 30 kg ha^{-1} seed rate could be used if there was minimal weed pressure in the previous crops. Seed rate could be increased up to 60 kg ha^{-1} to compensate for the damages by rats and birds, to partially overcome the adverse effects of herbicides, and to compensate for poor stand establishment if rains occurs immediately after sowing (Mahajan et al., 2010). Low plant density and high gaps encourage the growth of weeds, and in many cultivars, result in less uniform ripening and poor grain quality. On the other hand, very high plant stand should be avoided because it tends to have less productive tillers, increases lodging, prevents the full benefit of nitrogen application, and increases the chances of rat damage. In DSR, Chauhan et al. (2011) found a significant decrease in weed weight as the seeding rate increased from 15 to 125 kg ha^{-1} (Fig. 1). Thus, high seeding rate could partly control weeds. An increase in grain yield was observed in the untreated check plot but not in the weeded plot as a result of increased seeding rate (Chauhan et al., 2011). Increased panicle numbers with increased seed rate was offset by decrease in panicle length and panicle grain weight. Higher seeding rates would be beneficial if no weed control is planned or if only partial weed control is expected. In conclusion, different studies suggest that increasing seeding rates of the crop suppresses weed growth and reduced grain yield losses from weed competition. However, it is not necessary to use high seeding rates to suppress weeds in DSR if effective herbicides are used.

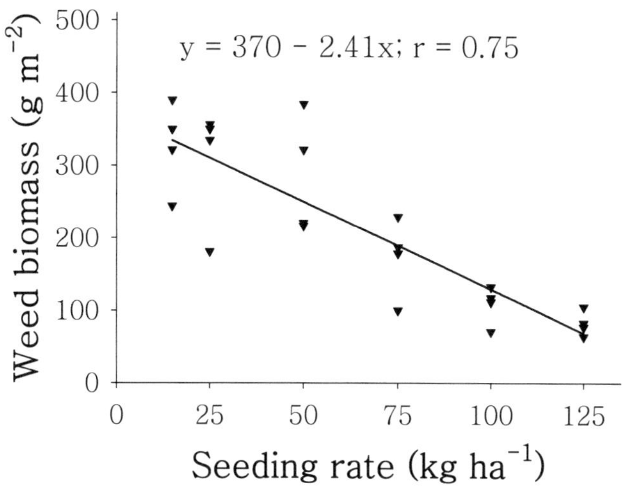

Fig. 1. The relation of weed biomass and rice seed rate at crop harvest (Chauhan et al., 2011).

2.6 Row spacing

The effect of crop geometry on weed incidence is a different issue from seeding rate. Mahajan & Chauhan (2011a) observed that paired row planting pattern (15-30-15-cm row spacing) in DSR had a great influence on weeds as compared to normal row planting system (23-cm row spacing) (Table 1). Paired row planting greatly facilitates weed suppression by maintaining rice plant's dominant position over weeds through modification in canopy structure. Earlier, Chauhan & Johnson (2010) reported that cultural management techniques, such as reduced crop row spacing, can increase the rice ability to compete against weeds for light.

Row spacing	Cultivar		
	Punjab Mehak 1		PR 115
	---------Dry matter (g m^{-2})---------		
35 days after sowing			
15-30-15-cm row spacing	131		136
23-cm row spacing	145		191
LSD (P = 0.05)		20	
Flowering stage			
15-30-15-cm row spacing	131		183
23-cm row spacing	140		244
LSD (P = 0.05)		21	

Table 1. Effect of planting pattern and cultivar on weed dry matter (g m^{-2}) at 35 days after sowing and at flowering stage (Mahajan & Chauhan, 2011a).

2.7 Water Management

Water management is an important component of any weed control program, whether any herbicide is used or not. Herbicides which give excellent control when applied into water may perform poorly in the absence of standing water (Kumar et al., 2009). There should be enough moisture in the field during the application of pre-emergence herbicides in DSR. In case of post-emergence application of herbicide, fields should be drained at the time of herbicide application and should not be irrigated immediately after its application. Good water management together with chemical weed control offers an unusual opportunity for conserving moisture and lowering the cost of rice production (Rao et al., 2007; Singh et al., 2009).

2.8 Nitrogen management

The proper management of N in DSR reduces the weed competition, and hence should be applied as per the requirement of the crop. The application of excess amount of N fertilizer, on the other hand, encourages weed growth and reduces yield. Recently, Mahajan & Timsina (2011) reported that when weeds were controlled, rice crop responded to higher amount of N application but under weedy and partially-weedy conditions, grain yield reduced drastically with higher amount of N fertilization.

2.9 Hand weeding

Many farmers do not realize that weed control is a limiting factor in crop production. Traditionally, they depend on manual labour to control weeds. By the time weeds are large enough to be removed by hand, damage has already occurred, and hand weeding alone cannot revert the yield loss. Maximum yields can only be achieved if weeds are controlled early because most damage takes place when crop plants are small. Hand weeding can only be done at a time when labour is available but this may not coincide with the best time for weed control. Improving weed management by alleviating labour constraints has repercussions for all aspects of crop production, the sustainability of cropping systems, and the social conditions of farming families.

Hand weeding in DSR is more time consuming and not as thorough as in transplanted rice. The weeders damage the rice plants as they move through the fields, especially during early crop growth. They fail to remove some of the grassy weeds or remove rice plants by mistake because of the difficulty in distinguishing some grassy weeds from rice. Also hand weeding is at least five times more expensive than herbicides for weed control in DSR, especially under labour-scarce or high labour cost environments (Rao et al., 2007). Hand weeding in DSR should only be done when there are typical weeds that cannot be controlled by either pre- or post-emergence herbicide application, or when labours are cheap and are in abundant supply.

2.10 Interrow cultivation

Recently, there have been many reports about the adaptation of DSR using a seed drill. The time taken for drilling rice seeds in an unpuddled field using a manually-operated planter is less than that required for seedling transplanting. Thus, this practice has potential to replace transplanting without any reduction in yield (Dixit et al., 2010). In India, under lowland conditions, it takes about 200-250 hours ha^{-1} for hand weeding in a DSR field, depending on the weed infestation. In a row-seeded or transplanted rice field, where weeds can be controlled by the use of mechanical weeders, it takes only about 50-60 hours ha^{-1} depending upon weed infestation and soil conditions. When the rice plants are 10-14 days old, a rotary weeder is passed through the field incorporating strips of rice seedlings into the soil so that the rice that is left appears as a row sown crop.

2.11 Herbicides

The success of DSR is dependent upon weed control with herbicides. However, herbicides should not be regarded as replacements for other weed control practices but should be used in conjunction with them. Herbicide use should be justified to coincide with the presence of sufficient weeds so as to warrant treatment, and should be used when weeds are most vulnerable. The optimum herbicide rate depends on factors such as cultural practices, soil type, and environmental conditions. Factors which must be considered when developing a herbicide program are the availability of herbicides, incidence of weed flora, application method and time, crop tolerance level, and cost effectiveness. The use of herbicides ensures effective weed control during periods of labour shortage when weeding coincides with other farm work. Temporary rice injury manifested as leaf chlorosis and inhibition of plant growth frequently occurs, but the crop usually recovers after two to three weeks and produces desired grain yield. Recently, sequential spray of pre-emergence application of pendimethalin (1 kg ha^{-1}) followed by bispyribac sodium (30 g ha^{-1}) at 15 days after sowing was found best for the control of weeds in DSR (Mahajan et al., 2009). Other herbicides that are found effective in DSR are pyrazosulfuron and oxadiragyl as pre-emergence and azimsulfuron, penoxsulam, cyhalopfop-butyl, and ethoxysulfuron as post-emergence (Rao et al., 2007). It must never be overlooked that all pesticides are toxic; they must be handled safely so as to reduce or avoid excessive and costly wastes, environmental concerns, crop damage, damage to adjacent crops by spray drift, injury to the applicator, excessive contamination and residues, and injury to beneficial organisms. It is advisable to rotate the herbicide combination in each year for delaying the development of herbicide resistance in weeds.

2.12 Integrated weed management

Herbicide use moves the agroecosystem to low species diversity with the possibility of appearance of new problem weeds, so there is a need for an ecological approach to control weeds instead of relying totally on chemical control methods. It was noted that reliance on a single herbicide could result in quantitative changes in the structure of the weed population in as few as five years (Singh, 2008). As weed populations are shifted by herbicides, weeds formerly of secondary importance emerge as primary weed problems. Such problems may be avoided by an integrated system of weed management, possibly by rotation of chemicals as well as rotation of crops. Many scientists advocate the alternative usage of herbicides to prevent the emergence of tolerant weeds.

Cultural and mechanical control methods have been the cornerstone of many pest control practices, in agriculture throughout the world. They remain the most widely used control practices in industrial and developing countries, even though many such controls have been eroded by pesticide substitution. Until the advent of herbicides, cultural practices through crop rotation and mechanical or hand weeding were virtually the only control mechanisms available against weeds. They remain vitally important but much more serious consideration is needed in establishing how they can be integrated with judicious herbicide use in order to minimize the constraint of untimely availability of labour for crop production. Weed control, whether done consciously or not by farmers, is often achieved by a combination of crop production practices and specific weed management activities. Integrated weed and crop management is not a new concept so, in theory, improved techniques need not be alien to farmers. However, farmers tend to be conservative and reluctant to change traditional practices, especially if they perceive risks. Weed problem in rice has been observed to be reduced by planting cowpea during dry season, rather than keeping the field fallow. Planting mungbean in dry season in northern India also reduced weed growth and weeding time and increased herbicide performance. Direct drilling under zero tillage is advantageous as far as weed control is concerned, but severe yield penalties were also noted by many workers (Brar et al., 2002). The practice of zero-till allows the retention of previous crop residues in the field and it is well known that mulches are a good tool for weed management. Mulches normally exclude light penetration and serve as a physical barrier to weed seedling emergence. However, efforts are underway to further improve this technique for better weed control and higher productivity. "Brown manuring" technique is also being utilized by many farmers for controlling weeds in DSR in northwest India. In this technique, sesbania and rice are grown together for 25-30 days and thereafter, sesbania is knocked down with the use of 2,4-D herbicide (Singh et al., 2007). The brown leaves of sesbania after the herbicide application would serve the purpose of mulch and hence smother the weed flora in rice. However, there are also reports that sesbania suppresses the rice crop too (Kumar & Ladha, 2011).

3. Conclusion

There is an increasing trend in NW-IGP to replace transplanting of rice by DSR. In effect, farmers are substituting capital in the form of seed and herbicides for labour. The results also suggest that the labour cost advantage of DSR more than compensates for the increase in herbicide cost. The adoption of DSR, being the farmers' friendly technology, is increasing even though herbicide costs are rising. DSR is expected to become increasingly more important in the next decade. However, the problems of weed control may become more important due to emergence of new weeds. Appropriate weed management strategies and technologies are needed to maintain yield stability and reduce the cost of production. Further studies are needed on weed species, population dynamics, and control practices by innovative cultural practices in conjunction with biological and chemical weed control. A dynamic research program is needed to develop the innovative and effective weed management strategies for DSR to maintain yield stability. There is also a need for more economical herbicides with reduced handling and environmental hazards, without reduction in herbicidal efficacy, for DSR. Worldwide concerns over safe handling of pesticides, environmental issues, and sustainable agriculture need to be addressed.

4. References

Bouman, B.A.M. & Tuong, T.P. (2001). Field water management to save water and increase its productivity in irrigated rice. *Agricultural and Water Management*, Vol. 49, pp. 11–30.

Brar, L.S., Mahajan, G. & Sardana, V. (2002). Weed management practices for basmati rice planted by different methods. *Proceeding of Second International Conference on Sustainable Agriculture for Food, Energy and Industry,* Beijing, China, 2002.

Chauhan, B.S., Singh, V.P., Kumar, A. & Johnson, D.E. (2011). Relations of rice seeding rates to crop and weed growth in aerobic rice. *Field Crops Research,* Vol. 21, pp. 105-115.

Chauhan, B.S. & Johnson, D.E. (2011). Row spacing and weed control timing affect yield of aerobic rice. *Field Crops Research,* Vol. 121, pp. 226-231.

Chauhan, B.S. & Johnson, D.E. (2010). The role of seed ecology in improving weed management strategies in the tropics. *Advances in Agronomy,* Vol. 105, pp. 221-262.

Dhillon, B.S., Kataria, P. & Dhillon, P.K. (2010). National food security vis-a-vis sustainability of agriculture in high crop productivity regions. *Current Science,* Vol. 98, pp. 33-36.

Dixit, A., Manes, G.S., Singh, A., Singla, C. & Mahajan, G. (2010). Evaluation of direct-seeded rice drill against Japanese manual transplanter for higher productivity in rice. *Indian Journal of Agricultural Sciences,* Vol. 80, pp. 884–887.

GOI (Government of India). (2011). The Mahatma Gandhi National Rural Employment Guarantee Act 2005, In: *Ministry of Rural Development, Government of India,* Accessed on 28 March 2011, Available from http://nrega.nic.in/netnrega/home.aspx.

Hira, G.S. (2009). Water management in northern states and the food security of India. *Journal of Crop Improvement,* Vol. 23, pp. 136-157.

Humphreys, E., Kukal, S.S., Christen, E.W., Hira, G.S., Singh, B., Yadav, S. & Sharma, R.K. (2010). Halting the groundwater decline in north-west India—Which crop technologies will be winners? *Advances in Agronomy,* Vol. 109, pp. 155-217.

Jat, M.L., Gathala, M.K., Ladha, J.K., Saharawat, Y.S., Jat, A.S., Kumar, V., Sharma, S.K., Kumar, V. & Gupta, R.K. (2009). Evaluation of precision land levelling and double zero-till systems in the rice-wheat rotation: Water use, productivity, profitability and soil physical properties. *Soil and Tillage Research,* Vol. 105, pp. 112–121.

Kumar, V. & Ladha, J.K. (2011). Direct seeding of rice: Recent developments and future research needs. *Advances in Agronomy,* Vol. 111, pp. 299-413.

Kumar, V., Ladha, J.K. & Gathala, M.K. (2009). Direct drill-seeded rice: A need of the day, In: *Annual Meeting of Agronomy Society of America,* Pittsburgh, November 1–5, 2009.

Mahajan, G., Chauhan, B.S. & Johnson, D.E. (2009). Weed management in aerobic rice in north western Indo-Gangetic Plains. *Journal of Crop Improvement,* Vol. 23, pp. 366-382.

Mahajan, G., Gill, M.S., Singh, K. (2010). Optimizing seed rate for weed suppression and higher yield in aerobic direct seeded rice in North Western Indo-Gangetic Plains. *Journal of New Seeds,* Vol. 11, pp. 225-238.

Mahajan, G. & Chauhan, B.S. (2011a). Effects of planting pattern and cultivar on weed and crop growth in aerobic rice system. *Weed Technology,* doi: 10.1614/WT-D-11-00025.1.

Mahajan G. & Chauhan, B.S. (2011b). Weed management in direct drilled rice. *Indian Farming,* April, 2011 pp: 6-9.

Mahajan, G., Ramesha, M.S. & Rupinder-Kaur. (2011). Screening for weed competitiveness in rice - way to sustainable rice production in the face of global climate change. *Proceedings of International Conf. on Preparing Agriculture for Climate Change,* Ludhiana, Feb 6-8, 2011.

Mahajan, G & Timsina, J. (2011) Effect of nitrogen rates and weed control methods on weeds abundance and yield of direct-seeded rice. *Archives of Agronomy and Soil Science*, Vol. 57, pp. 239-250.

Mahajan, G., Timsina, J. & Singh, K. (2011b). Performance and water use efficiency of rice relative to establishment methods in northwestern Indo-Gangetic Plains. *Journal of Crop Improvement*, In press.

Moody, K. (1982). Weed control in dry-seeded rice. In: *Report of a workshop on cropping system research in Asia*, International Rice Research Institute, Los Banos, Philippines.

Prasad, R. (2011). Aerobic rice systems. *Advances in Agronomy*, Vol. 111, pp. 207-233.

Rao, A.N., Johnson, D.E., Sivaprasad, B., Ladha, J.K. & Mortimer, A.M. (2007). Weed management in direct-seeded rice. *Advances in Agronomy*, Vol. 93, pp. 153–255.

Rodell, M., Velicogna, I. & Famiglietti, J.S. (2009). Satellite-based estimates of groundwater depletion in India. *Nature*, Vol. 460, pp. 999-1002.

Sharma, P.K., Bhushan, L., Ladha, J.K., Naresh, R.K., Gupta, R.K., Balasubramanian, V. & Bouman, B.A.M. (2002). Crop–water relations in rice–wheat cropping systems and water management practices in a marginally sodic, medium-textured soil, In: *Water-wise rice production*, B.A.M. Bouman et al., pp. 223–235, International Rice Research Institute, Los Banos, Philippines.

Singh, G. (2008). Integrated weed management in direct-seeded rice. In: *Direct Seeding of Rice and Weed Management in the Irrigated Rice-Wheat Cropping System of the Indo- Gangetic Plains*, Y. Singh, V. P. Singh, B. Chauhan, A. Orr, A. M. Mortimer, D. E. Johnson, and B. Hardy, pp. 161–176, International Rice Research Institute, Los Banos, Philippines and Directorate of Experiment Station, G.B. Pant University of Agriculture and Technology, Pantnagar, India.

Singh, S., Chhokar, R.S., Gopal, R., Ladha, J.K., Gupta, R.K., Kumar, V. & Singh, M. (2009). Integrated weed management: A key to success for direct-seeded rice in the Indo-Gangetic Plains. In: *Integrated Crop and Resource Management in the Rice-Wheat System of South Asia*, J.K. Ladha, Y. Singh, O. Erenstein, and B. Hardy, pp. 261–278, International Rice Research Institute, Los Banos, Philippines.

Singh, S., Ladha, J.K., Gupta, R.K., Bhushan, L., Rao, A.N., Sivaprasad, B. & Singh, P.P. (2007). Evaluation of mulching, intercropping with Sesbania and herbicide use for weed management in dry-seeded rice (*Oryza sativa* L.). *Crop Protection*, Vol. 26, pp. 518–524.

Sudhir-Yadav, Gill, G., Humphreys, E., Kukul, S.S. & Walia, U.S. (2010). Effect of water management on dry seeded and puddle transplanted rice. Part 1. Crop performance. *Field Crops Research*, Vol. 120, pp. 112-122.

Timsina, J. & Connor, D.J. (2001). Productivity and management of rice–wheat cropping systems: issues and challenges. *Field Crops Research*, Vol. 69, pp. 93–132.

12

Weed Management Challenges in Fairtrade Banana Farm Systems in the Windward Islands of the Caribbean

Wendy-Ann P. Isaac[1], Richard A.I. Brathwaite[1] and Wayne G. Ganpat[2]
[1]Department of Food Production, Faculty of Science and Agriculture, The University of The West Indies, St. Augustine,
[2]Department of Agricultural Economics and Extension, Faculty of Science and Agriculture, The University of The West Indies, St. Augustine, Trinidad

1. Introduction

The banana (*Musa* sp.) is the foundation of the agricultural and rural-based community life of the Windward Islands where about 8000 farmers are involved in its production. Banana is primarily grown on small farms in hilly areas averaging two hectares in size, usually owned by local family farmers and exported mainly to Britain and Europe. These farmers have limited financial resources, farm part-time and grow other crops and/or livestock in their system of farming. With the loss of preferential European market arrangements and higher production costs than Latin America, many banana growers have turned to alternative marketing arrangements such as Fairtrade to maintain their profitability.

Fairtrade is an organized social movement and market-based approach that aims to assist small-scale and other disadvantaged producers in developing countries to improve their quality of life through better trading conditions and sustainability (Moberg, 2005). The movement advocates the payment of a higher price to producers as well as social and environmental standard (Moberg, 2005). These disadvantaged producers receive a price premium of about 12 % and a social premium that is returned to support community projects such as street lights, bus shelters, community centres, school equipment and building, and scholarships (Moberg, 2009; Rodriguez, 2008). Fairtrade also stipulates the need for more sustainable production systems, which use fewer or no chemicals, and, in particular, restricts the use of herbicides (Moberg, 2005).

It is estimated that weed control accounts for approximately 50% of the total cost of banana production (Hammerton, 1981). There is no detailed information available about the amount of crop yield losses due to weeds in the Windward Islands. However, certain weeds associated with banana are known to harbour pests, which cause major losses in production. Among them, *Commelina diffusa* Burm. F (watergrass), which harbours the root-burrowing (*Radophilus similis*) (Queneherve et al., 2006), reniform (*Rotylenchulus reniformis*) and the banana lesion (*Pratylenchus goodeyi*) (Robinson et al., 1997) nematodes. It also harbours the soil borne fungus, *Fusarium oxysporum* (Waite & Dunlap, 1953) which causes Fusarium wilt. These nematodes all contribute to a significant reduction in banana production, particularly *R. similis* which may reduce yields by more than 50 % and decrease the productive life of banana fields by feeding on the secondary and tertiary roots of banana feeder roots and at high populations, cause severe necrosis and toppling of plants (Queneherve et al., 2006; Isaac et al., 2007a).

Bourdôt et al. (1998) noted that weed science should focus on those species that are causing, or are likely to cause, significant damage and loss of income in a particular industry or region. Myint (1994) described *C. diffusa* as a herbaceous, shade tolerant, tufted, rhizomatous perennial weed which is often spreading with stems growing up to 100 cm tall, creeping and rooting at the nodes. Fournet (1991) and Myint (1994) indicated that the weed is propagated mainly by seed, stem cuttings and rooting from nodes. Plants may also arise asexually when buds grow into autonomous, adventitiously erect leafy shoots, which later become separated from each other and grow into erect shoots directly without undergoing a period of inactivity (Duke, 1985). It is this type of reproductive capability and its long persistence in cultivation that Holm et al. (1977) attributes to one of the difficulties in controlling this weed.

Commelina diffusa has proliferated as the dominant weed in banana fields throughout the Windward Islands of the Caribbean because of several factors: i) *Commelina diffusa* was once encouraged as a groundcover to minimize soil erosion (Edmunds, 1971; Simmonds, 1959) and is recommended for use as a ground cover in plantation crops (Bradshaw & Lanini, 1995); ii) growers have for decades relied extensively on the use of herbicides such as 2,4-D, paraquat and glyphosate (Feakin, 1971; Hammerton, 1981), which non-selectively removed vegetation creating disturbances within the ecosystem and causing suspected herbicide-resistant biotypes of *Commelina* species; iii) banana growers recent adoption of the Fairtrade system, which restricts the use of herbicides, causing the spread of *Commelina* species to reach an all time high in the Windward Islands. Farmers sole use of cutlass and/or rotary trimmers as the only alternative strategy have further intensified the problem by spreading plant propagules (Isaac et al., 2007a; 2007b); iv) More importantly, these islands, which are characterized by hilly landscapes in a multitude of valleys have ideal moist conditions for the proliferation of *Commelina* species; and v) Crop rotations and recommended tillage practices have not been done on these banana fields for many years which have resulted in the stabilization of *Commelina* species populations.

Generally, the presence of weed populations in banana fields is the result of ecological reactions to previous management practices, soil characteristics of the site, and the climatic and weather conditions (Froud-Williams et al., 1983; Milberg et al., 2000). In addition, cultural, chemical and mechanical weed control activities can have a strong influence on weed populations. Knowledge of weed community structure therefore, is an important component of the development of an effective weed management programme.

There is a paucity of information on the management of weeds, including *C. diffusa,* by other means than herbicides. Banana growers in the Windward Islands urgently need to find alternatives to herbicides in order to sustain their export capacity and remain Fairtrade compliant.

This chapter gives an overview of weed management in Fairtrade banana systems in the Windward Islands of the Caribbean and addresses the myriad challenges faced by small producers in tropical climates as they seek to move to pesticide free production (PFP) systems. It also focuses on experiments carried out over 4 years in the Windward Islands and highlights the challenges faced with managing *C. diffusa.*

2. Methods of weed control in banana plantations

No single method of control has been identified for the effective management of *C. diffusa* in plantation crops such as banana (Terry, 1996). Wilson's review on the control of these *Commelina* species was directed towards finding suitable chemicals for their control in the early stages of growth, summarizing results of trials from difference parts of the world. However, he suggested that since dense mats of plant material make chemical weed control of older plants difficult, and recommended removal by hand as the only effective control at the mature stage (Wilson, 1981).

Hand removal increases cost of production, which for small farm systems operated by resource poor farmers, is hardly appropriate. Consequently, chemical control is still generally considered the only practical means of controlling large infestations of *Commelina* species (Ferrell et al., 2006; Webster et al., 2005; Webster et al., 2006). The management challenge associated with *C. diffusa* is intricately linked to its ability for regeneration after attempted management even by cultural, mechanical or chemical control. Further, current concerns worldwide about the environmental impacts of pesticide use in agriculture require the adoption of alternative cropping systems that are less pesticide based. An Integrated Management Strategy has been recommended as the best control of this weed species. Webster et al. (2006) suggested a multi-component approach, which included an effective herbicide for successful management.

The several components of such a multi-component approach include chemical, mechanical and biological strategies as well as the use of living and non-living mulches. These are now discussed.

2.1 Chemical weed control

Herbicides are not usually effective against *C. diffusa.* CABI (2002) indicated that control using herbicides is variable depending on the herbicide, accuracy of leaf coverage and environmental conditions. Spraying with a selective or non - selective herbicide may work but repeated treatments are required for preventing regrowth. Plants should not be under moisture stress when sprayed. Surfactants will improve penetration into the waxy-coated leaves. *Commelina elegans* has shown resistance to growth - regulator type herbicides (Ivens, 1967). The first resistance verified was registered in 1957, when *Commelina diffusa* biotypes were identified in the United States (Hilton, 1957; Weed Science, 2005).

Wilson (1981) indicated that many standard herbicides have relatively low activity on species of *Commelina.* These include 2,4-D, propanil, butachlor, trifluralin and pendimethalin. Treatment with 2,4-D or MCPA at the pre-emergent stage has been shown to be ineffective and although a reasonable kill of very young seedlings can be obtained with post-emergent spraying the plants develop a rapid resistance with age (Ivens, 1967). Research has shown that particular biotypes resistant to 2,4-D may be cross resistant to other Group O / 4 herbicides (Weed Science, 2005) which are mainly growth regulator herbicides.

It has been found that one biotype of *Commelina diffusa* could withstand five times the dosage of a susceptible biotype (WeedScience.org, 2005).

Resistance to residual herbicides has also been reported and relatively high doses of simazine and diuron appear to be necessary to achieve control (Ivens, 1967).

In the Windward Islands of the Caribbean, farmers started using paraquat around 1989 and noticed that it was ineffective. Gradually they started using Gramocil (paraquat + diuron) at high doses and this too was not effective and resistance in *Commelina* spp. began to show (Paddy Thomas, Personal Communication June 2002). Reglone (diquat + agral), Roundup (Glyphosate) and Talent (paraquat + asulam) have also been used with little success for the control of *C. diffusa* in the Windward Islands (Paddy Thomas, Personal Communication, April, 2002).

Studies were conducted on the efficacy of Basta (glufosinate ammonium) for weed control in coffee plantations and it was found that it did not effectively control *Commelina* spp. at a rate of 0.3-0.6 kg a.i. / ha. However, Paracol (paraquat + diuron) and Gardoprim (terbuthylazine) suppressed this perennial weed better (Oppong et al., 1998). Flex (fomesafen) and Cobra (lactofen) were shown to be two products with good potential for control of this broadleaf weed (Carmona, 1991).

2.2 Mechanical control

Mechanical weed control using tillage is not widely practised in these banana systems in the Windward Island where the terrain is undulating and sloping. Simmonds (1959) notes that tillage tends to damage the root systems of the banana plant and in general should be avoided. Terry (1996) indicated that alternatives to tillage are desirable. Kasasian and Seeyave (1968) cautioned that the most common method of weed control, slashing, will not be good enough to secure optimum yields because as Feakin (1971) pointed out, this practice may damage the banana stems and suckers if done carelessly. A real risk in small farm systems where temporary or casual labour is employed to slash weeds and payment is for work done in a day, usually much less care is taken by these workers than if the owner is doing the weed control. Feakin (1971) noted that a typical practice is to slash weeds 3-4 times a year, leaving a weed mulch on the surface to help avoid soil erosion to delay fresh weed growth. However, Terry (1996) indicated that this cannot prevent weed competition and eliminate weeds as it will encourage weeds with a prostrate habit such as *Cynodon dactylon* or *C. diffusa*.

Commelina diffusa is particularly difficult to control by tillage practices, partly because broken pieces of the stem readily establish roots and underground stems with pale, reduced leaves and flowers are often produced (Ivens, 1967). The plant is easy to rake, roll or hand pull and very small infestations can be dug out. Bagging and baking in the sun is also an effective destruction strategy. However, follow-up work is essential as any small fragment of the stem remaining will regrow and need to be removed and destroyed off - site. The use of the mechanical string trimmer has become popular in recent years because of the amount of acreage that could be covered compared to manual methods in the same time. Labour costs are reduced. However, this practice has contributed to to the spread of stem cuttings in addition to damaging the banana root system as much of that system lies within the top 15 cm of the soil (SVG farmers, Personal Communication 2004).

2.3 Living and non-living mulches for weed control

Agricultural environments in banana farms in the Windward Islands are characterised by high rainfall which makes it difficult to maintain soil organic matter and to retain residue on the soil surface on steep slopes where the crop is planted. Since soil is exposed to high levels of erosion from heavy rainfall after the removal of weeds, living mulches such as cover crops can play an important role in erosion reduction. It is critically important that the areas around each banana plant should be kept free of weedy vegetation particularly in the early stages of growth and development. A potential solution to overcoming weed infestations is by intercropping the banana with a fast, low-growing shade tolerant cover crop. This can be done by intercropping with melons, *Mucuna pruriens* (negra and ceniza), tropical alfalfa, *Cajanus cajan,* mung bean (*Vigna radiata*), cowpea (*V. unguiculata*), *Crotalaria juncea, Indigofera endecaphylla, Phaseolus trinervius,* Carioca beans and sweet potato. All of these have rapid canopy coverage which can suppress the establishment of weeds.

Studies in the Windward Islands by Rao & Edmunds (1980a-d) indicated that intercropping banana with cowpeas, corn, sweet potatoes and peanut could significantly suppress weed infestations. There was an increase in banana yield when intercropped with corn compared to pure stands in trials conducted and this was probably due to adequate fertilization of both crops.

Non-living mulches also offer another viable option for weed control in banana plantations. Mulching with rice straw, cut bush, grass, water hyacinth or even the dead or senescent banana leaves, pruned suckers and old stems can significantly suppress weed growth. Non-living mulches provide benefits which include retention of soil moisture, prevention of leaching, improved soil structure, disease and pest control, improved crop quality and weed control (Grundy & Bond, 2007). Synthetic mulches such as black plastic mulch also provide good weed control as it stifles weed seedling growth and development when light penetration is blocked. Research has shown that clear plastic mulch, which is used in soil solarization, a hydrothermal process of heating moist soil, can successfully disinfect soil pests and control weeds (Benjamin & Rubin, 1982; Ragone & Wilson, 1988; Abu-Irmaileh, 1991; Elmore & Heefketh, 1983). Soil solarization by covering with plastic sheeting for 6 weeks in the warmer months will weaken the plant. After removing the plastic any regrowth can be dug out or sprayed. However, this method will not be effective in full shade. Solarization can be used alone or in combination with other chemicals or biological agents as the framework for an Integrated Pest Management (IPM) programme for soil-borne pests in open fields. Mudalagiriyappa et al. (2001), recommended an integrated weed management approach using soil solarization with transparent polyethylene (TP) at 0.050 mm + Gliricidia (*Gliricidia sepium*) as reducing the weed count of *Commelina benghalensis* and other weeds by 77 and 74 % over the control at 90 days after sowing (DAS) and at harvest, respectively in a groundnut (*Arachis hypogaea*) -French bean (*Phaseolus vulgaris*) intercropping trial. They also found that yield and yield components were highest in the crop residue + soil solarization treatments. The highest yields of 20.8 t/ha (for groundnut) and 7.7 t/ha (for French bean) were obtained with Pongamia (*Pongamia pinnata*) + TP at 0.05 mm.

2.4 Biological control

There have not been many reports on biological control of *Commelina* species. *Commelina diffusa* is commonly grazed by small ruminants, pigs and cows. Holm et al. (1977) reported that because this species is very fleshy and has high moisture content, it is difficult to use as fodder for domestic stock. However, recent research has indicated that *C. diffusa* compared well with many commonly used fodder crops and could contribute as a protein source for ruminants on smallholder farms (Lanyasunya et al., 2006). There have also been reports of foraging of this weed by chickens (*Gallus domesticus*) (Anton Bowman, Personal Communication, August, 2005). Growers in Georgia will autumn graze beef cattle in fields infested with *C. benghalensis* following agronomic crop harvest (Theodore Webster, Personal Communication, November, 2006). As a forage crop, *C. benghalensis* was rated as 102 relative forage quality (RFQ) [10.5% crude protein (CP), 61% total digestable nitrogen (TDN)], comparable to bermudagrass (*Cynodon* species) 116 RFQ (12.1% CP, 59% TDN) and perennial peanut (*Arachis glabrata* Benth.) 133 RFQ (15.1% CP, 66% TDN) (Theodore Webster, Personal Communication, November, 2006).

There are no reports of promising insect candidates for biological control of *Commelina* spp. in the USA (Standish, 2001). In Korea and China, Zhang et al. (1996), reported *Lema concinnpennis* and *Lema scutellaris* (Coleoptera: Chrysomelidae), two leaf-feeding species, on *C. benghalensis*. Another leaf-feeding species, *Noelema sexpunctata* (Coleoptera: Chrysomelidae), occurred on *C. benghalensis* in the USA (Morton & Vencl, 1998).

In Central Virginia, USA, *Pycnodees medius* (Hemiptera: Miridae) was found to cause tissue necrosis on Asiatic dayflower (Johnson, 1997). Hill & Oberholzer (2000) also recorded feeding and nymphal development (up to 3rd and 4th instar) of *Cornop aquaticaum* (grasshopper) on *Commelina africana* L., and *Murdannia africana* (Vahl.). They observed that *Rhaphidopalpa africana* beetles fed more on *Commelina* species than on other weeds.

There are records of agromyzid leaf miners which may be promising sources of candidate biological control agents (Smith, 1990). *Liriomyza commelinae* (Diptera: Agromyzidae), a leaf-miner, was reported on *C. diffusa* in Jamaica (Smith, 1990). *C. diffusa* is the main food plant of *L. commelinae,* however, the leaf-miner is susceptible to predation by the formicid *Crematogaster brevispinosa* as well as competition by and exposure to the sun (high temperatures) which causes high mortality (Smith, 1990).

There are prospects for the management of alien invasive weeds in Latin America using co-evolved fungal pathogens of selected species from the genus *Commelina* (Ellison & Barrreto, 2004). Pathogens recorded in the native range of *Commelina* species include: *Cercospora benghalensis* Chidd., *Cylindrosporium kilimandscharium* Allesch. (Hyphomycete), *Kordyana celebensis* Gaum, (Exobasidiales: Brachybasidiaceae), *Phakopsora tecta* H.S. Jacks and Holw (Uredinales: Phakopsoraceae), *Septoria commelinae* Canonaco (Coelomycete), *Uromyces commelinae* Cooke (Uredinales: Pucciniaceae) (Ellison & Barreto, 2004). These mycobiota would appear to be good potential agents for classical biological control (CBC) (Ellison & Barreto, 2004). Although some of the most promising (e.g. the rusts *Phakopsora tecta* and *Uromyces commelinae*) are already present in the New World, they are restricted to certain regions and could be redistributed (Ellison & Barreto, 2004). It should be noted that the release of a phytopathogen in a new area could result in disastrous effects for the ecosystem, if it is not done under very strict control. The uredinal state of a rust was found widespread on spreading dayflower in Hawaii (Gardener, 1981) sometimes causing death of parts above ground. Studies aimed at identifying mycoherbicidal biocontrol agents have been conducted in Brazil on three endemic pathogens of tropical spiderwort which were: a bacterium (*Erwinia* sp.) and two fungi (*Corynespora cassiicola* and *Cercospora* sp.) (Lustosa & Barreto, 2001).

The use of biological control measures for weed control and more specifically for *C. diffusa* in banana has been largely unexplored in the Caribbean.

3. Alternative strategies for weed control

Alternative weed management strategies were compared in established banana fields under irrigated and non-irrigated regimes in the rainy and dry season, 2003 to 2004. The performance of these weed management strategies were compared to a reference standard system using herbicides. The treatments consisted of:

- Two herbicide treatments: (i) glufosinate-ammonium and (ii) fomesafen which were applied at early post-emergence (at the weed 3-5 leaf stage). These herbicides were applied with a backpack sprayer which delivered 269 L ha^{-1} at kPa pressure using a fan-nozzle (TJ-8002).
- Three non-living mulch treatments: (i) banana mulch (traditional practice) applied at a depth of 5-6 cm using fully green and senescing leaves, (ii) coffee hulls applied at a depth of 3-5 cm and (iii) clear plastic mulch using high - density, transparent polyethylene tarp at 0.5 mils (50 gauge) thick for a 6-week period.
- Three live mulch treatments: (i) *Arachis pintoi* Krapov & W.C. Greg. (wild peanuts) planted by seed and stem cuttings drilled into the soil in rows 16 cm apart with 5 seeds per hole; (ii) *Mucuna pruriens* (L) DC (velvet beans) drilled into the soil in rows 30 cm apart with 3 seeds per hole; (iii*) Desmodium heterocarpon* var *ovalifolium* (L) DC (CIAT 13651) broadcast at a rate of 5 kg/ha.
- Two organic treatments: (i) Corn gluten meal, a pre - emergent weed blocker and a slow release fertiliser (9-1-0), which controls emerging weeds and provides nutrients to the crop, applied at a rate of 10 kg/ha and (ii) Concentrated vinegar and acetic acid, a post-emergent contact herbicide, sprayed directly to weeds at the 3-5 leaf stage.
- Two control treatments: (i) Hand weeded control which was hand weeded once every 4 weeks and (ii) an unweeded control which was left unweeded from week 1 to week 16. The experimental design was a randomised complete block design with two replicates at each site. Treatments were arranged under banana planted 4.5 m x 4.5 m.

All treatments reduced *Commelina diffusa* and other weeds compared to the unweeded control (UWC) up to 49 days after treatment (DAT). At 21 DAT, the herbicide and non-living mulch treatments were as effective at suppressing weed growth as the hand weeded control (HWC) and at 35 and 49 DAT were actually more effective. Of the other treatments, only the *D. heterocarpon* (DH) live mulch gave such good weed control, being similar to the HWC at 35 and 49 DAT, although not quite as effective as the non-living mulches or fomesafen. All three non-living mulches gave excellent suppression of weed growth for 49 DAT, even better than the HWC by 35 DAT. The weed control efficiency (WCE) of the two dead mulches, banana mulch and coffee hulls, at 49 DAT was around 95 %.

Of the 3 cover crop treatments, *D. heterocarpon* suppressed the highest number of weeds at all 3 dates after application. The two organic treatments which included the vinegar and acetic acid and corn gluten meal were not as effective as other treatments in suppressing weed populations. Weed density under the corn gluten meal treatment increased from 53 to 62 % from 35 to 49 DAT, respectively, which was similar to the increase in the unweeded control from 55 to 66 % in the same DAT. Weed control scores at 63 DAT also showed similar trends to the weed density at 49 DAT (Figure 1).

The two dead mulches (banana mulch, 96 %, and coffee hulls, 95 %) showed excellent weed control. Weed control efficiency was also high (from 87 % to 72 %) in *D. heterocarpon*, followed by fomesafen, glufosinate-ammonium and clear plastic mulch. The hand-weeded treatment at 67 % was similar to the latter two treatments. *Arachis pintoi,* concentrated vinegar, *M. pruriens* and corn gluten meal were less efficient (52 % to 43 %) in controlling weeds, but still better than the unweeded control at 27 %.

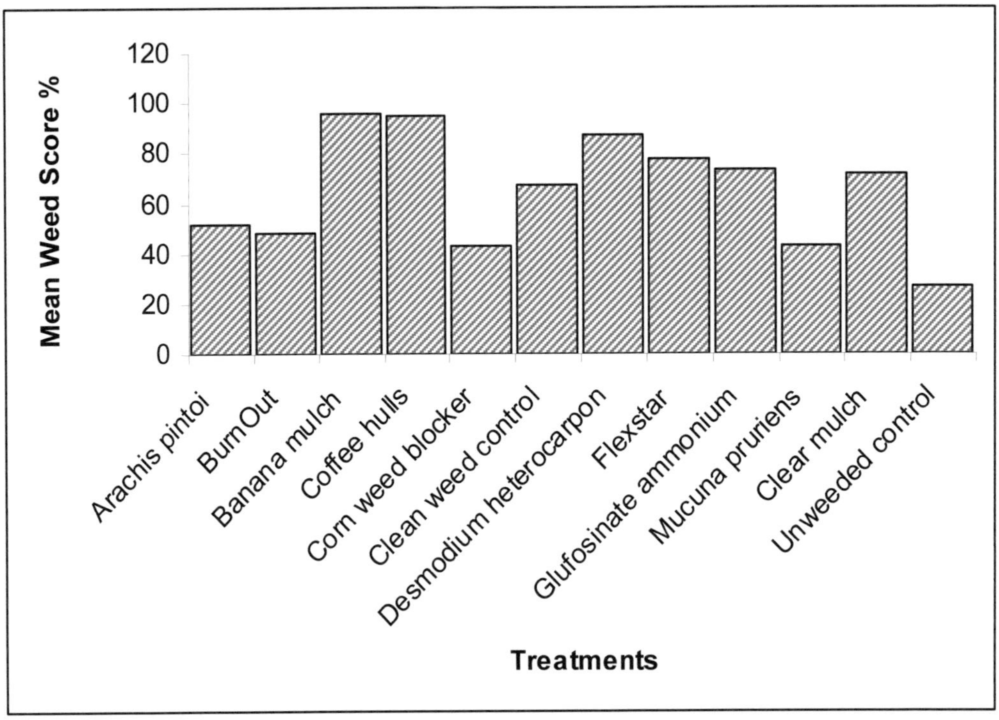

Fig. 1. Effect of treatments on mean weed control efficiency (WCE) at 63 DAT.

The non-living mulches, banana mulch and coffee hulls, as well as the clear plastic mulch, were the best weed management alternatives as they gave the highest levels of control. Coffee hulls significantly suppressed weed seed germination and seedling growth. This may have been due to the exclusion of light or from exudates released from the decaying plant material. It is possible that WCE of the decomposing coffee hulls is not only due to the amount of material applied on the soil surface but also to exudates released from the decaying material. Relating this to the caffeine found in coffee, Rizvi et. al. (1980) described caffeine as a natural herbicide selectively inhibiting germinating seed of *Amaranthus spinosus* L. After the clear plastic mulch was removed at 42 DAT, the stressed and etiolated weeds, which had germinated under the plastic recovered, causing an increase in weed density. Marenco & Lustosa (2000) reported an increase in seed germination of *C. benghalensis* L. when clear plastic was removed in a trial using clear plastic mulch for soil solarization in Brazil. The cost of clear plastic will be an issue for small resource poor farmers. In addition, to maintain the 5-6 cm thickness of the banana mulch, leaves had to be frequently replaced. The practicality of this is a concern on a commercial scale, as there are unlikely to be sufficient leaves available, as discussed by Cintra & Borges (1988) in studies conducted in Brazil. The availability of sufficient quantities of coffee hulls would be a similar concern.

Of the living mulches however, *D. heterocarpon* gave better coverage and was therefore more competitive than *Arachis pintoi* and *Mucuna pruriens*. The rapid establishment of *D. heterocarpon* quickly suppressed emerging weeds which is contrary to findings by Bradshaw & Lanini (1995) who noted that this cover crop required more than 2 months for establishment and control of weeds in Nicaraguan coffee. *Arachis pintoi* fell prey to predatory *Rattus novegicus* (rats) and *Gallus domesticus* (domestic chickens). Weed density in this treatment was high for most of the monitoring period; it was not until 12 weeks that any significant weed suppression occurred. Although *M. pruriens* vigorously established itself, weed density was also high for this treatment as most of its growth was in climbing the banana plant as reported by Buckles et al. (1998). The vines had to be removed often from banana and their effectiveness in suppressing weeds was low. Additionally, the vines began senescing after producing pods allowing further weed establishment. Buckles et al. (1998) noted that for effective weed control with *M. pruriens* fields should be planted over a 3-year period as exudates from senescing leaves have a herbicidal effect.

As a follow-up, a Participatory Approach which involved farmers in the design, conduct and evaluation of three potential cover crops (*Mucuna pruriens, Desmodium heterocarpon* var *ovalifolium* and *Arachis pintoi*) was conducted (Ganpat et al., 2009). Thirty six (36) participating farmers applied treatments using the paired-treatment design with three replicates. Weed data were collected weekly and farmers subjected data to the Overlap test (Ooi et al., 1999) to evaluate differences in treatments, which revealed that *Desmodium heterocarpon* was the most effective of the cover crops. Further analysis (F-test) confirmed that the most promising cover crop was *D. heterocarpon* as weed levels were significantly lower under this treatment ($p< 0.05$). Farmers and Researchers agreed that *D. heterocarpon* var *ovalifolium* could significantly reduce weed levels and provide a sustainable non-chemical approach to improved weed management of *C. diffusa* in banana fields.

4. Conclusions

The *Commelina* species are very persistent, noxious weeds which must be managed using an integrated approach. Weed management strategies that are narrowly focused will ultimately cause shifts in weed populations to species that no longer respond to the strategy resulting in adapted species, tolerant species or herbicide resistant biotypes as cautioned by Owen (2000). *Commelina* species in cropping systems falls into this category and this has been the case with *C. diffusa* in banana systems in the Windward Islands of the Caribbean. The integrated approach should utilize alternative strategies such as cover crops in addition to cultural and mechanical control and with a minimum and judicious use of herbicides. Such combinations should provide significant management levels of *Commelina* species for both conventional as well as organic growers using a PFP approach.

The integrated approach must begin very early as once an infestation is really entrenched it presents several difficulties because of the pernicious growth habit of this weed. As Webster et al. (2006) suggested for the successful management of *C. benghalensis,* a multi-component approach including an effective herbicide that provides soil residual activity is required. Studies on the management of *Commelina* species have focused primarily on effective herbicides and herbicide mixtures for their control despite hard evidence of the development of herbicide resistant biotypes. Additionally the adoption within recent years of genetically modified (GM) crops particularly herbicide - resistant crops presents serious issues involving their negative ecological impact as already there are reports of *Commelina* species prominence in some agroecosystems due to simple and significant selection pressure brought to bear by these herbicide - resistant crops and the concomitant use of the herbicide (Owen & Zelaya, 2004).

Perhaps the best way to control *Commelina* species for small holders in developing countries would be by implementing an integrated approach that embraces a variety of options which should be attuned to the individual farmer's agronomic and socio - economic conditions (soil type, climate, costs, local practices and preferences). The extent of his financial resources and whether he is a part-time or full-time farmer and is involved in mixed farming systems also needs to be considered. For example in banana growing areas in the Windward Islands, the growth of the weed can be suppressed by a single application of a herbicide or weed whacking very early before extensive spread of the weed followed by planting a competitive cover crop like *Desmodium heterocarpon* that would not only prevent re-invasion but improve soil fertility.

In the Windward Islands more than 70 % of the banana farmers still adhere to their traditional practices of chemical use. Adoption of Fairtrade practices is growing, yet many farmers still remain unconvinced of the benefits of integrated crop management to reduce or eliminate the use of certain pesticides. In the absence of herbicides, infestations of weeds such as the most prevalent, *Commelina* spp. have only served to dissuade farmers from adopting a more organic approach. Although the social and economic advantages have been elucidated by Fairtrade based on acceptance of produce into international markets with conformity to certain standards, further research into alternative agricultural practices is needed. Studies on the biology, ecology and dynamics of *Commelina diffusa* and strategies for their management in banana fields are therefore justified as they will provide valuable information for incorporation into an integrated weed management system for banana growers. Moreover, modern communication strategies have to be used to extend these findings if farmers are to be fully convinced. Appropriate research and Extension hold a key to meeting the challenges associated with weed management in Fairtrade banana systems in the Windward Islands of the Caribbean.

5. Acknowledgements

The authors thank the Association of Caribbean Farmers (WINFA)/Fairtrade Unit, the cooperating farmers of St. Vincent and the Grenadines and the School of Graduate Studies and Research, The University of the West Indies, St. Augustine, Trinidad for funding the research.

6. References

Abu-Irmaileh, B.E. (1991). Weed control in vegetables by soil solarization, In: *FAO Plant Production and Protection Paper*, pp. 155-165, ISBN 92-5-103057-X, Amman, Jordon

Benjamin, A., & Rubin, B. (1982). Soil solarization as a means of weed control. *Phytoparasitica,* Vol.10, pp. 4,280, ISSN 1876-7184

Bourdôt, G.W., Hurrell, G.A. & Saville, D.J. (1998). Weed flora of cereal crops in Canterbury, New Zealand. *New Zealand Journal of Crop and Horticultural Science,* Vol.26, pp. 233-247, ISSN 1175-8783

Bradshaw, L. & Lanini, W.T. (1995). Use of perennial cover crops to suppress weeds in Nicaraguan coffee orchard. *International Journal of Pest Management,* Vol.41 No.4, pp. 185-194, ISSN 1366-5863

Buckles, D., Triomphe, B. & Sain, G. (1998). Cover crops in hillside agriculture: Farmer Innovation with Mucuna. Mexico City, Mexico: International Development Research Centre (IDRC), Ottawa, Canada and International Maize and Wheat Improvement Centre (CIMMYT), ISBN 0-88936-841-4 p. 218

CAB International (CABI). (2002). *Crop Protection Compendium, 2006 edition.* Wallingford, UK: CABI, Accessed: March 14, 2006, Available from <http://www.cabi.org/compendia/cpc/index.htm>

Carmona, A. (1991). Flex (fomasafen) and cobra (lactofen): Two products with potential for broadleaf weed control for leguminous covers in oil palm plantations. *ASD Oil Palm Papers,* Vol.4, pp. 1-5, ISSN 1019-1100

Cintra, F.L.D., & Borges, A.L. (1988). Use of a legume and a mulch in banana production systems. *Fruits* Vol.43, No.4, pp. 211-217, ISSN 0248-1294

Duke, S.O. (1985). *Weed Physiology Vol. 1: Reproduction and Ecophysiology,* CRC Press, Boca Raton, Florida, ISBN 10: 0-8493-6313-6; ISBN 13: 978-0849363139

Edmunds, J.E. (1971). Association of *Rotylenchulus reniformis* with 'Robusta' banana and *Commelina* sp. Roots in the Windward Island. *Tropical Agriculture,* Vol.48, No.1, pp. 55-61, ISSN 0041-3216

Ellison, C.A., & Barreto, R.W. (2004). Prospects for the management of invasive alien weeds using co-evolved fungal pathogens: a Latin American Perspective. *Biological Invasions,* Vol.6, No.1, pp. 23-45, ISSN 1573-1464

Elmore, C., & Heefketh, K.A. (1983). Soil solarization an integrated approach to weed control, *Proceedings of the 35th Annual California Weed Conference,* pp. 143, Davis, California

Feakin, S.D. (1971). *Pest Control in Bananas.* PANS Manual No.1. Centre for Overseas Pest Research. 2d ed. College House, UK, PANS

Ferrell, J.A., MacDonald, G.E. & Brecke, B.J. (2006). *Tropical spiderwort (Commelina benghalensis L.), identification and control.* University of Florida Institue of Food and Agricultural Science SS-AGR-223, Accessed October 22, 2006, Available from:<http://edis.ifas.ufl.edu/AG230.htm>

Fournet, J., & Hammerton, J.L. (1991). *Weeds of the Lesser Antilles.* Paris France: Institute of National Research Agronomy (INRA), ISBN 2-7380-02090-9; ISSN 1150-3912

Froud-Williams, R.J., Drennen, D.S.H. & Chancellor, R.J. (1983). Influence of cultivation regime on weed floras of arable cropping systems, *Journal of Applied Ecology,* Vol.20, pp. 187-197, ISSN 1365-2664

Ganpat, W.G., Isaac, W.P., Brathwaite, R.A.I., & Bekele, I. (2009). Farmers' Attitude towards a Participatory Research and Development method used to evaluate weed management strategies in Bananas. *Journal of Agricultural Education and Extension,* Vol.15, No.3, pp. 235-244, ISSN 1389-224X

Gardener, D. E. (1981). Rust on *Commelina diffusa* in Hawaii. *Plant Diseases,* Vol.65, Vol.8, pp. 690-691 ISSN 0191-2917

Grundy, A.C. & Bond, B. (2007). Use of Non-living Mulches for Weed Control, In: *Non-Chemical Weed Management: Principles, Concepts and Technology,* Upadhyaya, M.K. and Blackshaw, R.E. (eds.), pp. 135-154, CAB International, ISBN 978-1-84593-290- 9

Hammerton J.L. (1981). Weed Problems and Weed Control in the Commonwealth Caribbean, *Tropical Pest Management,* Vol. 27, No.3, pp. 379-387, ISSN 0143-6147

Hill, M.P., & Oberholzer, I.G. (2000). Host specificity of grasshopper, *Cornops aquaticum,* a natural enemy of water hyacinth, *Proceedings of the X International Symposium on Biological Control of Weeds,* PP. 349-356 *4-14 July, 1999,* N.R. Spencer (ed.), Montana State University, Bozeman, Montana, USA, July 4-14, 1999

Hilton, H W. (1957). *Herbicide Tolerant Strains of Weeds,* Hawaiian Sugar Planters Association Annual Report. Honolulu, Hawaii, pp. 69-72

Holm, L.G., Pluknett, D.L., Pancho, J.V. & Herberger, J.P. (1977). *The World's Worst Weeds: Distribution and Biology,* The University Press, Honolulu, Hawaii

Isaac, W.A.P., Brathwaite, R.A.I., Cohen, J.E. & Bekele, I. (2007a). Effects of alternative weed management strategies on *Commelina diffusa* Burm. Infestations in Fairtrade banana (*Musa* spp.) in St. Vincent and the Grenadines, *Crop Protection,* Vol.26, pp.1219-1225, ISSN 0261-2194

Isaac, W.A.P., Brathwaite, R.A.I., Ganpat, W.G. and Bekele, I. (2007b). The impact of selected cover crops on soil fertility, weed and nematode suppression through Farmer Participatory Research by FairTrade banana growers in St. Vincent and the Grenadines, *World Journal of Agricultural Sciences* Vol.3, No.1, pp.1-9, ISSN 1817-3047

Ivens, G.W. (1967). *East African Weeds and Their Control,* Oxford University Press, ISBN 0196440173, 9780196440170, Nairobi, Kenya

Johnson, S.R. (1997). *Commelina communis* (Commelinacea) as host to *Pycnoderes medius* Knight (Hemiptera: Miridae) in central Virginia, USA, *The Entomologist* Vol.116, pp. 205-206, ISSN 0013-8878

Kasasian, L., & Seeyave, J. (1968). Chemical weed control in bananas - a summary of eight years' experiments in the West Indies, *Proceedings of the 9th British Weed Control Conference,* pp. 768-773, ISBN 10: 0901436119; ISBN 13: 9780901436115, Brighton, England, November 18-21, 1968

Lanyasunya, T.P., Wang Rong, H., Abdulrazak, S.A. & Mukisira, E.A. (2006). The potential of the weed, *Commelina diffusa* L., as a fodder crop for ruminants, *South African Journal of Animal Science,* Vol.36, No.1, pp. 28-32, ISSN 0375-1589

Lustosa, D. C., & Barreto, R.W. (2001). Primeiro relato de Cercospora commelinicola Chupp em *Commelina benghalensis* L. no Brasil, *34 Congresso Brasileiro de Fitopatologia, 2001, São Pedro. Fitopatologia Brasileira,* Vol.26, pp.364-364, ISSN 0100-4158

Marenco, R.A., & Castro-Lustosa, D. (2000). Soil solarization for weed control in carrot. *Pesquisa Agropecuaria Brasilia,* Vol.35, No.10, pp. 2025-2032, ISSN 0100-203X

Milberg, P., Anderson, L. & Thompson, K. (2000). Large-seeded species are less dependent on light for germination than small seeded ones, *Seed Science Research,* Vol.10, pp. 99-104, ISSN 0960-2585

Moberg, M. (2005). Fair Trade and Eastern Caribbean Banana Farmers: Rhetoric and Reality in the Anti-Globalization Movement. *Human Organization,* Vol.64, pp.4-15, ISSN 0018-7259

Moberg, M. (2009). *Slipping Away: Banana Politics and Fair Trade in the Eastern Caribbean,* Berghan Books, ISBN 1845451457, New York, USA

Morton, T.C., & Vencl, F.V. (1998). Larval beetles form a defence from recycled host-plant chemicals discharged as faecal waste, *Journal of Chemical Ecology,* Vol.24, pp.765-785, ISSN 0098-0331

Mudalagiriyappa, Nanjappa, H.V. & Ramachandrappa, B.K. (2001). Integrated weed management in groundnut-French bean cropping system. *Karnataka Journal of Agricultural Sciences,* Vol. 14, No.2, pp. 286-289, ISSN 0972-1061

Myint, A. 1994. *Common weeds of Guyana,* National Agricultural Research Institute, ISBN 976-8099-01-1

Ooi, P.A.C., van den Berg, H., Hakin, A.L., Ariawan, H. & Cahyana, W. (1999). Farmer Field Research. Jakata: The FAO Programme for Community IPM in Asia. 8

Oppong, F.K., Osei-Bonsu, K., Amoah, F.M. & Opoku-Ameyaw, K. (1998). Evaluation of Basta (gluphosinate ammonium) for weed control in Coffee, *Journal of the Ghana Science Association* Vol.1, No.1, pp.60-68, ISSN 0855-3823

Owen, M.D.K. (2000). The value of alternative strategies for weed management, (abstract) *Proceedings of the III international Weed Science Congress, Foz do Iguassu, Brazil,* pp.50 International Weed Science Society, Oregon, June 6-11, 2000

Owen, M.D.K., & Zelaya, I.A. (2004). Herbicide - resistant crops and weed resistance to herbicides, *Paper presented at the Symposium 'Herbicide – resistant crops from biotechnology: current and future status' held by the Agrochemicals Division of the American Chemical Society at the 227th National Meeting,* Anaheim, California, March 29 -30, 2004

Queneherve, P., Chabrier, C., Auwerkerken, A., Topart, P., Martiny, B. & Martie-Luce, S. (2006). Status of weeds as reservoirs of plant parasitic nematodes in banana fields in Martinique, *Crop Protection* Vol.25, pp. 860-867, 0261-2194

Ragone, D., & Wilson, J.E. (1988). Control of weeds, nematodes and soil born pathogens by soil solarization, The University of Hawaii, Honolulu, *Alafua Agricultural Bulletin* Vol.13, pp. 13-20, ISSN 1015-8499

Rao, M., & Edmunds, J.E. (1980a). Intercropping bananas with corn. *Windward Islands Banana Growers Association (WINBAN) Research and Development Division Advisory Bulletin* 14

Rao, M., & Edmunds, J.E. (1980b). Intercropping bananas with cowpeas. *Windward Islands Banana Growers Association (WINBAN) Research and Development Division Advisory Bulletin* 15

Rao, M., & Edmunds, J.E. (1980c). Intercropping bananas with peanuts. *Windward Islands Banana Growers Association (WINBAN) Research and Development Division Advisory Bulletin* 16

Rao, M., & Edmunds, J.E. (1980d). Intercropping bananas with sweet potatoes. *Windward Islands Banana Growers Association (WINBAN) Research and Development Division AdvisoryBulletin* 17

Robinson, A.F., Inserra, R.N., Caswell-Chen, E.P., Vovlas, N. & Troccoli, A. (1997). *Rotylenchulus* species: identification, distribution, host ranges and crop plant resistance, *Nematropica,* Vol.15, pp.165 - 170, ISSN 0099-5444

Rodriguez, K. (2008). Fairtrade, Dominica's shining light, Accessed January 16, 2011, Available from <http://dominica.tamu.edu/student%20projects/Dominica%20Projects%20pdf%20copy/Rodriguez_Kristen.pdf>

Rizvi, S.J.H., Jaiswal, V., Muierji, D. & Mathur, S.N. (1980). 1,3,7- trimethylxanthine--a new natural herbicide: its mode of action, *Plant Physiology,* Vol.65, No.6, pp. 99, ISSN 0032-0889

Simmonds, N.W. (1959). *Bananas,* Longmans, ISBN 0 582 06355 6, London, U.K.

Smith, D. (1990). Impact of natural enemies on the leaf mining fly *Liriomyza commelinace* (abstract), *Proceedings of the Interamerican Society for Tropical Horticulture, 1990,* Vol. 34, pp. 101-104, ISSN 0254-2528

Standish, R.J (2001). *Prospects for biological control of Tradescania fluminensis Vell. (Commelinaceae).* New Zealand Department of Conservation: Doc Science Internal Series 9, ISSN 1175-6519; ISBN 0-478-22156-8

Terry, P.J. (1996). Weed Management in Bananas and Plantains. *Weed Management for Developing Countries, edited by R. Labrada, J.C. Caseley and C. Parker,* pp. 311-315. FAO Plant Production and Protection Paper 120, ISBN 92-5-105019-8, ISSN 0259-2517

Waite, B.H., & Dunlap, V.C. (1953). Preliminary host range studies with *Fusarium oxysporum* f. *Cubense, Plant Disease Reporter,* Vol.37, pp. 79-80, ISSN 0032-0811

Webster, T.M., Burton, M.G., Culpepper, A.S., York, A.C. & Prostko, E.P. (2005). Tropical Spiderwort (*Commelina benghalensis*): A tropical invader threatens agroecosystems of the Southern United States, *Weed Technology,* Vol. 19, No. 3, pp. 501-508, ISSN 1939-7291

Webster, T.M., Burton, M.G., Culpepper, A.S., Flanders, J.T., Grey, T.L. & York, A.C. (2006). Tropical Spiderwort (*Commelina benghalensis* L.) Control and Emergence Patterns in Preemergence Herbicide Systems, *Journal of Cotton Science,* Vol. 10, pp. 68 -75, ISSN 1524-3303

WeedScience.org, (2005). *Group O/4 resistant spreading dayflower* (*Commelina diffusa*), USA: Hawaii, Accessed: February 15, 2006, Available from: <http://www.weedscience.org/Case/Case.asp?ResistID=394.htm.>

Wilson, A.K. (1981). Commelinaceae - a review of the distribution, biology and control of the important weeds belonging to this family, *Tropical Pest Management,* Vol. 27, No. 3, pp. 405-418, ISSN 0143-6147

Zhang, XiuRong, Ma Shu, Ying, Dai BingLi, Zhang, X.R., Ma, S.Y. & Dai, B.L. (1996). Monophagy of *Lema scutellaris* on *Commelina communis, Acta Entomologica Sinica,* Vol. 39, pp. 281-285, ISSN 0454-6296

13

Genetic Diversity in Weeds

Claudete Aparecida Mangolin[1], Rubem Silvério de Oliveira Junior[2] and Maria de Fátima P.S. Machado[1*]
[1]*Departamento de Biologia Celular e Genética;*
[2]*Departamento de Agronomia, Universidade Estadual de Maringá, Maringá,PR, Brasil*

1. Introduction

Weeds growing in cultivated areas are usually characterized as having high phenotype plasticity and genetic adaptability. They are frequently well-adapted to disturbance and often seed prolifically (Adahl et al., 2006). Thus, improving the current knowledge on the genetic diversity of weed populations is a challenge for management, primarily because this variability may be an important tool in determining the adoption and efficiency of weed control methods. Genetic markers may be important to improve current knowledge about important aspects of weeds, and may provide needed information to understand patterns of weed invasion, heritability of traits (e.g. herbicide resistance), taxonomic relationships, point of origin, and gene flow. Studies on genetic diversity both at population and species levels are important for weed management, and represent a source of information about genetic bottleneck effects, fitness, and the number of input events that contributed to a successful introduction (Goolsby et al., 2006; Hufbauer, 2004; Sterling et al., 2004). Improving the knowledge about genetics of weeds can provide vital information for the development of innovative control options (Slotta, 2008).

In natural habitats, plant populations often have greater genetic variability as compared to populations of exotic weed species, since very rarely all possible genotypes are introduced in a new environment. Thus, native weeds commonly represent major challenges for weed management programs, because of their wider genetic variability (Sterling et al., 2004). Assuming that the genetic variation among populations impacts the effectiveness of weed management tools, native weeds would represent the most difficult species to be controlled, as compared to exotic species. On the other hand, exotic species can prove way more aggressive in colonizing a habitat than natural weeds, since they might have no natural competitors or predators that control the population. Therefore, they can be as difficult to control as are the native species of weeds. Investigating the effectiveness of different weed management strategies in populations of native or exotic weeds may be important to provide information about the role of genetic variability on weed management success.

The magnitude of genetic variability may be estimated by using molecular or biochemical markers and studying nucleic acid or isozyme sequences. Isozyme markers have been used

* Corresponding Author

to study accessions of *Setaria glauca*, *S. geniculata* and *S. faberii*, worldwide weeds of tropics and temperate regions (Wang et al., 1995). By elucidating the genetic diversity and genetic structure of populations, the authors found low genetic variability for populations of *S. faberii*, *S. glauca*, and *S. geniculata*, although there was a substantial genetic differentiation between *S. glauca* and *S. geniculata*. Native populations also exhibited higher genetic diversity than exotic species. Patterns of genetic organization in *S. glauca* and *S. geniculata* may have also been influenced by several factors, including genetic bottleneck effect associated to the founder effect, random genetic drift and natural selection (Wang et al., 1995).

The genetic variability among six accessions of *Baccharis myriocephala* was evaluated by multivariate methods using isozyme and morphological descriptors (Castro et al., 2002). As regards the isozyme analysis, only the esterase system provided satisfactory resolution and two groups were formed. Use of morphological descriptors at 145 days after transplantation provided an efficient method to discriminate four groups of accessions. Isozymes were used as genetic markers to discriminate these accessions in *Baccharis myriocephala*, allowing its use in the characterization of varieties to complement morphological characteristics.

The analysis of products of different loci in plant tissues of weeds may be useful to estimate the genetic variability within each population and among different populations (Park, 2004). In recent literature addressing the detection of polymorphic loci to estimate the genetic diversity in plants through biochemical markers, the esterase system has been usually and primarily adopted as a genetic-biochemical marker. Many studies (Mangolin et al., 1997; Resende et al., 2004; Souza et al., 2004; Pereira et al., 2001; Orasmo et al., 2007) have demonstrated that the α- and β-esterases isozymes are produced by several and different loci, that mostly show co-dominant inheritance, being, therefore, an enzyme system suitable to estimate the genetic diversity and to analyze the genetic structure of plant populations.

For weeds, the analysis of α- and β-esterases isozymes using polyacrylamide gel electrophoresis (PAGE) was first established by Frigo et al. (2009), and was considered effective to analyze the genetic diversity and structure of populations of wild poinsettia (*Euphorbia heterophylla*). Frequencies of allelic variants found in this study were estimated for *Est-1*, *Est-2*, *Est-3*, *Est-4*, *Est-5*, *Est-6*, and *Est-7* loci, and the estimated proportion of polymorphic loci in populations was 87.5%. The positive value of F_{IS} (0.1248) indicated deficit of heterozygous or excess of homozygous plants. A relatively high level of differentiation was found among descendents of all 12 populations, indicating a reduced gene exchange among populations. Values of F_{ST} indicated that 16.63% of the total variance for allele frequency in populations of *E. heterophylla* occurred due to genetic differences among populations. Analysis of genetic diversity of 40 populations of *E. heterophylla* resistant to ALS herbicides growing in southern plateau of Brazil led to the conclusion that those populations present 60% of genetic variability. Such level of variability allowed grouping resistant plants in seven distinct groups (Winkler et al., 2003). In another study with different accessions of *Euphorbia spp.* from North America and Eurasia, the most divergent accession was the one collected in Austria, followed by accession from Italy and Russia. Accessions of *E. heterophylla* from USA were most intimately related to each other and also related to Russian accession (Nissen et al., 1992). A high degree of genetic variability was also described for the *Euphorbia esula* accession from USA (Rowe et al., 1997). The variability was attributed to multiple introductions of the species as well as to variability within native populations. For *Bidens pilosa* (beggartick) complex, a study on the genetic variability using 10 enzyme systems revealed 16 loci, but only three of them were polymorphic. For this complex, genetic diversity was low and characterized as being of 3.2% (Grombone-Guaratini, 2005).

Factors that can lead to or accelerate the development of herbicide resistance include weed characteristics, chemical properties and cultural practices. One of the characteristics related to weeds that may favor the selection of individuals resistant to herbicides is the wide genetic variability, due to the increased probability to find within the population a resistance allele to the herbicide in use. This fact leads to the general understanding that prolific, high density weeds are more likely to develop selection for herbicide resistance (Vidal and Meroto Jr., 2001; Winkler et al., 2002).

So far, the main methods used for weed control include biological, chemical, physical and cultural approaches, but the achievements of weed management programs may also be associated to the genetic variability of those plants. Some important questions that will be addressed and hopefully answered by increasing knowledge of genetic variability of weeds are as follow: *1)* Is the selection of weeds being imposed by management? *2)* Is the intensity of selection pressure imposed by agricultural practices more powerful than that imposed by natural ecosystems? *3)* Have weeds evolved and will keep evolving in response to agricultural management practices? *4)* Is the success of management programs associated to genetic variability of weeds?

Chances are that colonizing species, including exotic weeds, will have low genetic variability, since they have been through a genetic bottleneck (that is, they have lost alleles) after their introduction. However, despite low genetic variability, many exotic weeds will successfully colonize new geographic areas. In this case, as long as weed species have high fitness and great potential for colonization, they will persist.

The wider genetic variability and the genetic structure of populations have an impact on success of weed management. To expand knowledge and provide answers to unsolved questions, much more is needed to know about genetic variability of these populations. To seek for those answers, our research group has developed standard procedures to analyze esterase, malate dehydrogenase, and acid phosphatase isozymes. These isozymes were used to estimate the genetic diversity and the level of differentiation among populations of *Conyza spp.* and *Euphorbia heterophylla* collected from farms with a history of frequent use of herbicides and from sites with little or no herbicide use. Findings in these studies will provide help to guide weed management and control strategies of these two important species, which have recently evolved resistance to herbicides under field conditions.

2. *Euphorbia heterophylla*

The Euphorbiaceae family (also known as spurge family) includes more than 290 genera and 7500 species distributed all over the world. Plants in this family are mostly herbs, but some may also be shrubs or trees. *Euphorbia* is one of the genera of the Euphorbioideae subfamily and includes around 2300 species of wide morphological variety. Most species of the spurge family are shrubs, growing between 40 and 60 cm in height. A milky sap that oozes from damaged stems and leaves (latex) is also characteristic of some subfamilies such as Euphorbioideae within the spurge family (Joly, 1998).

Common names of *Euphorbia heterophylla* L. in Brazil include amendoim-bravo, leiteiro, parece-mas-não-é, flor-de-poeta, adeus-brasil, café-de-bispo, leiteira, café-do-diabo or mata-brasil (Suda, 2001). In English, it is usually referred as wild poinsettia, milkweed or mexican fire plant. In Brazil, it is widespread along Southern, Southeastern and Midwest regions and it is considered a native species in tropical and subtropical regions of the Americas (Cronquist, 1981). As a weed, it is a highly competitive, fast-growing, annual herb erect. Stems may be simple or branched; ovate to rhomboid leaves (4-7 cm long by 1.5-3 cm wide) occur on stems or branches as opposite, alternate or whorled, 2–12 cm long, leaf stalk 0.5–4 cm long; flowers are male or female in terminal clusters, each flower-head (cyathium) with a solitary terminal female flower surrounded by male flowers enclosed in a cup-shaped involucre with a solitary conspicuous gland; seeds with three longitudinal ridges. The fruits are small, segmented capsules and spread usually happens by seeds that are released explosively from ripe fruits (Cronquist, 1981; Kissmann and Groth, 1992) (Figure 1).

Fig. 1. Flowers of *Euphorbia heterophylla* in terminal clusters; each flower-head (cyathium) with a solitary terminal female flower surrounded by male flowers enclosed in a cup-shaped involucre with a solitary conspicuous gland.

Male flowers are constituted by one stamen, articulated in the pedicel, and the stamen surrounds the female flower (Kissmann and Groth, 1992). Reproduction of this species may be either by self or cross-fertilization (Cronquist, 1981; Barroso, 1984; Ingrouille, 1992) and is exclusively by seeds.

The fruits are small and contain trilocular segmented capsules, with one seed per locule. As fruit ripens, its color changes and at full maturity seeds are released explosively from ripe fruits, throwing seeds far away from mother-plants (Barroso, 1984). Seed is ovoid, with rough coat of variable color from light brown to almost black. Fruits are small, segmented capsules and have small bumps on the surface (mucilaginous cells); they have two cotyledons and dark coat (Cronquist, 1981; Barroso, 1984; Kissmann and Groth, 1992), and are produced in large quantities and with little to no dormancy. Causes of dormancy are not known, but light combined with alternating temperatures of 25-35 °C stimulate germination (Kissmann and Groth, 1992). Seeds germinate easily from a 4-cm depth, showing germination asynchrony along time during soybean growth season in southern states of Brazil, such as Rio Grande do Sul and Paraná (Kissmann and Groth, 1992).

E. heterophylla life cycle is short, allowing it to have two to three generations per year. The species develops well in almost all types of soil, but more prolific plants are found under fertile, well-drained soils. It is a representative of C4 photosynthetic pathway and the basic chromosome number is 2n = 32. The species is allogamous and produces up to 3000 seeds per plant (Kismann and Groth, 1992).

The center of origin of *E. heterophylla* is the Brazil-Paraguay area (Kissmann and Groth, 1992), and it is currently widely distributed in south-central Brazil and neighboring countries. Former field studies carried out at the Rio Grande do Sul State central plateau demonstrated that 74% of soybean fields were infested with this species (Vidal and Winkler, 2002). At a density of 10 plants m^{-2}, soybean yield was reduced by 7% by *E. heterophylla* competition during the whole crop cycle (Winkler et al., 2003; Chemale and Fleck, 1982). The presence of *E. heterophylla* at a 25 plants m^{-2} density caused a soybean yield daily loss of 5.15 kg ha^{-1}, whereas its absence provided a yield daily gain of 7.27 kg ha^{-1}. The critical period of interference for soybean in Brazil is considered to start as soon as 11 days after crop emergence (Meschede et al., 2002). Although reports of its relevance as a weed have been mostly related to soybeans, it is also an important competitor in other crops such as corn, peanuts, cotton, sugarcane, sorghum and beans.

3. *Euphorbia heterophylla* and resistance to herbicides

Among current agronomical techniques adopted to maximize crop growth and yield, weed management is considered one of the most influential on soybean grain yield. Weed management programs are set to minimize the effects of interference imposed by undesirable plants; this is important to maximize crop yield and also to reduce costs associated to crop production (Pitelli, 1985; Burnside, 1992).

The frequent use of a particular herbicide or of herbicides with the same mechanism of action may result in high selection pressure. Under high selection pressure the susceptible plants are killed while herbicide-resistant plants survive to reproduce without competition from susceptible plants, increasing, therefore, their frequency in the population (Ponchio, 1997; Mattielo et al., 1999).

Herbicide resistance is, by definition, the inherited ability of a plant to survive and reproduce following exposure to a dose of herbicide that would normally be lethal to the wild type. Resistance occurrence was hypothesized by Harper (1956) and documented by the first time in 1957 (Hilton, 1957; Switzer, 1957). Cases of herbicide resistance have been found in Brazil since 1996, and an average of one new case per year has been described so far (Winkler and Vidal, 2004).

In Brazil, the greatest concern about weed resistance is related to *E. heterophylla*, since its center of origin is located in Brazil-Paraguay region (Kissmann and Groth, 1992). It is usually found in high densities in field crops and imposes great impact to national agriculture. The most common herbicide alternatives in non transgenic crops include acetolactate synthase (ALS) and protoporphyrinogen oxidase (PROTOX) inhibitors (Vidal and Merotto Jr., 2001). Over the last decade, ALS-resistant biotypes have been identified in states such as Rio Grande do Sul, Paraná, São Paulo, Mato Grosso do Sul, Mato Grosso, Bahia, Tocantins and Minas Gerais, and in neighbor countries like Paraguay (Gazziero et al., 1998; Vidal and Winkler, 2002; Heap, 2010). Previous work has demonstrated that resistance to ALS herbicides is nuclear and dominant in *E. heterophylla* and that resistance is coded by a single dominant gene (Vargas et al., 2007).

According to the assumptions postulated by Winkler et al. (2003), soybean seeds commercialized in Rio Grande do Sul were all from the same commercial source and, therefore, seeds from weeds that are usually found in this crop may have spread geographically. Both species, soybean and wild poinsettia, would have evolved in a parallel process along the last years, favoring the selection of resistant biotypes to ALS-inhibiting herbicides in soybean fields, due to the intensive use of such mechanism of action. Weed resistance to these herbicides has already been described for six weeds in Brazil. The probability of selecting *E. heterophylla* biotypes with multiple resistance increases in the same order of magnitude as other herbicides with the same mechanism of action are used. Studies conducted by Trezzi et al. (2005) in southern states from Brazil (Paraná and Santa Catarina) have confirmed the presence of *E. heterophylla* biotypes with multiple resistance both to ALS and PROTOX inhibitor herbicides. In 2006, multiple resistance to ALS and EPSPs inhibitors was also reported (Vidal et al., 2007; Trezzi et al., 2009).

Weeds have a background of genetic diversity that gives them the ability to adapt to many different environments. The natural genetic diversity of weeds favors the selection of individuals resistant to herbicides most likely due to the highest probability of finding alleles that provide resistance to that particular herbicide (Winkler et al., 2003). Many studies have demonstrated that agronomic practices such as soybean mono cropping as well as the frequent and intensive use of herbicides with the same mechanism of action increase the selection pressure for resistant biotypes under such systems (Owen, 2001).

Several important points related to weed characteristics, herbicide properties, and cultural practices have been highlighted as key factors to explain the rapid occurrence and spread of weed resistance to herbicides (Vidal and Merotto, 1999). As concerned to weeds, annual growth habit, high seed production, relatively rapid turnover of the seed bank due to high percentage of seed germination each year (i.e., little seed dormancy), several reproductive generations in each growing season, extreme susceptibility to a particular herbicide (also called hypersensitivity of weeds to a particular herbicide) and the initial frequency of a resistant biotype in the population. Characteristics related to herbicides include a single site of action, dose violation (i.e. use of low or very high doses in relation to the optimum rates prescribed for a specific crop and situation), broad spectrum of weed control, frequency of use and long residual activity in soil. Cultural practices that may contribute in selecting for resistance include monocrop farming, reduced soil cultivation or zero tillage systems, failure to eliminate weeds that escape control by herbicides and use of a single herbicide or combinations that have same the mechanism of action in every season persistently.

4. Genetic diversity in *Euphorbia heterophylla*

Frigo et al. (2009) employed a non-denaturing PAGE system to identify *polymorphism* in α- e β-esterases loci in leaf tissues of *E. heterophylla* from seeds of 12 populations collected in states of Paraná and Mato Grosso – Brazil (Figure 2) to analyze the genetic diversity and structure of populations. Eight clearly defined loci were detected by this method (Figure 3). The α-preferential esterases, β-preferential esterases, and α/β-esterases were numbered in sequence, starting from the anode, according to their decreasing negative charge. The esterase with the slowest migration was named as Est-8 (Figure 3). These results confirmed a previous hypothesis that PAGE may be a powerful procedure for analysis of α/β-esterases isozymes from leaf tissues of wild poinsettia plants. Eight loci for isoesterases were simultaneously and clearly evident in the same electrophoresis, that is, using only one enzymatic system. Isozyme studies in other *Euphorbia* species have revealed only 11 loci from analysis of eight enzymatic systems (Park, 2004). The analysis of different enzymatic systems generally requires higher cost and time investments. Thus, α- and β-esterase isozymes analysis in PAGE system may be used in further studies to detect genetic diversity in other *Euphorbia* species

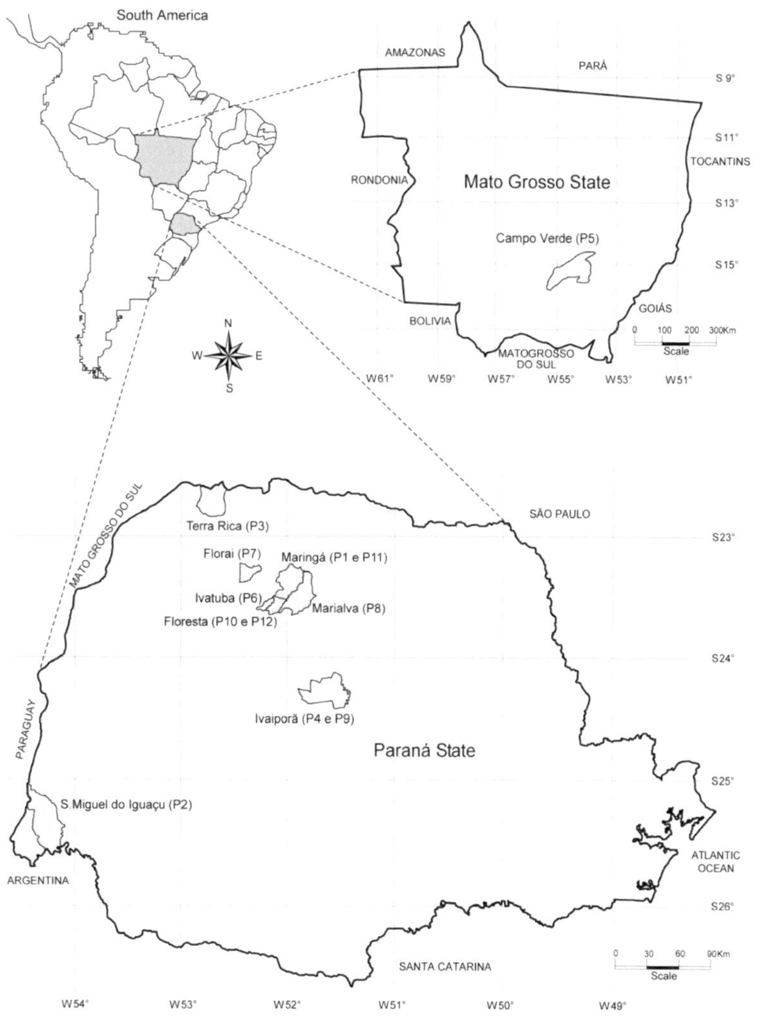

Fig. 2. Localities where seeds of wild poinsettia were collected at Mato Grosso, Brazil (MT) and Paraná, Brazil (PR) states: P1 and P11 (Maringá, PR), P2 (São Miguel do Iguaçu, PR), P3 (Terra Rica, PR), P4 and P9 (Ivaiporã, PR), P5 (Campo Verde, MT), P6 (Ivatuba, PR), P7 (Floraí, PR), P8 (Marialva, PR), P10 and P12 (Floresta, PR) populations. Source: Frigo et al. (2009).

High and low genetic diversity levels have been reported in different populations of wild poinsettia by DNA fragment analysis as molecular markers (Vasconcelos et al., 2000; Winkler et al., 2003). Deploying α- and β-esterase polymorphisms in the PAGE system (Frigo et al., 2009) indicates that genetic diversity of wild poinsettia has higher mean values for grades of genetic variation (number of alleles per locus, proportion of polymorphic loci, observed and expected proportion of heterozygous loci) when compared to other *Euphorbia* species (Park, 2004).

The genetic variation in wild poinsettia is nearly as high as the genetic variability in *Euphorbia ebracteolata*, a widespread species (Park et al., 1999). The proportion of polymorphic loci in 12 wild poinsettia populations is much higher than the mean proportion value (31%) reported for dicotyledons (Hamrick et al., 1979) and also for 16 species of *Euphorbia* (reviewed comparisons in Park, 2004).

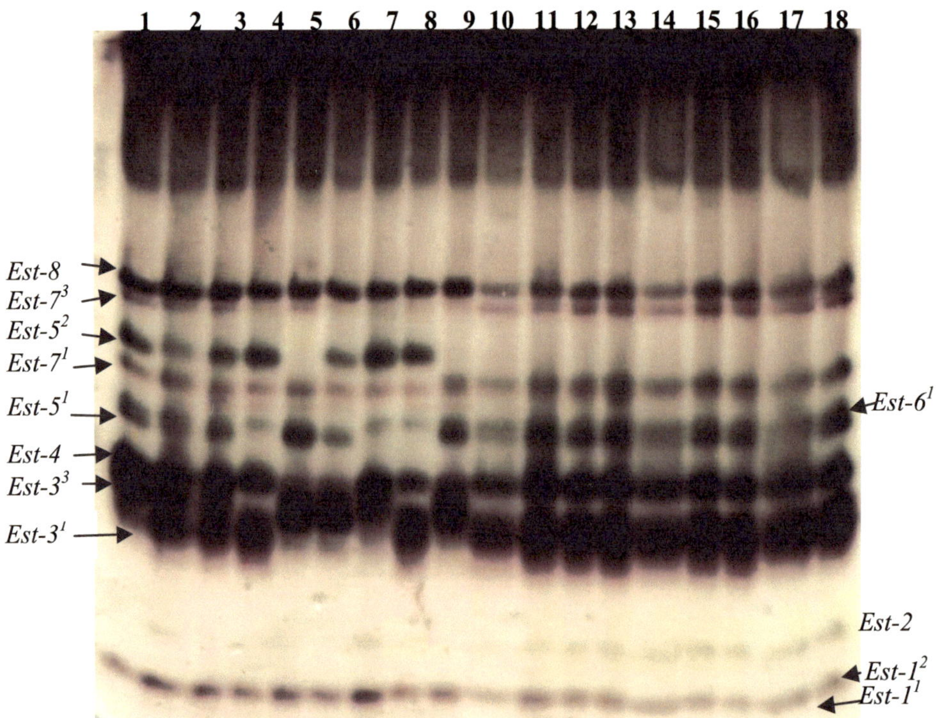

Fig. 3. Polymorphism of α- and β-esterases detected in eight loci of wild poinsettia plant descendants from 12 populations. Source: Frigo et al. (2009).

On the other hand, high and low values for observed (H_o) or expected (H_e) proportion of heterozygous loci in descendants from 12 different wild poinsettia populations sustain our preliminary hypothesis that wild poinsettia populations are genetically structured. Differential allele frequencies and proportions of heterozygous loci in different populations determined genetic divergence between the 12 populations (F_{ST} = 0.1663). According to Wright (1978), F_{ST} values between 0.15 and 0.25 indicate high interpopulational divergence level, or high genetic differentiation level between populations. A highest level of genetic differentiation between populations (F_{ST} > 0.25) has been described in 12 out of the 16 different *Euphorbia* species analyzed by Park (2004). The establishment of isolation and structuring mechanisms in populations has been reported in *Euphorbia nicaeensis* as a consequence of the inflorescence-architecture variability (Al-Samman et al., 2001). Substantial differences in the amount of genetic variation between different populations may indicate limited spatial dispersal or recent reduction in genetic variation caused by human action (Allendorf and Luikart, 2007).

Both limited spatial dispersal and populations frequently disturbed by human interference may determine high levels of spatial differentiation within wild poinsettia species. The explosive seed dispersal as a primary form of seed dispersal in *Euphorbia* species (Narbona et al., 2005) may explain the highly genetically structured populations. The seed dispersal of *Euphorbia* species may also occur by the activity of different ant species. In fact, the mean distance of seed dispersal has been positively correlated with size and species of ants (Gómez and Espadaler, 1998). Additionally, small-scale disturbances such as constant use of herbicides may create increased spatial heterogeneity. High selection pressure adopted in conventional weed management has caused selection of resistant biotypes (Holt and LeBaron, 1990) and may have determined highly structured populations within species.

High selection pressure imposed by the frequent use of herbicides in populations of *E. heterophylla* has not been detected through data obtained from α- and β-esterases. Parallel analysis comparing wild poinsettia plant descendants from seeds collected in organic culture of soybean (not exposed to herbicides) and plant descendants from seeds obtained in soybean culture frequently exposed to herbicides showed low genetic differentiation (F_{ST} = 0.03). Similarity between plants of *E. heterophylla* from organic and nonorganic fields was high (I = 0.9621) (Table 1). However, the mean observed and expected heterozygosity was higher in wild poinsettia plants from organic crops (Ho = 0.3526; He = 0.3980) than in plants from nonorganic crops (Ho = 0.2569; He = 0.3641). A comparison of organic and nonorganic populations suggests that frequent herbicide exposure may lead to increased homozygosity.

The heterozygous deficiencies in 12 populations of wild poinsettia may be evident by the positive value of F_{IS} (F_{IS} = 0.1248). Positive F_{IS} value indicates heterozygous deficit (12.48%) or excess of homozygous plants, which could be the result of human selection pressure (frequent herbicide application) in soybean areas and/or the result of self pollination. In consonance with the significant F_{IT} value (F_{IT} = 0.2703), overall inbreeding or nonrandomized breeding did play a major role in shaping the population's genetic structure. Increased homozygosity in wild poinsettia populations is important, because it leads toward a great number of deleterious recessive alleles in inbred plants, with a subsequent lowering of their fitness. Reduced heterozygosity reduces the fitness of inbred individuals at loci in which the heterozygous entities have a relative advantage over homozygous specimens (Allendorf and Luikart, 2007). Alternatively, a high number of heterozygous plants in populations of wild poinsettia may result in differential reactions and prevent uniform plant responses. High heterozygosity would indicate that the plant population has probably a substantial amount of adaptive genetic variations to escape the effects of a control agent.

The level of interpopulational genetic divergence in wild poinsettia species is revealed in the dendrogram through the genetic identity values (I) of 12 populations. Results in the dendrogram provide evidence that genetic divergence is independent of geographic distance (Figure 4). Lack of concordance between the geographic-distribution pattern and genetic identity for descendants from 12 populations may also be the result of the differential selection pressure or of the heterogeneity of environmental factors. Major understanding of the meaning of identity values could lead to important evidence related to differential tolerance to herbicides in field conditions and to development and spread of resistance. This in turn could lead to development of more effective policies of wild poinsettia control. For populations with higher identity values it may be possible to adopt similar strategies and processes for their control.

In subsequent studies carried out in our laboratory, the polymorphism for the α- and β-esterases loci of *E. heterophylla* plants from three distinct populations (organic population, herbicide-susceptible and herbicide-resistant populations) was evaluated in order to characterize diversity and genetic differentiation among these populations. The proportion of esterases polymorphic loci was 85.71%. Allelic frequencies were analyzed for *Est-1, Est-3, Est-4, Est-5, Est-6, Est-7,* and *Est-8* loci (unpublished results).

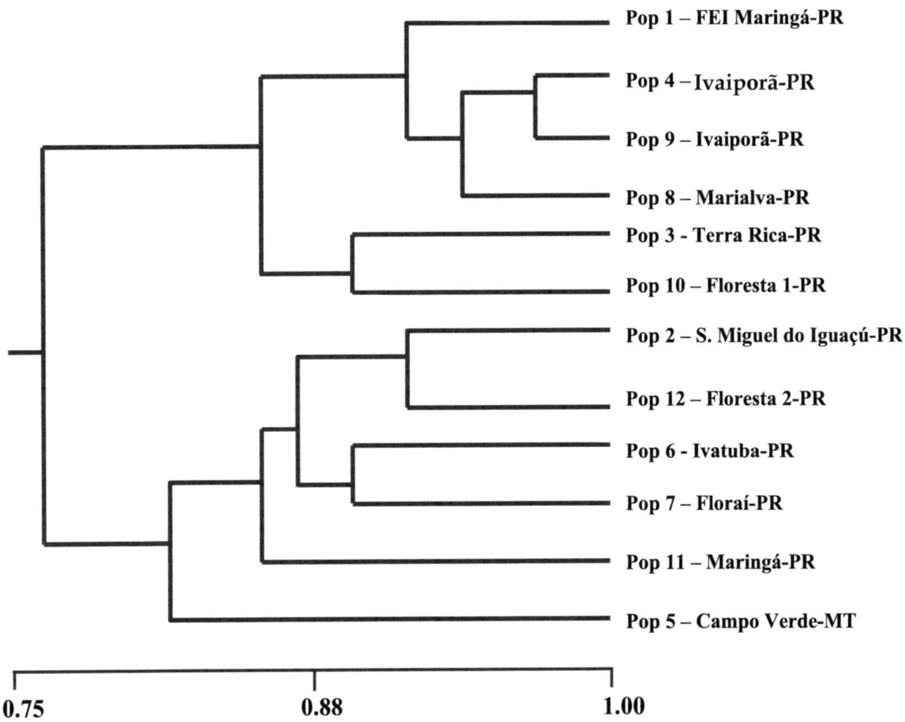

Fig. 4. The dendrogram represents the relationship between the plant descendants from 12 populations of wild poinsettia based on UPGMA cluster analysis of the allele polymorphism at *Est-1, Est-3, Est-4, Est-5, Est-6,* and *Est-7* loci, by Jaccard's similarity coefficient.

As seen in Table 1, exclusive alleles and alleles with different frequencies were found for the three populations, suggesting that these enzymes may be involved with the differential metabolism of herbicides. Two alleles were detected in tissues from leaves of plants from organic and herbicide-resistant populations for *Est-1, Est-4* and *Est-5* loci. Locus *Est-4* had three alleles in the susceptible population and the allele *Est-4*[3] has a low frequency in population (0.0014). For *Est-3*, *Est-6* and *Est-7* loci, three alleles were found for the three populations in this study (Table 1). The EST-2 esterase encoded by the locus *Est-2* was found in 71.39% of plants of *E. heterophylla* and was absent in 28.61% of plants. In the research carried out by Frigo et al. (2009), EST-2 was not found for 100% of herbicide-resistant plants analyzed (Figure 5), suggesting that this enzyme may be also involved with the differential herbicide metabolism.

Locus	*Est-1*	*Est-3*	*Est-4*	*Est-5*	*Est-6*	*Est-7*	*Est-8*
Alleles							
Organic population							
1	0.4454	0.4055	0.2143	0.7815	0.6324	0.6681	1.0000
2	0.5546	0.3613	0.7857	0.2185	0.3487	0.0021	
3		0.2332			0.0189	0.3298	
Herbicide-susceptible population							
1	0.6320	0.4105	0.1832	0.5289	0.8416	06956	1.0000
2	0.3680	0.4270	0.8154	0.4711	0.1556	0.0110	
3		0.1625	0.0014		0.0028	0.2934	
Herbicide-resistant population							
1	0.3467	0.4315	0.7200	0.9667	0.6333	0.0533	1.0000
2	0.6533	0.5342	0.2800	0.033	0.3267	0.7333	
3		0.0342			0.0400	0.2133	

Table 1. Allelic frequencies for *Est-1, Est-3, Est-4, Est-5, Est-6, Est-7,* and *Est-8* loci observed in *E. heterophylla* from organic, herbicide-susceptible and herbicide-resistant populations.

A moderate level of genetic differentiation (F_{ST} = 0.1410) was found for all three populations, suggesting a reduced genetic exchange between them (N_m= 1.5231). High selection pressure imposed by the use of herbicides on *E. heterophylla* populations has been detected in data from α- and β-esterases. Similarity between plants of *E. heterophylla* from organic and herbicide-susceptible populations was high (I = 0.9670), however, the mean observed and expected heterozygosity was higher in wild poinsettia plants from organic crops (H_o = 0.3529; H_e = 0.3923) than in plants from nonorganic crops (H_o = 0.2597; H_e = 0.3693), and the lowest values of heterozigosity were found for the herbicide-resistant population (H_o = 0.2070; H_e = 0.3360). A comparison of organic and herbicide-susceptible populations suggests that frequent herbicide exposure may lead to decreased heterozygosity and that the selection process of resistant biotypes further reduces heterozigosity. Because of the difference in allele frequency and heterozigosity, the three populations formed a group consisting of organic and herbicide-susceptible demonstrating greater similarity between them, while the herbicide-resistant population was isolated from this group, being the most divergent.

The dendrogram based on the genetic distances calculated by the UPGMA method (Figure 6), provided evidences of a group constituted by herbicide-susceptible and organic populations, demonstrating that these two populations present greater similarity, while the herbicide-resistant population was isolated from the other two populations. As regards the other two populations in this study, descendants of herbicide-resistant population had the highest level of differentiation observed. Nei's identity values (I) ranged from 0.7623 (descendants from herbicide-susceptible and herbicide-resistant populations) and 0.9670 (descendants from herbicide-susceptible and organic populations).

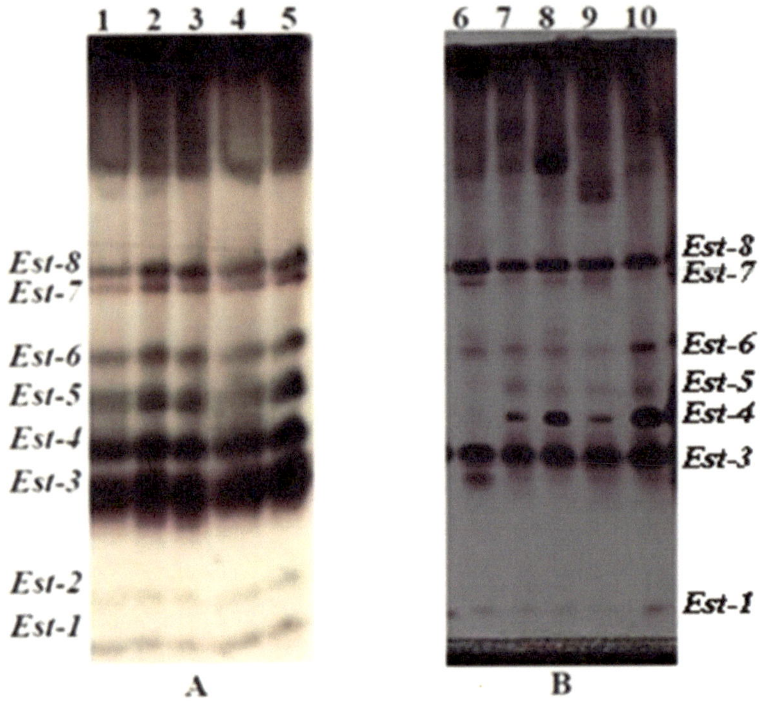

Fig. 5. Polymorphism of α- and β-esterases detected in plants of *Euphorbia heterophylla* descending from herbicide-susceptible (samples 1-5; **gel A**) and herbicide-resistant (samples 6-10; **gel B**). **Gel A**, samples from plants susceptible to ALS inhibitor herbicides, where Esterase-2 is present. **Gel B**, samples from ALS-resistant plants, where Esterase-2 is absent.

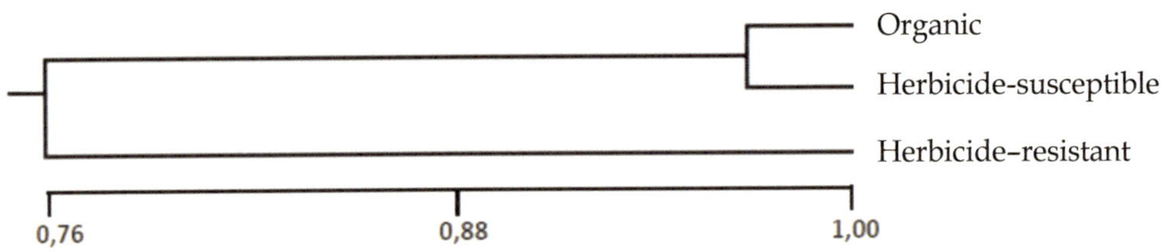

Fig. 6. Dendrogram representing the relationship between plants from organic, herbicide-susceptible and herbicide-resistant populations of *Euphorbia heterophylla,* based on similarity measures by UPGMA and cluster analysis for the alleles polymorfism from *Est-1*, *Est-3*, *Est-4*, *Est-5*, *Est-6*, and *Est-7* loci by Jaccard's similarity coefficient.

Data from the studies evaluating α- and β-esterases provide evidences that populations of *E. heterophylla* have been under high selection pressure imposed by herbicide use. This has been verified by the differentiation between organic, herbicide-susceptible and herbicide-resistant populations. Exclusive alleles and different frequencies for alleles in different loci of esterases found for the three populations suggest that these enzymes may be involved with differential metabolism of herbicides. Frequent use of a single herbicide or mechanism of action may exert a high selection pressure, reducing the susceptible populations, and, therefore, resulting in herbicide-resistant biotypes dominance, which already were found in natural populations, but in very low frequencies.

5. Biology and ecophysiology of *Conyza spp.*

The genus *Conyza* includes around 50 species, distributed all over the world (Kissmann and Groth, 1999). The species that stand out by their negative effects are *Conyza bonariensis* (L.) (fleabane, hairy fleabane) and *Conyza canadensis* (L.) (horseweed, marestail); both from Asteraceae family. The first is native to South America and abundant in Argentina, Uruguay, Paraguay and Brazil. In Brazil, its dispersion is more evident in South, Southeast and Midwest regions. It can also be found in coffee plantations in Colombia and Venezuela (Kissmann and Groth, 1999). *Conyza canadensis,* however, is native to North America (Frankton and Mulligan, 1987) and is one of the most widely distributed species globally (Thebaud and Abbott, 1995). It can be predominately found in Northern hemisphere temperate regions (Holm et al., 1997) and in subtropical regions of Southern hemisphere (Kissmann and Groth, 1999). *C. canadensis* is also present in Canada (Rouleau and Lamoureus, 1992), Western Europe (Thebaud and Abbott, 1995), Japan and Australia (Holm et al., 1997).

Propagation of both *Conyza bonariensis* and *C. canadensis* occurs through the seed only. The fruit is an achene with pappus, a simple one-seed fruit which has an apical structure of radiating fine light bristles (pappus) that serves as a means for seed dispersion by wind (Andersen, 1993), as well as by water (Lazaroto et al., 2008). Both *Conyza* species are self-compatible and seem not to be pollinated by insects (Thebaud et al., 1996), although insect visits to open flowers of *C. canadensis* have been reported (Smisek, 1995).

Seeds are able to disperse by wind in distances over 100 m (Dauer et al., 2006). The average number of seeds found in *C. canadensis* and *C. bonariensis* ranges from 60 to 70 per achene (Smisek, 1995; Thebaud and Abbott, 1995) and from 190 to 550 per capitulum (Wu and Walker, 2004), respectively. *C. canadensis* densities of 10 plants m^{-2} growing in areas with no soil disturbance may produce as much as 200,000 seeds per plant (Bhowmik and Bekech, 1993). About 80% of them germinate next to mother-plants (Loux et al., 2004). With increasing densities, the number of flowering plants, the individual plant size and the number of seeds per plant decreases, but the global seed production per area remains very similar (Lazaroto et al., 2008).

Conyza seeds have no dormancy and germinate when ever favorable conditions of temperature and moisture are present (Wu and Walker, 2004). Minimum temperature required for germination of *C. canadensis* was estimated in 13 °C (Steinmaus et al., 2000). Seeds from both *C. canadensis* and *C. bonariensis* germinate after exposition to temperatures between 10 and 25 °C (Zinzolker et al., 1985). In a study carried out in Australia, the optimum temperature for *C. bonariensis* was 20 °C, but minimum and maximum temperatures were estimated in 4.2 °C and 35 °C (Rollin and Tan, 2004).

Germination of *C. canadensis* seeds was higher during a period of light, but under lab conditions, these seeds also germinate either under no light or when submitted to alternate periods of light/dark (13/11 h) (Nandula et al., 2006). Aggregation of mulching as soil covers such as those propitiated by no tillage cropping systems may delay or prevent germination, allowing the crop to establish and suppress later fluxes of weeds that eventually will emerge (Lazaroto et al., 2008). Germination of *C. canadensis* seeds occurs preferentially in neutral to alkaline soils (Nandula et al., 2006). Therefore, soil liming should be planned and balanced to meet the crop needs of crops and not to promote favorable conditions to *Conyza* germination (Lazarotto et al., 2008). *Conyza* is also able to grow and reproduce under more limited soil resources (rough, stony areas) (Hanf, 1983), as well as flat, poorly drained areas , provided that there is no flooding (Smith and Moss, 1998).

Conyza canadensis is an annual or biennial species, depending on environmental conditions (Regehr and Bazzaz, 1979; Holm et al., 1997) and *C. bonariensis* is considered a typical annual (Kissmann and Groth, 1999). Studies in Australia demonstrated that plants emerge throughout the year, but maximum emergences occur in spring (Walker et al., 2004). Other studies have shown that *Conyza* is able to grow under a different set of climate types, but *C. canadensis* is rare under tropical conditions (Holm et al., 1997). In Canada, seed production and invading potential of *Conyza* as weed tend to be limited by latitude 52 °N (Archibold, 1981). However, the two species of *Conyza* are tolerant to water stress conditions and use of irrigation is considered as an alternative to improve crop competitive ability against those weeds (Lazaroto et al., 2008).

6. Management and herbicide resistance in *Conyza* species

Both species *Conyza bonariensis* and *C. canadensis* are typical colonizers of abandoned areas, perennial and annual crops (soybeans, maize, cotton and wheat) (Thebaud and Abbott, 1995). Bruce and Kells (1990) demonstrated that the interference imposed by *C. canadensis* decreased soybean grain yield in 83% under no tillage conditions and weed densities around 150 plants m^{-2}. Leroux et al. (1996) also demonstrated the effects of *Conyza spp.* in other crops such as onions and carrots, concluding that, in carrots, negative effects on crop harvest may be even more important than those found in crop yield. In Indiana (USA), *Conyza* infestations have been detected in about 63% of soybean areas cropped with soybeans for two consecutive years, in 51% of soybean areas with no crop rotation and in 47% of areas cultivated under soybean/corn rotations (Barnes et al., 2004). The inclusion of barley as a successional winter crop decreased *C. canadensis* populations in onion and carrots over the next summer (Leroux et al., 1996).

In Brazil, *Conyza* most prolific growth usually is found between winter crops harvest and summer crops sowing. Farmers have related poor control of *Conyza* with herbicides, especially those used for burndown prior to summer crops. Problems are mostly related to tolerance of adult plants to herbicides and also to resistance to glyphosate. In several field experiments carried out in the last years, we found that a fall application (usually one to two weeks after corn harvest in July/August) including tank mixtures of burndown herbicides and residual herbicides provide an excellent alternative for these areas. Residual herbicides added to these treatments improve control of emerged *Conyza* and provide residual control, so that at the point that next crop is about to be sowed, seed bank is adequately controlled or, when emerged, is still within a range of size (≤ 10 cm) that permits control with a regular burndown treatment (Oliveira Jr. et al., 2010).

Increases in soil disturbance reduce the densities of *C. canadensis* by 50% or more (Buhler and Owen, 1997). Seedlings of *C. canadensis* were detected in 61% of the crops that were not submitted to soil tillage, as compared to 24% under minimum soil tillage (Barnes et al., 2004). Thus, as survival rate is drastically reduced when these species are submitted to soil tillage, this has been a strategy to limit infestation in agricultural areas. The impediment to periodical soil tillage, like that imposed by no-tillage cropping areas, and the fact that under that cropping system the seeds of weeds are deposited in the soil surface or buried very shallow may be used as management tools to obtain a more uniform emergence of these plants in the field. Uniform emergence of weeds favor the efficiency of herbicides and tend to allow the use of non chemical alternatives of weed control, like mowing (Lazaroto et al., 2008).

Therefore, weed management practices as regards the *Conyza* species require the combination of multiple actions like increased intensity of soil management, adoption of routine crop rotations and cultural strategies (Lazaroto et al., 2008). In addition, the correct identification of *Conyza* species is important so that a suitable control method may be chosen.

The frequent use of a particular herbicide or of herbicides with the same mechanism of action in *Conyza* species may also result in high selection pressure. Glyphosate has been safely used for over 40 years in weed management. It is considered as a non-selective herbicide and is a very useful tool to promote soil protection by plant residues that are obtained from natural vegetation or a cover crop cultivated during the intercropping season in no-tillage areas. The growing dependence and overreliance on glyphosate to control weeds is a major concern for the maintenance of long-term viability of such valuable tool in weed managements, since the repeated use of one single herbicide molecule may select preexisting weed resistant biotypes, leading to increased densities of these biotypes in field (Powles et al., 1994). In general, species or biotypes of a species best adapted to a particular practice are selected and multiply rapidly (Holt and Lebaron, 1990). Evidences suggest that emergence of herbicide resistance in a plant population is due to the selection of preexisting resistant genotypes, which, because of the selection pressure exerted by repeated applications of a single herbicide, find conditions for multiplication (Betts et al., 1992).

Weed resistance to herbicides is not a new phenomenon. Plants of field bindweed (*Convolvulus arvensis*) resistant to glyphosate were identified in Indiana (USA) in the mid-80's in fields that had been sprayed repeatedly with glyphosate (Degennaro and Weller, 1984). However, weed resistance to glyphosate has become a major concern a few years after the release of the first Roundup Ready® soybean varieties in USA in 1996. Species that are currently considered as of greatest concern include those from the genus *Conyza*. The first reported case of glyphosate-resistant *Conyza* was found in Delaware (USA) in 2000 (Van Gessel, 2001).

Currently, *Conyza* resistant biotypes are distributed in over 20 U.S. states and in over 40 countries worldwide (Heap, 2010; Alcorta et al., 2011). In Brazil, the first sites of resistance were reported in Rio Grande do Sul in 2005, and thereafter these biotypes have rapidly dispersed in all southern states and, more recently, in Midwest and Southeast. All sites of reported detections of glyphosate-resistant *Conyza spp*. share the frequent use of this herbicide in weed control, little or no use of alternative herbicides that provide adequate control of *Conyza spp*., and long-term, no-till agricultural practices (Loux et al., 2009).

Resistance of *Conyza spp*. in relation to other herbicides has also been previously described. In 1980, Japanese scientists detected a resistant biotype of *C. canadensis* to the herbicide paraquat (Heap, 2010). Increased activity of detoxification enzymes such as superoxide dismutase or the compartimentalization of herbicide molecules at cellular organelles were related to the mechanism of resistance to paraquat (Ye et al., 2000). In Hungary, herbicide-resistant populations of *Conyza* were simultaneously found for paraquat and atrazine (Lehoczki et al., 1984). In Israel and U.S. populations resistant to atrazine and chlorsulfuron (a ALS inhibitor) were also found (Heap, 2010).

Among the recommended measures to manage weed resistance to herbicides, the frequent monitoring of crops in field is essential, in order to identify eventual suspected plants, which should be systematically eliminated (Lazaroto et al., 2008).

7. Genetic diversity in *Conyza bonariensis* and *C. canadensis*

To estimate the level of genetic diversity and the level of differentiation among populations of *Conyza bonariensis* and *C. canadensis*, we have developed routine lab procedures to analyze esterase isozymes as well as malate dehydrogenase and acid phosphatase. We assume that this information may serve as a guideline for weed management of both species in view of the growing concern related to the spread of cases of resistance to herbicides.

Fig. 7. Plants of *Conyza bonariensis* (A) and *C. canadensis* (B) used for analyze isozymes esterase, malate dehydrogenase and acid phosphatase.

For analyze esterase isozymes (EST; EC 3.1.1._), malate dehydrogenase (MDH; EC 1.1.1.37) and acid phosphatase (ACP; EC 3.1.3.2) leaves of plants of *C. bonariensis* e *C. canadensis* were used (Figure 7). Leaf fragments (200 mg) were homogenized with 60 μL of a extraction solution prepared with phosphate buffer 1.0 M, pH 7.0 containing PVP-40 5%, EDTA 1.0 mM, 0.5% β-mercaptoetanol, and glicerol 10%; extraction was performed in an ice bath using 2.0 mL microcentrifuge tubes. After homogenization, samples were centrifuged (centrifuge Juan 23 MRi, at 14,000 rpm - 48,200 *g*) for 30 minutes, at 4 °C, and the supernatant (50 μL) of each sample was used in electrophoresis.

Electrophoresis and α- and β-esterase identification was previously established by Frigo et al. (2009). To analyze enzymes malate dehydrogenase and acid phosphatase, after centrifugation supernatants were absorbed in strips of paper (Whatman nº 3; 5 x 6 mm), and these were vertically inserted into 16% starch gel, following the protocols described by Machado et al. (1993). Visualization (staining procedures) and identification of the malate dehydrogenase and acid phosphatases isozymes were also performed according to the protocols described by Machado et al. (1993).

Analysis of esterase isozymes from plant leaves of *Conyza*, through the non-denaturing PAGE method and using as substrates α-naphthyl acetate and β-naphthyl acetate disclosed seven esterase loci clearly defined. The α/β-esterases were listed starting from anode as EST-1 and following the order of decreasing negative charges. The slowest-migration esterase was named as EST-7 (Figure 8).

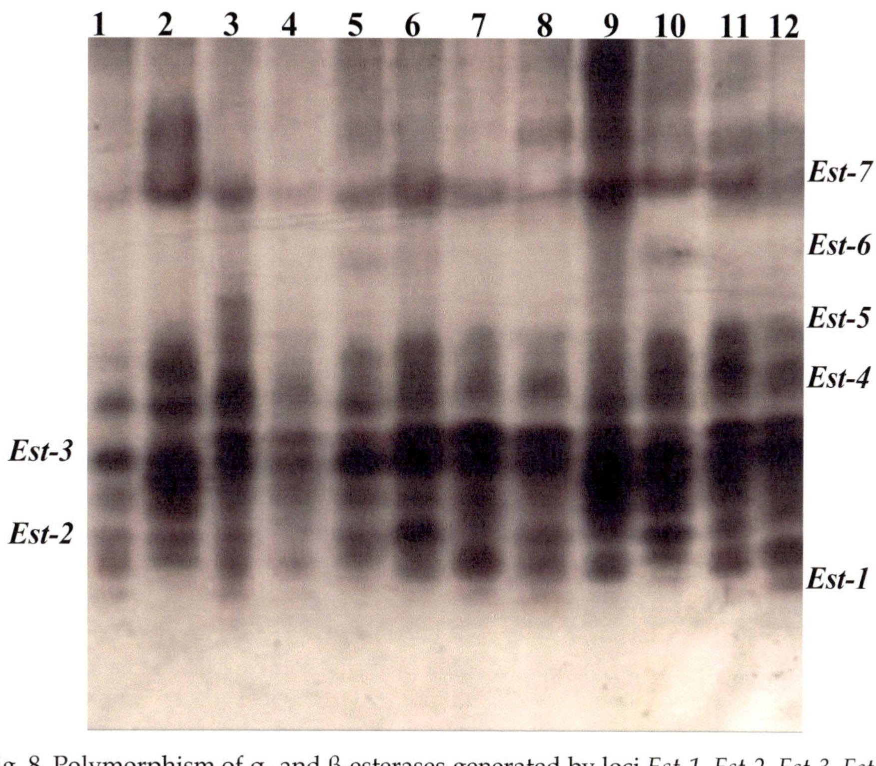

Fig. 8. Polymorphism of α- and β-esterases generated by loci *Est-1*, *Est-2*, *Est-3*, *Est-4* and *Est-5* detected in 12 plants of *Conyza canadensis*.

Electrophoresis for malate dehydrogenase in starch gel demonstrated the presence of three different groups of malate dehydrogenase isozymes: the microbodies MDH (mbMDH), mitochondrial MDH (mtMDH), and cytosol or soluble MDH (sMDH) (Figure 8). Four mbMDH isozymes were evident (mbMDH-1, mbMDH-2, mbMDH-3, and mbMDH-4) at the *mbMdh* locus; two other loci were evident for mtMDH isozymes (*mtMdh-1* and *mtMdh-2*) and another two loci for sMDH isozymes (*sMdh-1* and *sMdh-2)* (Figure 9). Due to the complex structural organization of these enzymes, which includes the ability to produce heterodimers between products from alleles of the *mtMdh* and *sMdh* loci, studies related to genetic diversity were only performed for the analysis of the *mbMdh* locus. At the *mbMdh* locus the presence of four alleles was evident (Figure 9).

For acid phosphatase isozymes, two loci *(Acp-1* e *Acp-2*) were evident, but only the *Acp-2* locus was analyzed, with four alleles for ACP-2^1, ACP-2^2, ACP-2^3 and ACP-2^4 isozymes (Figure 10).

Considering the three enzymatic systems, 14 loci were detected; from those, five that code for esterases were analyzed, as well as one locus coding for malate dehydrogenase and one locus for acid phosphatase, totalizing seven loci.

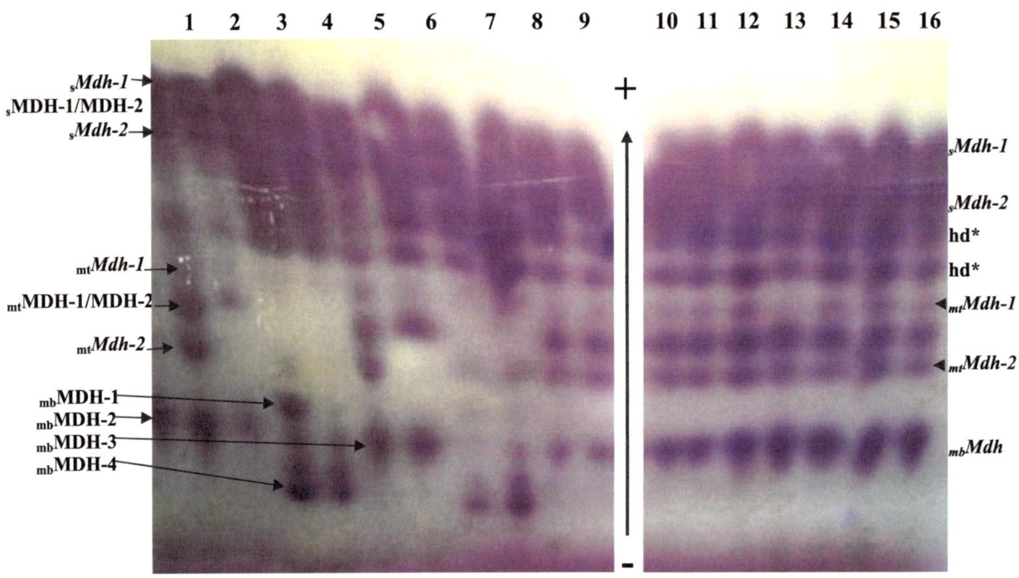

Fig. 9. Malate dehydrogenase from plants of *Conyza canadensis* (1-7) and *C. bonariensis* (8-16). Polymorphism for $_{mb}$*Mdh* locus showing the four isozymes coded by their alleles ($_{mb}$MDH-1, $_{mb}$MDH-2, $_{mb}$MDH-3, and $_{mb}$MDH-4). Evidence of two loci ($_{mt}$*Mdh-1* and $_{mt}$*Mdh-2*) for mitochondrial MDH and two loci ($_{s}$*Mdh-1* and $_{s}$*Mdh-2*) for the soluble MDH isozyme. The $_{mt}$MDH-1/MDH-2 is the heterodimer between the product of the loci $_{mt}$*Mdh-1* and $_{mt}$*Mdh-2;* $_{s}$MDH-1/MDH-2 is the heterodimer between the product of the $_{s}$*Mdh-1* and $_{s}$*Mdh-2* loci; hd* are heterodimers between the products of the loci $_{mt}$*Mdh-2* and $_{s}$*Mdh-1* and loci $_{mt}$*Mdh-1* and $_{s}$*Mdh-1*.

The genetic diversity found in our studies with *C. bonariensis* and *C. canadensis* based on MDH, ACP and α-/β-esterases can be considered high, since polymorphism was detected for 7 out of 14 loci analyzed (50%). That value may still be underestimated, because the polymorphism for the loci of the soluble and mitochondrial malate dehydrogenase, as well as for the *Acp-1* locus, was not reported here. Hence the proportion of polymorphism for *C. bonariensis* and *C. canadensis* is much higher than mean values reported for dicotyledons (31%) (Hamrick et al., 1979) and also for other weed species such as *Euphorbia* (Park, 2004; Frigo et al. 2009).

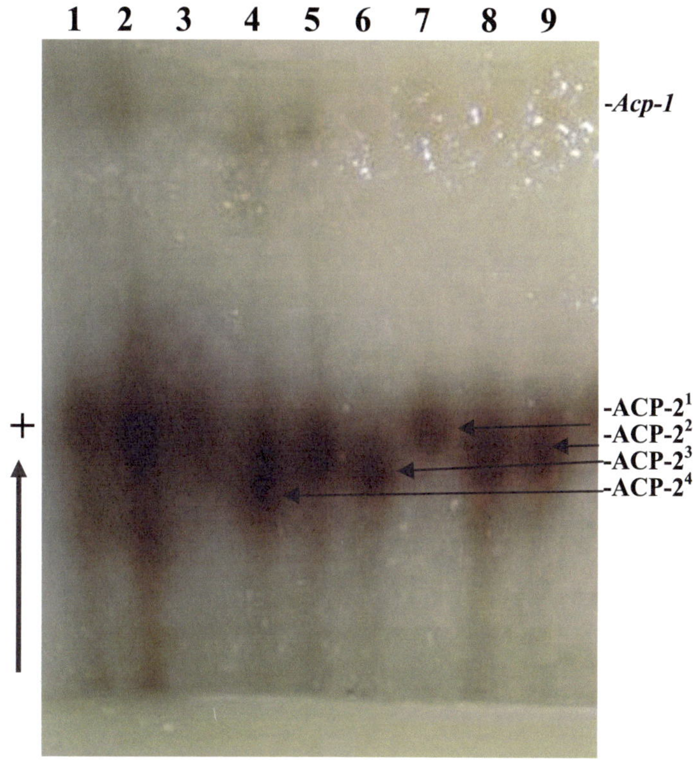

Fig. 10. Polymorphism of the acid phosphatase isozymes produced by locus *Acp-2* coding for isozymes ACP-2^{1}, ACP-2^{2}, ACP-2^{3} and ACP-2^{4}. Samples from 1 to 9 are *Conyza bonariensis* plants.

Observed heterozygosity (H_o) for α- and β-esterases from *C. canadensis* and *C. bonariensis* was 0.4310 and 0.4293, respectively; these values are lower than values for expected mean heterozygosity (H_e), which was found to be 0.5125 and 0.4978. The deficit of heterozygotes was higher when plants of *Conyza* were evaluated for the *mbMdh* and *Acp-2* loci. Observed heterozygosity (H_o) for *C. canadensis* was 0.4410 compared to an expected value (H_e) of 0.6699. For *C. bonariensis*, the observed mean heterozygosity was 0.4333 compared to an expected value of 0.6149.

For all 14 tests performed to analyze the Hardy-Weinberg equilibrium in both populations, equilibrium was found in only six of them (42.86%). The lack of equilibrium for the analyzed loci is result of a deficit of heterozygous individuals. The fixation index (F_{IS}) was positive for the *Est-1, Est-2, Est-4, mbMdh* and *Acp-2* loci. For both populations, the H_o was lower than expected, indicating the lack of heterozygous plants in populations of *Conyza*.

A low level of genetic differentiation was found for the two populations of *Conyza*, both when α- and β-esterase (F_{ST} = 0.0137), and malate dehydrogenase and acid phosphatase isozymes (F_{ST} = 0.0239) were evaluated. According to Wright (1978), values of $F_{ST} < 0.05$ indicate a low genetic differentiation between populations. That finding suggests an extensive genetic exchange (N_m) between populations of *C. canadensis* and *C. bonariensis*, which was estimated to be of 18.008 and 10.203 for the α-/β-esterases loci and malate dehydrogenase/acid phosphatase loci, respectively. According to this, the estimates of gene flow were high (N_m = 18.008 and 10.203). The pattern of allele migration or the exchange of alleles between populations may have contributed to maintain homogeneity between the two populations of *C. canadensis* and *C. bonariensis*.

The frequency of alleles in both populations is very homogenous, i.e., the distribution of alleles for all loci analyzed showed no preferential distribution in any of the evaluated populations. Estimates calculated in the present study lead to the proposition of no reproductive isolation between species of *C. bonariensis* and *C. canadensis*. Lack of genetic differentiation may indicate exchange of alleles between both populations, what is reasonable to occur as long as plants share the same space for long periods of time.

When reproductive aspects of these species are considered, a greater differentiation should be expected, since both species of *Conyza* are self-compatible and apparently are not pollinated by insects, suggesting the occurrence of autogamy or wind pollination (Thebaud et al., 1996). *Conyza canadensis* is self-compatible (Mulligan and Findlay, 1970); pollen is released before capitulum full opening, suggesting self-pollination, although insects visit open flowers (Smisek, 1995). However, when paraquat-resistant plants were used as markers, the average level of self-crossing within a population of *C. canadensis* was only 4% (ranging from 1.2 to 14.5%). Therefore, besides self-crossing not to be the most frequent reproduction form for these species, a second way to explain the low differentiation between them is the ability to develop hybridization between *C. canadensis* and other species in *Conyza* genus, specially *C. sumatrensis* and *C. bonariensis*, since they usually grow in associated populations and occur on a frequent basis (Thebaud and Abbott, 1995). Small differences for genetic variation among different populations may indicate a large spatial dispersion or a recent disturbance in genetic variation associated to human action (Allendorf and Luikart, 2007).

Concurrent occurrence of spatial coexistence for long periods of time, ability of hybridization and populations frequently disturbed by human interference may determine the low level of genetic differentiation between the two species of *Conyza*. Small-scale disturbances such as the continuous use of herbicides can promote an increase in spatial homogeneity. The increased selection pressure imposed by traditional weed management tools has contributed to selection of herbicide-resistant biotypes (Holt and LeBaron, 1990) and may be an important component to determine how the populations are genetically structured.

The deficiency of heterozygous plants in both populations of *Conyza* was also evident by positive values of F_{IS} (F_{IS} = 0.1484 for α- and β-esterases and F_{IS} = 0.32 for MDH and ACP). Positive values of F_{IS} indicate a deficit of heterozygous or excess of homozygous plants. This event could be the result of selection pressure imposed by frequent application of herbicides in soybean fields or of self pollination, which is described for these species. Significant values of F_{IT} (F_{IT} = 0.1607 for α- and β-esterases and F_{IT} = 0.3363 for MDH and ACP) indicate that frequent self-crossing or nonrandom-crossing should have a fundamental role in shaping the genetic structure of *C. bonariensis* and *C. canadensis* populations. On the other hand, the high heterozygosity found in populations of *C. bonariensis* and *C. canadensis* may indicate that these plant populations have a substantial amount of adaptive genetic variations, and that these variations may be enough for them to escape of the eventual effects of a control agent.

Levels of interpopulational genetic diversity in species *C. bonariensis* and *C. canadensis* may be evaluated based on values of Nei's genetic identity (I) of both populations. I values also demonstrated a very small genetic differentiation between the two species of *Conyza*. The degree of genetic divergence may be used to develop crop and weed management policies to provide more effective control of *C. bonariensis* and *C. canadensis*. For populations of these two species with high genetic identity values, similar weed control strategies may be adopted.

This research on genetic diversity in species of *Conyza* was important since no studies could be found in the scientific literature. So far, most available information have been focused on reports of resistant populations, on aspects related to herbicide efficiency in its control and on elucidating mechanisms that promote herbicide resistance in this genus. Such studies are important since herbicide resistance has already been reported in six countries for *C. bonariensis* and in 13 countries for *C. canadensis* (Heap, 2010). Biotypes of both species have also been reported as resistant to glyphosate in Brazil by Christoffoleti et al. (2006), Montezuma et al. (2006), Moreira et al. (2007), Vargas et al. (2007), and confirmed by Lamego and Vidal (2008). Developing knowledge on genetic diversity and population structure is important to orientate weed management programs, leading to differential or specific forms of control, making them more effective to control species in *Conyza* genus.

The analysis of α- and β-esteases, malate dehydrogenase and acid phosphatase isozymes revealed high genetic diversity in *C. bonariensis* and *C. canadensis* species, and limited genetic differentiation between them, indicating that it may be possible to develop similar weed control mechanisms and strategies (type and doses of herbicides, for instance) for both species. Based on our results, there is an expectation that weed control approaches developed for one species should also be effective for the other species, considering the limited genetic differentiation detected between them.

8. References

Adahl, E.; Lundberg, P. & Jonzan, N. (2006). From climate change to population change: the need to consider annual life cycles. *Global Change Biology*, v. 12, n. 9 (September), p. 1627–1633. ISSN: 1365-2486.

Al-Samman, N.; Martin, A. & Puech, S. (2001). Inflorescence architecture variability and its possible relationship to environment or age in a Mediterranean species, *Euphorbia nicaeensis* All. (Euphorbiaceae). *Botanical Journal of the Linnean Society*, v. 136, n. 2 (February), p. 99-105. ISSN: 1095-8339.

Alcorta, M.; Fidelibus, M.W.; Steenwerth, K.L. & Shrestha, A. (2011). Effect of vineyard row orientation on growth and phenology of glyphosate-resistant and glyphosate-susceptible horseweed (*Conyza canadensis*). *Weed Science*, v. 59, n. 1 (January-March), p. 55–60. ISSN: 1550-2759.

Allendorf, F.W. & Luikart, G. (2007). *Conservation and the genetics of populations*. Blackwell Publishing Maden, Massachusetts, 642p. ISBN: 9781405121453.

Andersen, M.C. (1993). Diaspore morphology and seed disperal in several wind-dispersed Asteraceae. *American Journal of Botany*, v. 80, n. 5 (May), p. 487-492. ISSN: 0002-9122.

Archibold, O.W. (1981). Buried viable propagules in native prairie and adjacent agricultural sites in central Saskatchewan. *Canadian Journal of Botany*, v. 59, n. 5 (May), p. 701-706. ISSN: 1916-2804.

Barnes, J.; Johnson, B.; Gibson, K. & Weller, S. (2004). *Crop rotation and tillage system influence late-season incidence of giant ragweed and horseweed in Indiana soybean*. In: <http://www.plantmanagementnetwork.org/pub/cm/brief/2004/late/>. April, 19, 2010.

Barroso, G.M. (1984). *Sistemática de angiospermas do Brasil*. Universidade Federal de Viçosa, Viçosa, vol. 2, 377p. ISBN: 85-7269-127-8.

Betts, K.J.; Ehlke, N.J.; Wyse, D.L.; Gronwald, J.W. & Somers, D.A. (1992). Mechanism of inheritance of diclofop resistance in italian ryegrass (*Lolium multiflorum*). *Weed Science*, v. 40, n. 2 (April-June), p. 184-189. ISSN: 1550-2759.

Bhowmik, P.C. & Bekech, M.M. (1993). Horseweed (*Conyza canadensis*) seed production, emergence, and distribution in no-tillage and conventional tillage corn (*Zea mays*). *Agronomy*, v. 1, n.1, p. 67-71. ISSN: 0065-2113.

Bruce, J.A. & Kells, J. (1990). Horseweed (*Conyza canadensis*) control in no-tillage soybeans (*Glycine max*) with preplant and preemergence herbicides. *Weed Technology*, v. 4, n. 4 (October-December), p. 642-647. ISSN: 0890-037X.

Buhler, D.D. & Owen, M.D.K. (1997). Emergence and survival of horseweed *(Conyza canadensis)*. *Weed Science*, v. 45, n. 1 (January- March), p. 98-101. ISSN: 1550-2759.

Burnside, O.C. (1992). Rationale for developing herbicide-resistant crops. *Weed Technology*, v. 6, n. 4 (October-December), p. 621-625. ISSN: 0890-037X.

Castro, H.G.; Silva D.J.H.; Ferreira, F.A. & Ribeiro Júnior, J.I. (2002). Estabilidade da divergência genética em seis acessos de carqueja. *Planta Daninha*, v. 20, n.1 (January-March), p. 33-37. ISSN: 0100-8358.

Chemale, V.M. & Fleck, N.G. (1982). Avaliação de cultivares de soja (*Glycine max* (L.) Merrill) em competição com *Euphorbia heterophylla* L. sob três densidades e dois períodos de ocorrência. *Planta Daninha*, v.5, n. 2 (April-June), p. 36-45. ISSN: 0100-8358.

Cronquist, A. (1981). *An integrated system of classification of flowering plants*. New York, Columbia University Press. 1262p. ISBN 0-231-03880-1.

Dauer, J.T.; Mortensen, D.A. & Humston, R. (2006). Controlled experiments to predict horseweed *(Conyza canadensis)* dispersal distances. *Weed Science*, v. 54, n. 3 (May), p. 484-489. ISSN: 1550-2759.

DeGennaro, F.P. & Weller, S.C. (1984). Differential susceptibility of field bindweed (*Convolvulus arvensis*) biotypes to glyphosate [Indiana]. *Weed Science*, v. 32, n. 3 (May), p. 472-476. ISSN: 1550-2759.

Frankton, C. & Mulligan, G.A. (1987). *Weeds of Canada*. Toronto: NC, Pr. 217p. ISBN-10: 1550210165.

Frigo, M.J.; Mangolin, C.A.; Oliveira, R.S.J. & Machado, M.F.P.S. (2009). Esterase polymorphism for analysis of genetic diversity and structure of wild poinsettia (*Euphorbia heterophylla*) populations. *Weed Science*, v. 57, n. 1 (January), p. 54-60. ISSN: 1550-2759.

Gazziero, D.L.P.; Brighenti, A.M.; Maciel, C.D.G.; Christoffoleti, P.J.; Adegas, F.S. & Voll, E. (1998). Resistência de amendoim-bravo aos herbicidas inibidores da enzima ALS. *Planta Daninha*, v. 16, n. 2 (December), p. 117-125. ISSN: 0100-8358.

Gómez, C. & Espadaler, X. (1998). Seed dispersal curve of a Mediterranean myrmecochore: Influence of ant size and the distance to nests. *Ecological Research*, v. 13, n. 3 (October), p. 347-352. ISSN: 1440-1703.

Goolsby, J.A.; De Barro, P.J.; Makinson, J.R.; Pemberton, R.W.; Hartley, D.M. & Frohlich, D.R. (2006). Matching the origin of an invasive weed for selection of a herbivore haplotype for a biological control program. *Molecular Ecology*, v. 15, n. 1 (January), p. 287–297. INSS: 1365-294X.

Grombone-Guaratini, M.T.; Silva-Brandão, K.L.; Solferini, V.N.; Semir, J. & Trigo, R. (2005). Sesquiterpene and polyacetylene profile of the *Bidens pilosa* complex (Asteraceae: Heliantheae) from southeast of Brazil. *Biochemical Systematics and Ecology*, v. 33, n. 10 (July-September), p. 479-486. ISSN: 0305-1978.

Hamrick, J.L.; Linhart, Y.B. & Mitton, J.B. (1979). Relationship between life history characteristics and electrophoretically-detectable genetic variation in plants. *Annual Review of Ecology and Systematics*, v. 10, p. 173-200. ISSN: 0066-4162.

Hanf, M. (1983). *The arable weeds of Europe with their seedlings and seeds*. BASF Aktiengesellschaft, Ludwigshafen, Germany. 494p. ASIN: B0007B15JC.

Harper, J.L. (1956). The evolution of weeds in relation to resistance to herbicides. *British Crop Protection Conference*, v. 1, n. 1, p.179-188. ISSN: 0144-1612.

Heap, I. (2010). International survey of herbicide-resistant weeds. In: <http://www.weedscience.com>. February, 20, 2010.

Hilton, H.W. (1957). Herbicide tolerant strains of weeds. *HSCPA Annual Report*, p.66.

Holm, L.; Doll, J.; Holm, E.; Pancho, J. & Herberger, J. (1997). *Conyza canadensis* (L.) *Cronq*. (*syn. Erigeron canadensis* L.). *World Weeds: Natural Histories and Distributions*. John Wiley and Sons, Inc. Toronto, (March), p. 226-235. ISBN: 978-0-471-04701-8.

Holt, J.S. & LeBaron, H.M. (1990). Significance and distribution of herbicide resistance. *Weed Technology*, v. 4, n. 2 (July), p. 141-149. ISSN: 0890-037X.

Hufbauer, R. (2004). Population genetics of invasions: Can we link neutral markers to management? *Weed Technology*, v. 18, n. 4 (December), p. 1522–1527. ISSN: 0890-037X.

Ingrouille, M. (1992). *Diversity and evolution of land plants*. London, Chapman & Hall, 340p. ISBN: 0412442302 9780412442308.

Joly, A.B. (1998). *Botânica: Introdução à taxonomia vegetal*. 12th. Ed. São Paulo. Companhia Editora Nacional, 778p. ISBN: 8504002314.

Kissmann, K.G. & Groth, D. (1992). *Plantas infestantes e nocivas*. São Paulo: Basf, 798p. ISBN: 8588299038.

Kissmann, K.G. & Groth, D. (1999). *Plantas infestantes e nocivas*. 2th. Ed. São Paulo: Basf, v. 2, 978p. ISBN: 85-88299-02-X.

Lamego, F.P. & Vidal, R.A. (2008). Resistência ao glyphosate em biótipos de *Conyza bonariensis* e *Conyza canadensis* no estado do Rio Grande do Sul, Brasil. *Planta Daninha*, v. 26, n.2 (April-June), p. 467-471. ISSN: 0100-8358.

Lazaroto, C.A.; Fleck, N.G. & Vidal, R.A. (2008). Biologia e ecofisiologia de buva (*Conyza bonariensis* e *Conyza canadensis*). *Ciência Rural*, v. 38, n. 3 (Maio-Junho), p. 852-860. ISSN: 0103-8478.

Lehoczki, E.; Laskay, G.; Pölös E. & Mikulás J. (1984). Resistance to triazine herbicides in horseweed *(Conyza canadensis)*. *Weed Science*, v. 32, n. 4 (Julio) p. 669-674. ISSN: 0043-1745.

Leroux, G.D.; Benoît, D-L. & Banville, S. (1996). Effect of crop rotations on weed control, *Bidens cernua* and *Erigeron Canadensis* populations, and carrot yields in organic soils. *Crop Protection*, v. 15, n. 2 (March), p. 171-178. ISSN: 0261-2194.

Loux, M.; Stachler, J.; Johnson, B.; Nice, G.; Davis, V. & Nordby, D. (2004). *Biology and management of horseweed.* In: <http://www.ipm.uiuc.edu/pubs/horseweed.pdf>. April, 17, 2010.

Loux, M.; Stachler, J.; Johnson, B.; Nice, G.; Davis, V. & Nordby, D. (2009). *Biology and management of horseweed. The glyphosate, weeds, and crop series*, Purdue University (Purdue Extension publication number ID-32), 11p. ISBN: 978-0-19-532694-9.

Machado, M.F.P.S.; Prioli, A.J. & Mangolin, C.A. (1993). Malate dehydrogenase (MDH; EC 1.1.1.37) isozymes in tissues and callus cultures of *Cereus peruvianus* (Cactaceae). *Biochemical Genetics*, v. 31, n. ¾ (April), p. 167-172. ISSN: 0006-2928.

Mangolin, C.A.; Prioli, A.J. & Machado, M.F.P.S. (1997). Isozyme variability in plants regenerated from calli of *Cereus peruvianus* (Cactaceae). *Biochemical Genetics*, v. 35, n. 5/6 (June), p. 189-204. ISSN: 0006-2928.

Mattielo, R.R.; Ronzeli Júnior, P. & Puríssimo, C. (1999). Mecanismos de resistência: fatores biológicos, agronômicos e genéticos. In: *Curso de manejo da resistência de plantas daninhas aos herbicidas, 2*. Ponta Grossa. Anais. Ponta Grossa: AECG, p. 27-40.

Meschede, D.K.; Oliveira Jr., R.S.; Constantin, J. & Scapim, C.A. (2002). Período crítico de interferência de *Euphorbia heterophylla* na cultura da soja sob baixa densidade de semeadura. *Planta Daninha*, v. 20, n. 3 (July-September), p. 381-387. ISSN: 0100-8358.

Moreira, M.S.; Nicolai, M.; Carvalho, S.J.P. & Christoffoleti, P.J. (2007). Resistência de *Conyza canadensis* e *C. bonariensis* ao herbicida glyphosate. *Planta Daninha*, v. 25, n. 1 (January-March), p. 157-164. ISSN: 0100-8358.

Mulligan, G.A. & Findlay, J.N. (1970). Reproductive systems and colonization in Canadian weeds. *Canadian Journal of Botany*, v. 5, n. 5 (May), p. 859-860. ISSN: 0008-4026.

Nandula, V.K.; Eubank, T.W.; Poston, D.H.; Koger, C.H. & Reddy, K.N. (2006). Factors affecting germination of horseweed *(Conyza canadensis)*. *Weed Science*, v. 54, n. 5 (September), p. 898-902. ISSN: 0043-1745.

Narbona, E.; Arista, M. & Ortiz, P.L. (2005). Explosive seed dispersal in two perennial mediterranean *Euphorbia* species (Euphorbiaceae). *American Journal of Botany*, v. 92, n. 3 (March), p. 510-516. ISSN: 1537-2197.

Nissen, S.J.; Masters, R.A.; Lee, D.J. & Rowe, M.L. (1992). Comparison of restriction-fragment-length-polymorphisms in chloroplast DNA of 5 leafy spurge (*Euphorbia* spp) accessions. *Weed Science*, v. 40, n. 1 (March), p. 63-67. ISSN: 0043-1745.

Oliveira Jr, R.S.; Constantin, J.; Blainski, E. & Oliveira Neto, A.M. (2010). Fall application as alternative for glyphosate-resistant biotypes of *Conyza* sp. control in Paraná (southern Brazil). In: *Pan-American Weed Resistance Conference*, 2010, Miami. *Annals*. Miami: Bayer Crop Science, v. 1 (January 2010).

Orasmo, G.R.; Oliveira-Collet, S.A.; Lapenta, A.S. & Machado, M.F.P.S. (2007). Biochemical and Genetic Polymorphisms for Carboxylesterase and Acetylesterase in Grape Clones of *Vitis vinifera* L. (Vitaceae) Cultivars. *Biochemical Genetics*, v. 45, n. 9-10 (October), p. 663–670. ISSN: 1573 4927.

Owen, M.D.K. (2001). World maize/soybean herbicide resistance. In: Powles, S.B.; Shander, D.L. *Herbicide resistance and world grains*. Boca Raton, CRC Press, p. 101-164. ISBN/ISSN: 0849322197.

Park, K.; Jung, H.; Ahn, B.; Lee, K. & Kim, J. (1999). Genetic and morphological divergence in Korean *Euphorbia ebracteolata* (Euphorbiaceae). *Korean Journal of Plant Taxonomy*, v. 29, p. 249-262. ISSN: 1225-8318.

Park, K.R. (2004). Comparisons of allozyme variation of narrow endemic and widespread species of Far East *Euphorbia* (Euphorbiaceae). *Botanical Bulletin of Academia Sinica*, v. 45, n. 3 (July), p. 221-228. ISSN: 1817-406X.

Pereira, A.J.; Vidigal-Filho, P.S.; Lapenta, A.S. & Machado, M.F.P.S. (2001). Differential esterase expression in leaves of *Manihot esculenta*, Crantz infected with *Xanthomonas axonopodis pv. manihotis*. *Biochemical Genetics*, v. 39, n. 9/10 (October), p. 289-296. ISSN: 0006-2928.

Pitelli, R.A. (1985). Interferência de plantas daninhas em culturas agrícolas. *Informativo Agropecuária*, v. 11, n. 126 (March), p. 16-27. ISSN: 0325-2116.

Ponchio, J.A.R. (1997). *Resistência de Bidens pilosa L. aos herbicidas inibidores da enzima acetolactato sintase*. Tese (Doutorado) - Escola Superior de Agricultura "Luiz de Queiroz", Universidade de São Paulo, Piracicaba, 138p.

Powles, S.B. & Holtum, J.A.M. (1994). *Herbicide resistance in plants: Biology and biochemistry*. Boca Raton, 353p. ISBN: 0873717139.

Regehr, D.L. & Bazzaz, F.A. (1979). The population dynamics of *Erigeron Canadensis*, a sucessional winter annual. *Journal of Ecology*, v. 67, n. 3 (November), p. 923-933. ISSN: 1365-2745.

Resende, A.G.; Machado, M.F.P.S. & Vidigal-Filho, P.S. (2004). Esterase polymorphism marking cultivars of *Manihot esculenta*, Crantz. *Brazilian Archives of Biology and Technology*, v. 47, n. 3 (June), p. 347-353. ISSN: 1516-8913.

Rollin, M.J. & Tan, D. (2004). Fleabane: first report of glyphosate resistant flax-leaf fleabane from western Darling Downs. In: <http://www.weeds.rc.org.au/documents/fleabane_preceedings%20_mar_04.pdf > June, 26, 2010.

Rouleau, E. & Lamoureux, G. (1992). *Atlas of the vascular plants of the island of Newfoundland and of the islands of Saint Pierre et Miguelon*. Quebec: Group Fleurbec, p. 777. ISBN: 2920174126 9782920174122.

Rowe, M.L.; Lee, D.J.; Bowditch, B.M. & Masters, R.A. (1997). Genetic variation in North American leafy spurge (*Euphorbia esula*) determined by DNA markers. *Weed Science*, v. 45, n. 4 (December), p. 446-454. ISSN: 0043-1745.

Slotta, T.A.B. (2008). What we know about weeds: insights from genetic markers. *Weed Science*, v. 56, n. 2 (September), p. 322–326. ISSN: 0043-1745.

Smisek, A.J.J. (1995). *The evolution of resistance to paraquat in populations of Erigeron Canadensis L.* Master Thesis. University of Western Ontario, London, Ontário, 102p.